THE YEAST TWO-HYBRID SYSTEM

Advances in Molecular Biology

The Yeast Two-Hybrid System, edited by Paul L. Bartel and Stanley Fields

THE YEAST TWO-HYBRID SYSTEM

Edited by
Paul L. Bartel
& Stanley Fields

New York Oxford • Oxford University Press 1997

Oxford University Press

Oxford New York
Athens Auckland Bangkok Bogota Bombay Buenos Aires
Calcutta Cape Town Dar es Salaam Delhi Florence Hong Kong
Istanbul Karachi Kuala Lumpur Madras Madrid Melbourne
Mexico City Nairobi Paris Singapore Taipei Tokyo Toronto Warsaw

and associated companies in
Berlin Ibadan

Published by Oxford University Press, Inc.
198 Madison Avenue, New York, New York 10016

Library of Congress Cataloging-in-Publication Data
The yeast two-hybrid system / edited by Paul L. Bartel and Stanley Fields.
p. cm. — (Advances in molecular biology)
Includes bibliographical references and index.
ISBN 0-19-510938-4
1. Protein binding—Research—Methodology. 2. Yeast fungi.
3. Plasmids. I. Bartel, Paul L. II. Fields, Stanley. III. Series.
QP551.Y4 1997
574.87'322—dc20 96-39040

9 8 7 6 5 4
Printed in the United States of America
on acid-free paper

Contents

Part II Analyzing Interactions with the Two-Hybrid System

Part III Variations on the Two-Hybrid Theme

Part IV Related Methodologies

Contributors

David C. Amberg
Department of Biochemistry and Molecular Biology
State University of New York Health Science Center
Syracuse, New York 13210

Richard Baer
Departments of Microbiology and Pediatrics
The University of Texas Southwestern Medical Center
Dallas, Texas 75235-9140

Chang Bai
Merck Research Laboratories
West Point, Pennsylvania 19486

Paul L. Bartel
Myriad Genetics,Inc.
Salt Lake City, Utah 84108

Robert O. Bash
Departments of Microbiology and Pediatrics
The University of Texas Southwestern Medical Center
Dallas, Texas 75235-9140

Vivian Berlin
Mitotix, Inc.
Cambridge, Massachusetts 02139

David Botstein
Department of Genetics
Stanford University School of Medicine
Stanford, California 94305-5120

Roger Brent
Department of Genetics
Harvard Medical School
Boston, Massachusetts 02114

Jayhong A. Chong
Department of Pharmacology
Mayo Clinic Foundation
Rochester, Minnesota 55905

Jonathan A. Cooper
Fred Hutchinson Cancer Research Center
Seattle, Washington 98104

Stephen J. Elledge
Verna and Marrs McLean Department of Biochemistry
Howard Hughes Medical Institute
Baylor College of Medicine
Houston, Texas 77030

Stanley Fields
Departments of Genetics and Medicine
Markey Molecular Medicine Center
University of Washington
Seattle, Washington 98195

Russell L. Finley, Jr.
Center for Molecular Medicine and Genetics
Wayne State University School of Medicine
Detroit, Michigan 48201

Erica A. Golemis
Fox Chase Cancer Center
Philadelphia, Pennsylvania 19111

David Gunn
CLONTECH Laboratories, Inc.
Palo Alto, California 94303

Gregory J. Hannon
Cold Spring Harbor Laboratory
Cold Spring Harbor, New York 11724

Stanley M. Hollenberg
Vollum Institute
Oregon Health Sciences University
Portland, Oregon 97201-3098

Larn-Yuan Hwang
Departments of Microbiology and Pediatrics
The University of Texas Southwestern Medical Center
Dallas, Texas 75235-9140

Ying Jin
Departments of Microbiology and Pediatrics
The University of Texas Southwestern Medical Center
Dallas, Texas 75235-9140

Nils Johnsson
Max-Delbrück-Laboratorium
Köln, Germany

Jarema P. Kochan
Department of Metabolic Diseases
Hoffmann-La Roche, Inc.
Nutley, New Jersey 07110

Brian Kraemer
Department of Biochemistry
University of Wisconsin
Madison, Wisconsin 53706

Sailaja Kuchibhatla
CLONTECH Laboratories, Inc.
Palo Alto, California 94303

Manuel Lubinus
Department of Metabolic Diseases
Hoffmann-La Roche, Inc.
Nutley, New Jersey 07110

Gail Mandel
Department of Neurobiology and Behavior
State University of New York at Stony Brook
Stony Brook, New York 11794

Mark A. Osborne
Genome Therapeutics Corporation
Waltham, Massachusetts 02154

Bradley A. Ozenberger
American Cyanamid Company
Princeton, New Jersey 08543-0400

John A. Printen
Department of Chemistry
Institute of Molecular Biology
University of Oregon
Eugene, Oregon 97403-1229

Dhruba SenGupta
Departments of Genetics and Medicine
Markey Molecular Medicine Center
University of Washington
Seattle, Washington 98195

George F. Sprague, Jr.
Department of Biology
Institute of Molecular Biology
University of Oregon
Eugene, Oregon 97403-1229

Julia Tsou Tsan
Departments of Microbiology and Pediatrics
The University of Texas Southwestern Medical Center
Dallas, Texas 75235-9140

Alexander Varshavsky
Division of Biology
California Institute of Technology
Pasadena, California 91125

Marc Vidal
Massachusetts General Hospital Cancer Center
Charlestown, Massachusetts 02129

Anne B. Vojtek
Department of Biological Chemistry
The University of Michigan Medical School
Ann Arbor, Michigan 48109-0606

Zhuowei Wang
Departments of Microbiology and Pediatrics
The University of Texas Southwestern Medical Center
Dallas, Texas 75235-9140

Michael A. White
Department of Cell Biology and Neuroscience
University of Texas Southwestern Medical Center
Dallas, Texas 75235-9030

Marvin Wickens
Department of Biochemistry
University of Wisconsin
Madison, Wisconsin 53706

Meijia Yang
CuraGen Corporation
Branford, Connecticut 06405

Kathleen H. Young
American Cyanamid Company
Princeton, New Jersey 08543-0400

Beilin Zhang
Department of Biochemistry
University of Wisconsin
Madison, Wisconsin 53706

Li Zhu
CLONTECH Laboratories, Inc.
Palo Alto, California 94303

THE YEAST TWO-HYBRID SYSTEM

1

The Two-Hybrid System

A Personal View

Stanley Fields
Paul L. Bartel

Origins of the Method

The two-hybrid system dates to early 1987, when one of us (S.F.) was a new assistant professor at the State University of New York at Stony Brook with a small National Science Foundation grant. The university had a seed grant program to fund ideas with commercial potential, and obtaining such a grant struck us as more likely than obtaining another Federal grant. Unfortunately, the lab was working on pheromone response in the yeast *Saccharomyces cerevisiae*, in particular the role of a protein implicated in transcriptional induction, and the intricacies of yeast mating behavior seemed unlikely to excite the seed grant panel. However, our research interests kept us familiar with the current findings in transcriptional regulation. Specifically, we knew of two key results: one was the work of Brent and Ptashne (1985) which demonstrated that a hybrid transcriptional activator could be generated from the *Escherichia coli* LexA repressor and the yeast Gal4 protein; the second was the work of groups such as that of Steve McKnight (Triezenberg et al. 1988) suggesting that transcriptional activators could function by binding to DNA-bound proteins rather than directly to DNA. We toyed with various notions in the hope of linking yeast transcription to commercial potential. Late one afternoon, the idea came to use two different hybrid proteins, one containing a DNA-binding domain and one a transcriptional activation domain, to detect protein-protein interactions. Thus was born the two-hybrid system, not as an incremental step in our continuing studies but in an instant. Along with the basic idea came the immediate realization of its most important consequence: it might be feasible to construct libraries of activation domain hybrids and search them to identify interacting proteins.

We submitted a grant on this idea in March 1987, and began gathering the necessary reagents. It did not take long to learn that the grant would not be funded; the review panel rated the research as having no possibilities for commercial development. Nevertheless, it seemed too persuasive an idea to abandon, and we continued both our experimental and fund-raising efforts, eventually getting the assay to work and the Procter and Gamble Company to support us. This story can be viewed in either of two lights: the competition for grants works well because it drives the generation of good ideas, or the system does not work as well as it might since good ideas often are not funded. In any event, the history of the two-hybrid system is one of many examples in which small laboratories have made contributions; in the current granting climate, it seems more critical than ever to keep these kinds of labs operational.

The original two-hybrid experiments were based on several suppositions for which there was then no experimental support. Specifically, we assumed the following: many proteins, in addition to transcriptional activators, were capable of maintaining their structural integrity as hybrid proteins; the typical affinities of protein-protein interactions that would be studied in this system would be sufficient to reconstitute a transcriptional signal; activation domains would be accessible to the transcriptional machinery when present as hybrid proteins; and nonnuclear proteins could be targeted to and function within the nucleus. In addition, the view of transcriptional activation back then was simpler than it is now. If we had been aware of the need in this process for not only site-specific activators and the basal transcription factors, but all the additional complexity of TATA-binding protein-associated factors, mediator complexes, chromatin-affecting proteins and the like, we might not have considered it as likely that an idea so simple as the two-hybrid system was workable. In fact, we ran into considerable skepticism, not only from protein chemists but also from molecular biologists.

The first two-hybrid test, the combination of the yeast proteins Snf1p and Snf4p, was assayed in July 1988. Although the results were somewhat encouraging, the transcriptional response due to the protein-protein interaction was barely above background. We considered this result to be likely due to low expression of the hybrid proteins, but after spending several months to swap promoters, we failed to get any increase in β-galactosidase expression. It was only when, in early 1989, we obtained a yeast strain from Grace Gill in Mark Ptashne's lab—GGY1::171—that the same Snf1p and Snf4p constructions yielded a significant signal.

By the time one of us (P.L.B.) came to the lab that August to interview for a postdoc position, the initial two-hybrid experiments had been published (Fields and Song 1989). It was an easy decision to join the lab to continue work on this system, and specifically, to use this approach to screen libraries for interacting proteins. In late 1989, we began to collaborate with Rolf Sternglanz, a colleague working in a lab upstairs, and Cheng-ting Chien, who was then a graduate student with Rolf. They had used the two-hybrid system to detect the homodimerization of a yeast protein, Sir4p, which provided us with a test case for a library search. We developed a set of activation domain vectors that allowed us to produce fusions in all three reading frames, and we used these vectors to generate libraries from yeast genomic DNA. We screened the first library we had made for Sir4p-binding proteins, and from just over 200,000 transformants we identified two positives that required the presence of the Sir4p fusion protein for a *lacZ* signal. Fortunately, one of these positives encoded Sir4p, demonstrating that a library approach was feasible (Chien et al. 1991). We later determined that the other positive was, in fact, our first false positive (but not the last one!). Thanks to the efforts of a number of researchers, two-hybrid searches are much easier now than they were in those days.

In the following years, the lab turned some of its attention to p53, using the two-hybrid system to screen for its protein partners and to identify p53 mutants that had lost the ability to associate with SV40 large T antigen. Like many other labs, we began to explore other uses of the two-hybrid system, including the study of antibody-antigen and protein-peptide interactions. More recently, we have been engaged in developing global approaches to study protein-protein interactions and in generating a related system to study RNA-protein interactions.

Why Is the System Popular?

The two-hybrid system addresses one of life's fundamental questions: how to find a meaningful partner. If a protein has a known function, new proteins that bind to it bring additional components into play, ultimately contributing to understanding the process under study. Alternatively, a protein's function may be obscure but the protein may be of obvious relevance; for example, its gene may be mutated in human disease. In this case, partners with known roles may turn up and provide some essential clues. Thus the method is a tractable and rapid form of genetics for organisms, for example mammals, that cannot be readily manipulated, and it can accomplish some of the tasks that suppressor screens and similar genetic strategies can do in simple organisms.

Among its strongest features, the two-hybrid system has the virtue of being easy to perform. Once the essential steps of putting plasmids into yeast and getting them back out again are learned, and distinguishing true positives from false ones, the method is no more difficult than other routine procedures in molecular biology. This simplicity means that once one protein has been successfully used in a search, it is possible to screen many more. In this sense, the method is largely insensitive to the individual properties of proteins that make them unique and interesting.

The popularity of the method owes tremendously to the contributions of labs such as those of Roger Brent, Steve Elledge, Dan Nathans, Richard Treisman, and Hal Weintraub, which early on began building vectors, libraries, and reporter strains (some of which are described in chapters 2, 3, and 4). The willingness of these and other labs to freely provide these reagents meant that bugs in the procedure were worked out fairly quickly, new innovations came into play, numerous combinations of proteins were tested, and diverse proteins were used in searches. In addition to the yeast community, investigators in some fields such as signal transduction and cell cycle control (chapters 10 and 11) quickly adopted the technology and spread it to nearby labs. A few early successes, including searches with the retinoblastoma protein (Durfee et al. 1993), the HIV gag protein (Luban et al. 1993), Ras (Vojtek et al. 1993), and cyclin-dependent kinases (Harper et al. 1993; Hannon et al. 1993; Gyuris et al. 1993) gave support to the idea that this could be a general method. The increasing availability of activation domain libraries (chapter 5) also facilitated spread of the technology.

Unintended Consequences

While the possibility of carrying out library searches may have been implicit in the original idea, what we did not forsee was the potential for so many permutations and variations on the two-hybrid theme. The advent of one-hybrid systems (chapter 17) brought similar methodology to the analysis of protein-DNA interactions as was becoming available for protein-protein interactions; more recently, a three-hybrid system (chapter 18) was developed that may have comparable uses in the analysis of protein-RNA interactions. Specific protein-protein interactions could be analyzed to identify mutations that affected binding (chapters 7 and 8); in other experiments, such mutations were correlated with structural information (chapter 6). The principle of using hybrid proteins to detect interaction was shown not to be limited to transcriptional activators; ubiquitin, for example, could also be split into two domains for reconstitution (chapter 19). Initial limitations to the two-hybrid assay could be circumvented by mammalian-based systems (chapter 13), the presence of a small molecule (chapter 15), or the addition of a protein modifying activity (chapter 14). The assay proved to be amenable, as well, to certain normally extracellular proteins (chapter 9) and to peptides (chapter 16). Finally, the ability to apply the two-hybrid method on a genome-wide scale, initially for a small genome like that of bacteriophage T7 (Bartel et al. 1996) but potentially for much larger ones, means that whole complexes and pathways may be amenable to this approach (chapter 12).

Phil Hieter has pointed out that yeast technologies, like yeast artificial chromosome (YAC) construction and the two-hybrid system, have brought labs working on all kinds of biological problems into contact with labs working on yeast. While the initial contact is for technical advice, a side effect is the exchange of the two labs' respective research ideas, sometimes resulting in collaborations. It is a delight to us to see that the two-hybrid system has brought together more than proteins.

References

Bartel, P.L., Roecklein, J.A., SenGupta, D., and Fields, S. (1996). A protein linkage map of *Escherichia coli* bacteriophage T7. Nat. Genet. 12:72–77.

Brent, R., and Ptashne, M. (1985). A eukaryotic transcriptional activator bearing the DNA specificity of a prokaryotic repressor. Cell 43:729–736.

Chien, C.-T., Bartel, P.L., Sternglanz, R., and Fields, S. (1991). The two-hybrid system: A method to identify and clone genes for proteins that interact with a protein of interest. Proc. Natl. Acad. Sci. USA 88:9578–9582.

Durfee, T., Becherer, K., Chen, R.-L., Yeh, S.H., Yang, Y., Kilburn, A.E., Lee, W.H., and Elledge, S.J. (1993). The retinoblastoma protein associates with the protein phosphatase type 1 catalytic subunit. Genes Dev. 7:555–569.

Fields, S., and Song, O.-K. (1989). A novel genetic system to detect protein-protein interactions. Nature 340:245–246.

Gyuris, J., Golemis, E., Chertkov, H., and Brent, R. (1993). Cdi1, a human G1 and S phase protein phosphatase that associates with Cdk2. Cell 75:791–803.

Hannon, G.J., Demetrick, D., and Beach, D. (1993). Isolation of the Rb-related p130 through its interaction with CDK2 and cyclins. Genes Dev. 7:2378–2391.

Harper, J.W., Adami, G.R., Wei, N., Keyomarsi, K., and Elledge, S.J. (1993). The p21 Cdk-interacting protein Cip1 is a potent inhibitor of G1 cyclin-dependent kinases. Cell 75:805–816.

Luban, J., Bossolt, K.L., Franke, E.K., Kalpana, G.V., and Goff, S.P. (1993). Human immunodeficiency virus type 1 Gag protein binds to cyclophilins A and B. Cell 73:1067–1078.

Triezenberg, S.J., Kingsbury, R.C., and McKnight, S.L. (1988). Functional dissection of VP16, the trans-activator of herpes simplex virus immediate early gene expression. Genes Dev. 2:718–729.

Vojtek, A.B., Hollenberg, S.M., and Cooper, J.A. (1993). Mammalian Ras interacts directly with the serine/threonine kinase Raf. Cell 74:205–214.

Part I

THE TWO-HYBRID SCREEN

Undoubtedly, the most valuable use of the two-hybrid system has been to identify protein partners for a protein of interest. Even without any specific knowledge of where or how a particular protein acts, a researcher can find its partners and potentially use this information to learn more about how that protein might function. In the following part, members of three of the labs that early on developed many of the vectors and strains that led to widespread use of the two-hybrid system describe their particular reagents and strategies. In chapter 2, Bai and Elledge describe vectors and strains based on the use of the *Saccharomyces cerevisiae* Gal4p DNA-binding and activation domains and in chapter 3, Vojtek, Cooper, and Hollenberg describe reagents for the use of the *Escherichia coli* LexA protein as the DNA-binding domain and the herpes virus VP16 protein as the activation domain. The activation domain vectors for the Gal4p- and VP16-based systems just described carry the same yeast nutritional marker, such that activation domain libraries created for either system are compatible with the other. A third system, described by Golemis and Brent in chapter 4, uses the *E. coli* LexA protein as DNA-binding domain and the *E. coli* B42 sequence as its activation domain, and this system features a number of useful reporter genes of different sensitivities.

The vectors, strains, and protocols required to conduct a two-hybrid search are now readily available, either commercially or from the labs that developed the reagents. As successful searches can be conducted using any of these and other systems, choice of reagents may depend on the particular advantages of that system or the availability of an appropriate activation domain library. Additionally, numerous activation domain libraries are available commercially and from individual researchers. However, in some cases, an appropriate library may not be available and must be constructed. In chapter 5, Zhu, Gunn, and Kuchibhatia present protocols for the construction of high-complexity cDNA libraries in activation domain vectors.

As Bai and Elledge point out, a protocol is never fully optimized. With this in mind, we have retained each group's version of some protocols (for example, for large-scale yeast transformation) and invite you to select the one that works best for you.

2

Searching for Interacting Proteins with the Two-Hybrid System I

Chang Bai
Stephen J. Elledge

Protein-protein interactions have attracted much attention because they form the basis of a wide variety of biochemical reactions. The identification of proteins that interact with a known protein is an essential aspect of the elucidation of the regulation and function of that protein. This interest has stimulated the development of a number of biochemical and genetic approaches to identify and clone genes encoding interacting proteins including coimmunoprecipitation, copurification, cross-linking, and direct expression library screening using proteins as probes. However, the development of the yeast two-hybrid system appears to have had the greatest impact on interaction cloning methodology.

The yeast two-hybrid system was devised to identify genes encoding proteins that physically associate with a given protein in vivo. This is a versatile and powerful method that is applicable to most, if not all, proteins once their genes have been isolated. In contrast to biochemical methods detecting protein-protein interaction, this system is based on a yeast genetic assay in which the interaction of two proteins is measured by the reconstitution of a functional transcription activator in yeast (Fields and Song 1989; Chien et al. 1991). This method not only allows identification of proteins that interact, but also can be used to define and/or test the domains/residues necessary for the interaction of two proteins (Li and Fields 1993). Since its development, a large number of genes from a variety of studies have been identified using this method, including many cell cycle regulators that have contributed significantly to our understanding of the eukaryotic cell cycle (Harper et al. 1993; Toyoshima and Hunter 1994; Fields and Sternglanz 1994; Allen et al. 1995; Serrano et al. 1993; Matsuoka et al. 1995; Bai et al. 1996; chapter 11).

The purpose of this chapter is to provide a practical guide to the yeast two-hybrid system emphasizing recent improvements of this method. There are several different designs for the actual hybrid proteins and selection utilized in two-hybrid screens. This chapter will focus on derivatives of the Gal4p-based system (for a description of LexA-based systems see Gyuris et al. 1993; Vojtek et al. 1993; and chapters 3 and 4). Due to the limitation of space, general methods of yeast genetics, for which several comprehensive manuals are available (Ausubel et al. 1994; Guthrie and Fink 1991) will not be covered in detail in this chapter. As with all protocols, the one presented here is not completely optimized and the readers are encouraged to try alternative methods to improve performance.

The basis of the two-hybrid system relies on the structure of particular transcription factors that have two physically separable domains: a DNA-binding domain and a transcription activation domain. The DNA-binding domain serves to target the transcription factor to specific promoter sequences (designated UAS for upstream activation sequence) whereas the activation domain serves to facilitate assembly of the transcription complex allowing the initiation of transcription. The fact that a functional transcription factor can be reconstituted through noncovalent interaction of two independent hybrid proteins containing either a DNA-binding domain or an activation domain constitutes the basis of the two-hybrid approach (Fields and Song 1989). The hybrid proteins are normally transcriptionally inactive alone or when coexpressed with a noninteracting hybrid protein. However, if when coexpressed they associate via the interaction between the two fusion protein partners, they become active causing the expression of a reporter gene driven by the specific UAS for the DNA-binding domain (Fields and Song 1989). A schematic representation of the two-hybrid genetic selection by activation of transcription of *HIS3* and *lacZ* genes is shown in Figure 2-1. The plasmid encoding the fusion protein with the Gal4p DNA-binding domain is often referred to as the bait plasmid.

Since the emergence of the two-hybrid approach in 1989, a number of improvements have been incorporated into this system that have greatly increased its usefulness, several of which are listed below. The most significant improvement has been the conversion of the assay from a color-based screen to a nutritional selection via inclusion of a gene such as *HIS3* as a transcription reporter in addition to *lacZ* (Durfee et al. 1993). This allows much larger libraries to be efficiently screened on many fewer plates. New methods for the elimination of false positives have also had a large impact

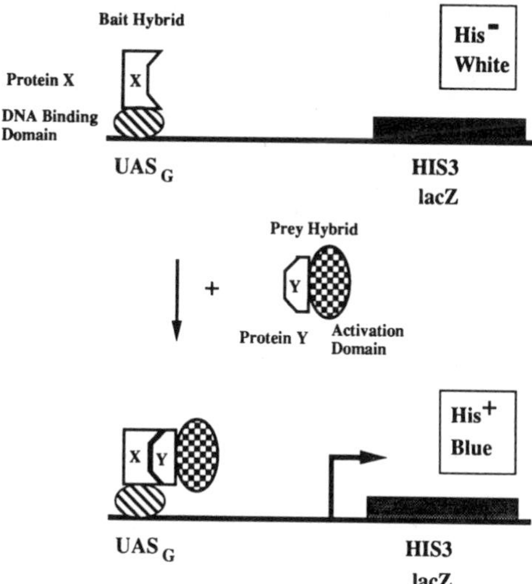

Fig. 2-1. A schematic representation of the two-hybrid system. Two hybrid proteins are generated. The bait hybrid consists of protein X fused to a DNA-binding domain while the prey hybrid consists of protein Y fused to an activation domain. Neither of these alone is able to activate transcription of *lacZ* or *HIS3*. However, if interaction of protein X and Y occurs, a functional transcription activator is generated and results in the transcription of reporter genes that confer the His^+ and blue phenotype on the host cells.

on the efficacy of the system. The mechanistic basis for false positives is not understood, but they are operationally defined as clones that activate multiple unrelated bait proteins. This has served as the basis of a mating assay to eliminate these clones as described in the following pages (Harper et al. 1993). Improvements in testing for specificity were also accomplished by the simultaneous use of two different reporters that share minimal sequence overlap in their promoters (Durfee et al. 1993; Bartel et al. 1993). Finally, advances have been made in the generation of large representative cDNA libraries by the development of highly efficient lambda cloning vectors, λACT (Durfee et al. 1993) and λACT2, that also allow direct conversion into plasmid via cre-lox site-specific recombination. λACT2 (S. Elledge, unpublished data) has an epitope tag and allows directional libraries using unique *Eco*RI and *Xho*I restriction sites. A commercially available kit exists for preparing *Eco*RI/*Xho*I cDNA (Stratagene).

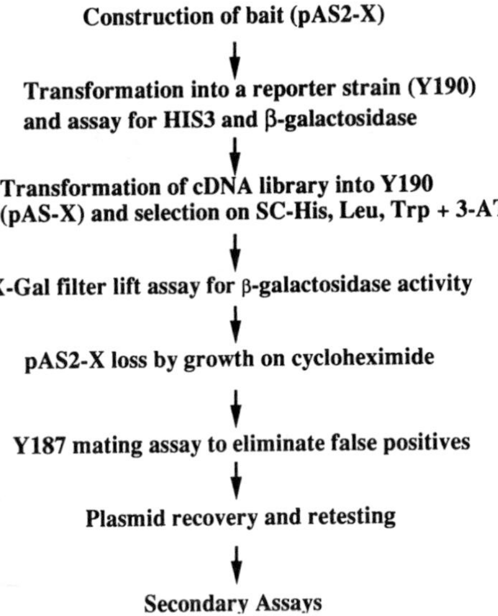

Fig. 2-2. A flow chart describing the steps employed in a typical two-hybrid screen and false positive detection assays.

Key Reagents

A flow chart for cloning by the two-hybrid system is outlined in Figure 2-2. A search for interactions with a known protein requires a yeast reporter strain, a DNA-binding domain vector, and a cDNA library in an activation domain vector, as well as yeast media, basic yeast manipulation techniques, and some patience.

Vectors

A number of different DNA-binding domain and transcription activation domain vectors have been successfully employed in this system (Chien et al. 1991; Durfee et al. 1993; Bartel et al. 1993; Gyuris et al. 1993; Vojtek et al. 1993; Fields 1993; Dalton and Treisman 1992). The most extensively used vectors are Gal4p-based. Examples are the Gal4p DNA-binding domain vectors pAS1 and pAS2 (see restriction map in Figure 2-3A) and the Gal4p activation domain vectors pACT1 and pACT2 (Figure 2-3B). However, it is possible to combine vectors and strains from different sources as long as they are compatible with each other with regard to the selectable markers and the UAS for the reporters.

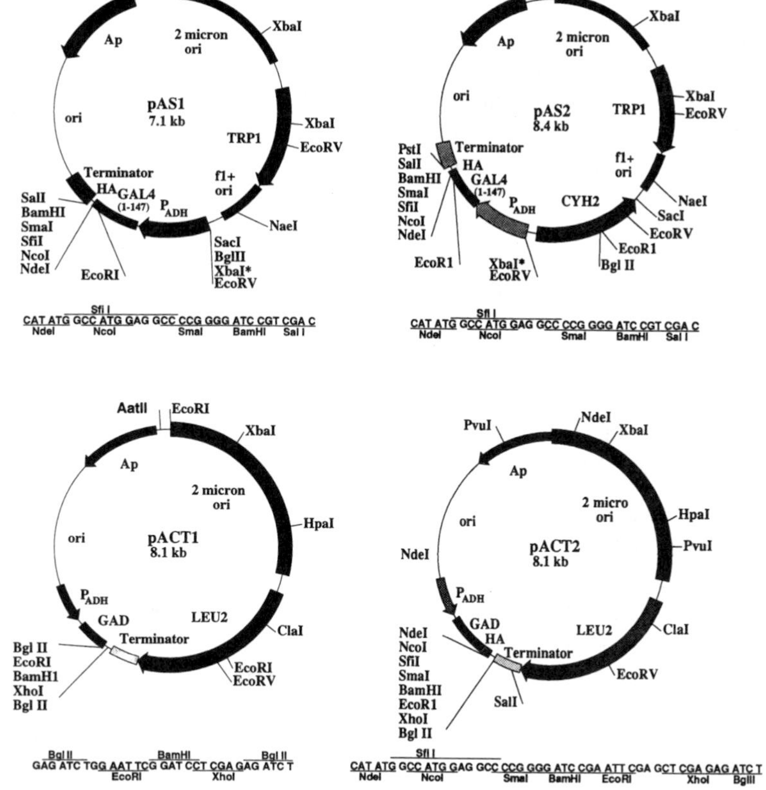

Fig. 2-3. The restriction maps of DNA-binding domain vectors pAS1 and pAS2, and activation domain vectors pACT1 and pACT2. The reading frames of the polylinkers are shown in coding triplets. *denotes methylation and blocks cutting. GAD: *G*al4p *a*ctivation *d*omain.

Strains

The reporter strain must contain a reporter gene that is under the control of the UAS corresponding to the DNA-binding domain vector. It should also carry nonreverting mutations in the genes corresponding to those present on both two-hybrid vectors. A number of reporter strains are available in which the bacterial gene *lacZ* is the reporter. Some strains, for example, Y190, also carry the yeast biosynthetic gene *HIS3* under the control of the *GAL* UAS so that the selection for interacting clones can be carried out by both nutritional selection for histidine and an assay for ß-galactosidase activity. Y190 has the genotype *MATa gal4 gal80 his3 trp1-901 ade2-101 ura3-52 leu2-3,-112 URA3::GAL-lacZ, LYS2::GAL(UAS)-HIS3 cyh.r*

cDNA Libraries

The most commonly used two-hybrid cDNA libraries are made as Gal4p transcription activation domain fusion libraries. A large number of libraries from different tissues and organisms have been published and several more are available commercially, but it may be necessary to construct your own library. A description of cDNA library construction is beyond the scope of this chapter, but has been covered previously (Mulligan and Elledge, 1994 and see chapter 5). Theoretically, any activation domain library can be used, but one must keep in mind that *GAL1-10* (the UAS region between the *GAL1* and *GAL10* genes) promoter directed libraries cannot be used in a system based on the Gal4p DNA-binding domain.

Establishing a Reporter Strain Expressing the Gal4p DNA-Binding Domain Fused to the Gene of Interest

First, the gene you wish to examine must be subcloned into pAS2 or a similar Gal4p DNA-binding domain fusion vector. Care must be taken to maintain the proper reading frame to create a hybrid protein. Often, an in-frame fusion can be generated by incorporation of an adapter oligonucleotide or by PCR. DNA sequencing or Western analysis can be used to verify that the proper fusion has been generated. The sequencing primer we used for pAS is 5'GAA TTC ATG GCT TAC CCA TAC. The pAS vectors also encode a hemagglutinin epitope tag (HA) that allows the protein, if present in sufficient amounts, to be visualized with commercially available antibodies. This construct (pAS-X) must be introduced into a reporter strain such as Y190 by transformation and nutritional selection using synthetic complete (SC) media lacking the appropriate amino acid. Mini-prep plasmid DNA is sufficient for this purpose.

Regents and Media

LiOAC refers to lithium acetate

LiAcTE=100 mM LiOAC, 10 mM Tris pH 8, 1 mM EDTA. Autoclave.

LiSORB=100 mM LiOAC, 10 mM Tris pH 8, 1 mM EDTA, 1 M Sorbitol. Autoclave.

70% PEG=70% *p*oly*e*thylene *g*lycol (MW 3350) in water. Can be briefly heated in a microwave to help the PEG go into solution. Store in a tightly sealed container to prevent evaporation after autoclaving.

YEPD= 10 g *y*east *e*xtract, 20 g *p*eptone, 20 g *d*extrose per 1 liter. Autoclave.

SC = synthetic media used for selection for yeast containing plasmids with specific nutritional markers. SC lacking a particular amino acid is referred to as dropout media.

10x YNB= 67 g *y*east *n*itrogen *b*ase without amino acids in 1 liter water, filter-sterilized. Store in the dark.

Dropout mixture components:

adenine	800 mg	arginine	800 mg
aspartic acid	4000 mg	histidine	800 mg
leucine	2400 mg	lysine	1200 mg
methionine	800 mg	phenylalanine	2000 mg
threonine	8000 mg	tryptophan	800 mg
tyrosine	1200 mg	uracil	800 mg

To make a dropout mixture, simply weigh out the different components leaving out the ones you are going to select for, combine them, and grind into fine powder with a mortar and pestle.

SC-Trp plates = 870 mg dropout mixture (minus tryptophan), 20 g dextrose, 1 ml 1N NaOH, 20 g agar, add water to 900 ml, autoclave. Add 100 ml 10x YNB just prior to pouring plates.

Procedure for Small-Scale Yeast Transformation

1. Inoculate Y190 in 10 ml YEPD and grow to OD600=1.
2. Wash with water and resuspend in 200 μl LiOAcTE, incubate at 30°C for 1 hour.
3. Add ~1 μg pAS-X DNA to 100 μl cells and 100 μl 70% PEG, incubate at 30°C for 1 hour.
4. Heat shock at 42°C for 5 minutes, add 1.2 ml water, pellet by spinning for 5 seconds in an eppendorf centrifuge.
5. Resuspend in 100 μl water and plate on a SC-Trp plate and incubate at 30°C for 3 days.

You can follow any of the commonly used yeast transformation protocols. The transformation mixture should be plated on SC-Trp plates in the case of pAS2. Colonies should be visible after about 48 hours incubation at 30°C. It is always advisable to include a no DNA control during the transformation to test the reversion frequency of the yeast selectable marker or the presence of contaminants. It should be noted that a few independent transformants need to be picked, and the expression of the fusion protein should be verified by Western blotting (with anti-HA antibodies which are commercially available if specific antisera are not available for your protein), since reversion and gene conversions do occur in the process of transformation. Occasionally, a functional fusion is made and cannot be detected with anti-HA. In this case, you may at least verify the presence of your plasmid in the strain by recovery into *Escherichia coli*, Southern blotting, or the polymerase chain reaction (PCR) with primers specific to the gene of your interest.

Checking the Feasibility of Using a Two-Hybrid Screen for Your Protein

Some proteins that are not transcription factors themselves activate reporter gene transcription when fused to the Gal4p DNA-binding domain. This activation will severely interfere with the two-hybrid system. Prior to beginning a screen, you must check transformants carrying the bait protein for *lacZ* activation and growth properties on SC-His plates containing differing concentrations of 3-AT (3-amino-1,2,4-triazole, SIGMA, A8056). 3-AT, an inhibitor of *HIS3*-encoded IGP-dehydratase, is used because the basal level of *HIS3* expression from the reporter construct, in which the *HIS3* UAS sequences have been replaced by the *GAL1-10* UAS, is sufficient to allow growth on SC-His media. We have found that 3-AT concentrations of 25 to 50 mM are typically sufficient to select against the growth of strains bearing pAS2 subclones that fail to activate transcription on their own. If your construct fails to activate transcription, like most fusions, you may proceed to the library transformation step. If it activates transcription alone (turns blue in the X-Gal assay or grows on SC-His + 50 mM 3-AT), it cannot be used in this version of the two-hybrid system. Usually the X-Gal assay is more sensitive (Breeden and Nasmyth 1985).

Reagents and Media

SC-His, Trp + 3-AT plates: 3-AT is dissolved in sterile water as a 2.5 M stock, heated to 50°C to get into solution if necessary, filter sterilized, and added to SC-His,Trp media just prior to pouring plates at a concentration of 50 mM. These plates should be thick (ten 15 cm plates per liter media).

Z Buffer = Na_2HPO_4 $7H_2O$ 16.1 g, NaH_2PO_4 H2O 5.5 g, KCl 0.75 g,

$MgSO_4$ $7H_2O$ 0.25 g, adjust pH to 7.0, add water to 1 liter, autoclave. 2.7 ml ß-mercaptoethanol should be added just before use.

X-Gal = 5-bromo-4-chloro-3-indolyl-ß-D-galactoside dissolved in DMF (N, N-dimethyl formamide) at 100 mg/ml as a stock. Stored at −20°C in the dark.

The 3-AT Histidine Prototrophy Assay

Y190 (pAS-X) colonies are streaked onto SC-Trp, His + 3-AT plates and incubated at 30°C for 72 hours to examine their ability to form colonies. It is advisable to include both positive and negative controls at this point.

The X-Gal Colony Filter Assay to Detect ß-Galactosidase Activity

1. Label Schleicher and Schuell BA85 45 μm circular nitrocellulose filters (cat. # 20440) or nytran filters (cat. # D868/4)* with a ball-point pen.
2. Lay the filter onto the plate of yeast colonies** and allow it to wet completely. Place asymmetric orientation markers on filters and plates with India ink and a needle.
3. Lift the filter off of the plate carefully to avoid smearing the colonies and place the filter in liquid nitrogen to permeabilize the cells. Five to 10 seconds is sufficient. Filters can either be submerged in the liquid nitrogen alone or be placed on an aluminum foil float and submerged, which minimizes handling.

*We observe that nitrocellulose filters become more brittle in liquid nitrogen than nytran filters. However, nitrocellulose filters adhere to colonies slightly better than nytran. Filters can be thoroughly washed with deionized water and reused.

**It is best to have at least medium size colonies with which to perform this assay.

4. *Carefully* remove the filters from the liquid nitrogen (frozen nitrocellulose filters become very brittle), allow to thaw, and place cell side up in a petri dish that contains 3MM chromatography paper circles soaked with 0.30 ml/square inch of Z Buffer containing 1 mg/ml X-Gal.
5. Incubate at 30°C for minutes to overnight for the development of color.

Library Transformation and Selection

If the Y190 reporter strain expressing your protein of interest does not activate *HIS3* or *lacZ*, this strain can now be used for screening an activation domain cDNA library. The library is introduced into the strain by transformation. Each transformant presumably contains an individual library plasmid in addition to your bait plasmid (pAS-X). Cells containing a library plasmid encoding a protein interacting with your bait will be selected on SC-His plates and assayed for ß-galactosidase activities. Since only one out of three (for directional cDNA libraries) or one out of six cDNAs are in the correct reading frame with the activation domain, a large number of independent yeast transformants need to be screened. We typically screen between 1 and 10 million colonies. The transformation procedures must be carefully followed to ensure a high transformation efficiency. The following protocols should give greater than 1 x 10^6 transformants with 50 µg library plasmid.

Reagents

LiAcTE
LiSORB
LiAcPEG = LiAcTE containing 40% polyethylene glycol (MW 3350).
Carrier ssDNA or total yeast RNA at 20 mg/ml
SC-Trp liquid media
YEPD liquid media
SC-Trp, Leu, His liquid media
SC-Trp, Leu, His + 3-AT = 15 cm dropout plates (minus tryptophan, leucine and histidine) with the addition of 25 mM or 50 mM 3-AT.

Procedure

1. Grow the recipient strain, Y190 (pAS-X), to mid-log (1×10^7 cells/ml) in SC-Trp.
2. Determine the OD_{600} of the above culture and inoculate 1 liter YEPD such that the cell density becomes 1×10^7 cells/ml (OD_{600} = 0.5 to 0.8) in two generations.*

*SC-Trp media is used to select for pAS-X but YEPD gives the best transformation efficiencies. The doubling time for Y190 (pAS-X) in YEPD varies between 2.0 and 3.5 hours depending on the plasmid it carries.

3. Pellet the cells and resuspend the pellets in LiAcTE; the volume is not critical because this is a wash step.
4. Pellet the cells once more and resuspend the cells in 5 ml of LiSORB for every 200 ml of starting culture.
5. Incubate the cells for 30 minutes at 30°C with shaking.
6. Pellet the cells as above and resuspend in 500 μl of LiSORB per 200 ml of culture. (We usually do large-scale transformations with a liter of cells so the final volume is 2.5 ml.)
7. After removing 100 μl of cells for a negative control, add 2 μg of pACT library DNA and 200 μg of yeast total RNA or ss DNA** for every 100 μl of cells remaining. (50 μg DNA and 5 mg of total RNA carrier for 1 liter worth of cells.)
8. Mix well, then incubate for 10 minutes at 30°C without shaking.
9. Add 900 μl of LiAcPEG for each 100 μl of cells and mix well. (This is 22.5 ml for the large scale.) We usually do this in 125 ml or 250 ml flasks. This aids the heat shock step later.
10. Heat shock the tubes or the flask in a 42°C water bath for 12 minutes.
11. At this point, cells can be plated out to check the transformation frequency. Five μl on a SC-Trp,Leu plate should give 1000 or more transformants with a good transformation. However, in order to plate a large number of cells per plate, we generally take our transformation mixture and add it to 500 ml of SC-Trp,Leu,His media for an initial 1 liter culture and allow it to recover at 30°C for 4 hours***. At this point, cells have been established as transformants. They can be pelleted, and resuspended in 6 mls of SC-His,Trp,Leu liquid media and plated 300 μl per 15 cm plate (SC-His,Trp,Leu + 3-AT) or frozen at −80°C with addition of DMSO to 9% final concentration and stored for future use. Storing small aliquots allows you to thaw them and plate them out later at an appropriate density. It is important not to have extra histidine around during plating since it interferes with the 3-AT selection.

Screening Using the X-Gal ß-Galactosidase Assay

Colonies that grow after 4 to 5 days (or longer: we have waited up to 10 days in some cases) are then tested for ß-galactosidase activity using the X-Gal colony filter assay (see earlier subsection). Blue colonies are taken for further study and they can often be recovered directly from the filters. However, when this does not work, they can be recovered from the original plate. All positives should be struck out to single colonies and retested for

**We have found that total yeast RNA works more reproducibly as a carrier but it is more work to prepare than DNA. Protocols for preparing ssDNA and yeast RNA for transformation can be found in Gietz and Woods 1994 and Kohrer and Domdey 1991, respectively.

***Cells in PEG are more fragile and often die when pelleted, so the recovery step is useful. Plating directly from the PEG without the recovery step also works but is more messy. Cells lose less than half their viability when frozen and can be stored indefinitely. This is useful because they can be thawed and plated at an optimal density for screening/selection at your leisure.

ß-galactosidase activity. In the case that only a few colonies formed from your library selections on SC-His,Trp,Leu + 3-AT plates, you might want to pick all these colonies and streak on a single plate before performing the X-Gal assay to save filters and labor. In case a larger number of plates are used, filters can be washed and reused.

The *HIS3*/3-AT selection sometimes functions as a tight selection but more often like an enrichment, depending on the bait employed. We often see many microcolonies on the original selection plates, although it has been reported that inclusion of 3-AT in the recovery step eliminates these. Occasionally, these microcolonies are 1% of the total Leu+Trp+ colonies. In most cases, true positives continue to grow into large colonies while the microcolonies stop growing. The secondary X-Gal screen eliminates the vast majority of the microcolonies. The majority of blue colonies are the large His+ colonies that grow out. An enrichment of 100-fold is very useful because it allows you to screen 100 times as many colonies on a single plate so that large libraries can be screened in only 20 large plates. We have also developed a *GAL-URA3* selection system that requires higher levels of two-hybrid activated transcription than the His selection. That strain, Y166, is available upon request.

We typically use pSE1111 (Snf4p fused to the activation domain in pACT) and pSE1112 (Snf1p fused to the DNA-binding domain of Gal4p in pAS1) as a positive control for our X-Gal and 3-AT resistance assays. It should be noted that pAS2 alone can activate *lacZ* weakly and is not a good negative control. pSE1112 is a better negative control. The weak activation observed with pAS2 may be due to residual transcriptional activation activity of amino acids 1–147 of the Gal4 protein used in pAS1 and pAS2 (S. Johnston, personal communication). This weak activation capacity appears to be eliminated when genes are cloned into pAS1 and pAS2.

Eliminating False Positives

The mechanistic basis of false positives is not well understood. A few false positives activate reporter transcription by themselves, but most require the DNA-binding domain plasmid for their activities. These false positives might result from nonspecific interaction with the DNA-binding domain hybrid or from interaction with the DNA or DNA-bound proteins at a particular promoter. Dual selection systems (Durfee et al. 1993; Finley and Brent 1994), the use of two different Gal4p-dependent promoters and reporters, for example *lacZ* and *HIS3* in Y190, with little sequence overlap decrease the frequency of these false positives. The best way to determine whether a particular library plasmid encodes a bona fide positive is to test its ability to activate in the presence of the original bait and several other unrelated bait plasmids as a negative control. This can be very cumbersome, requiring the recovery of many plasmids and transformation into several strains. A genetic assay has been developed to facilitate this process

(Harper et al. 1993). The basis of this assay is the loss of the original bait plasmid, leaving a strain containing only the library plasmid. The plasmids used in the two-hybrid system are lost at a low frequency during growth in the absence of nutritional selection. The mating strain can be made by screening for loss of the pAS derivative by replica-plating for inability to grow on SC-Trp or by the use of a negative selectable marker incorporated into pAS2. The strain Y190 is resistant to cycloheximide (2.5 µg/ml) due to a mutation in the *CYH2* gene. This is a recessive drug resistance marker. When a plasmid, for example pAS2, carrying the wild type *CYH2* gene is in the strain, cells become sensitive to cycloheximide. If you begin with pAS2-X, it is a good idea to streak the colonies out first on SC-Leu before streaking on cycloheximide media to allow plasmid loss and dilution of the *CYH2* gene product. Loss of the pAS2 derivative can be achieved by streaking colonies onto SC-Leu plates containing 2.5 mg/ml cycloheximide. The colonies that grow should be Trp-, but they should be checked for loss of the TRP1 marker, just to be safe, to avoid *CYH2* gene conversion events. This plasmid loss allows one to check for plasmid dependency of *lacZ* activation as well as generating a strain that contains only the library plasmid, facilitating plasmid recovery into bacteria (see next section). This strain is mated to strains of the opposite mating type (Y187) that harbor pAS2 clones containing unrelated fusions. Since the library plasmid-containing strain is Leu+Trp- and the tester strains are Leu-Trp+, the diploids resulting from the mating can be selected for on SC-Trp,-Leu plates. The genotype of Y187 is *MATα gal4 gal80 his3 trp1-901 ade2-101 ura3-52 leu2-3,-112 URA3::GAL-lacZ*. The unrelated fusions we generally use are in pAS1 or pAS2(Leu-Trp+) and include Cdk2, Snf1p(pSE1112), lamin, and p53 (a gift of Stan Fields), although you may want to use your own set of unrelated fusions. The resulting diploids, selected by growth on SC-Trp,Leu, can be immediately tested for ß-galactosidase activity in the filter screen assay. Colonies that activate *lacZ* expression significantly above background levels (pSE1112 alone) are likely to contain library plasmids encoding false positives that nonspecifically activate in the presence of your fusion and should, therefore, be discarded.

Media and Reagents

SC-Leu plates and liquid media

SC-Leu*Cyh* = SC-Leu plus 2.5 µg/ml cycloheximide plates. Cycloheximide (dissolved in water and filter sterilized) should be added just prior to pouring plates.

SC-Trp plates

YEPD plates

SC-Trp, Leu plates

Replica-plating velvets

Procedure for Selecting the Loss of the DNA-Binding Domain Plasmid

1. Streak your positive clones on an SC-Leu plate and incubate for 48 hours (optional).
2. Pick colonies from above plates and streak them on SC-Leu *Cyh* and incubate until colonies appear.
3. Replica-plate to SC-Trp and SC-Leu plate and incubate for 48 hours (optional).
4. Colonies that grow on SC-Leu only are to be picked for further analysis. Clones that activate *lacZ* by themselves are false positives and should be eliminated.

Procedure for the Mating Assay to Eliminate False Positives

1. Make a series of wide parallel streaks of individual Y190 derivatives from above in a straight line crossing the length of a SC-Leu plate. Six evenly spaced streaks can be accommodated on a single plate.
2. Do the same for the false positive detector Y187 derivatives on an SC-Trp plate.
3. Replica-plate the above two plates to a common YEPD plate at 90° angles so that cells from SC-Leu plate crossover with cells from SC-Trp, and incubate 4 hours to overnight to allow mating.
4. Replica-plate to a SC-Trp, Leu plate and incubate for 2 to 3 days. Colonies of diploid cells will grow in patches at the intersection of the two streaks. Care must be taken to properly label the mating plates to ensure that there is no confusion in the identity of the streak. Replica-plating can be confusing.
5. Perform the X-Gal filter assay as described earlier.

The more specificity fusions with which you test your positives, the more you can trust the significance of your positives. We encourage you to gather your colleagues' bait constructs as your false positive detectors. The above strains eliminate many but not all false positives.

Recovery of Plasmids from Yeast into *E. coli*

Once putative positive two-hybrid clones are obtained, the library plasmid can be recovered through bacterial transformation using DNA isolated from these clones.

Reagents

Yeast lysis solution: 300 mM NaCl, 10 mM Tris, 1 mM EDTA, pH 8, 0.1% SDS.
PCI: 49% Phenol + 49% $CHCl_3$ + 2% iso-amyl alchohol. Phenol saturated with 10 mM Tris, 100 mM NaCl, 1 mM EDTA, pH 8.
CHCl3.
TE: 10 mM Tris pH 8, 1 mM EDTA.
Acid washed glass beads 0.45 μm (Sigma # G8772).

Procedure for Small-Scale DNA Isolation from Yeast

1. Spin down 1 ml culture of Y190 (containing library plasmids) grown in SC-Leu or scrape a few colonies from an SC-Leu plate into an eppendorf tube.
2. Add 200 µl yeast lysis solution, vortex to resuspend cells.
3. Add glass beads until the total volume reaches 400 µl, vortex vigorously for 1 min.
4. Add 200 µl PCI and vortex vigorously for 1 min.
5. Spin for 5 min. in an eppendorf centrifuge.
6. Recover the aqueous phase and repeat steps 4 and 5 for the aqueous phase.
7. Extract once more with 200 µl $CHCl_3$.
8. Precipitate the DNA by addition of 2 volumes of ethanol and place at room temperature for 15 min.
9. Spin for 10 min. in an eppendorf centifuge at room temperature and wash the DNA pellet with 1 ml of 70% ethanol.
10. Dissolve DNA in 50 µl TE.

The plasmid can be easily recovered using several of the common *E. coli* strains with ampicillin selection. Electroporation gives us consistently high efficiency of transformation. Five µl of the above mini-prep DNA should give you hundreds of colonies. If you are not using the pAS2 cyclohexemide selection for plasmid loss and you are attempting to selectively recover the library plasmid that also has the bait vector, you may want to transform into a *leuB- E. coli* strain such as JA226. The yeast *LEU2* gene can complement the *leuB* mutation in *E. coli.* The selection must then be performed on *E. coli* minimal medium lacking leucine.

It is important to note that the library transformation into yeast often places more than one plasmid into a single yeast cell, and as many as 10% may be doubly transformed. Thus, several *E. coli* transformants must be recovered and checked for inserts and retested for transcriptional activation. Plasmid DNAs recovered into *E. coli* should be retransformed into Y190 (pAS-X) followed by a secondary X-Gal assay to confirm the interaction. Plasmids that give a positive signal can then have their 5' ends sequenced to determine if they are previously known genes. The sequencing primers we use for identifying the sequences for the 5' and 3' ends of the cDNA in pACT libraries are:

pACT forward 5'	5'CTATCTATTCGATGATGAAG
pACT reverse 3'	5'ACAGTTGAAGTGAACTTGCG

Characterization of Two-Hybrid Clones

It is always important to have an independent test for the interaction of the two proteins once genetically interacting clones are obtained, because false positives do occur even when clones meet all the criteria described in the previous section. In addition, secondary evidence of physical association is usually required for publication. We have used several different tests to detect in vitro binding of positives. One is to make a PCR primer to the library plasmid that has a T7 promoter placed in an appropriate position to place the insert of the library plasmid under T7 control. The PCR-derived fragments can then be directly added to a coupled transcription-translation system (TnT™ from Promega) and radio-labeled protein prepared. We usually add 6 μl of a robust 30 cycle PCR reaction to 25 μl of the translation mix (TnT). Five μl of this reaction mixture run on a SDS gel gives a readily detectable signal on an overnight exposure. This tells you the size of the fused protein and can be used to detect interaction in vitro with a column of your bait protein, usually as a GST fusion.

The sequences of the PCR primers we have used successfully are:

(1) 5' TAA TAC GAC TCA CTA TAG GGA GAC CAC *ATG* GAT GAT GTA TAT AAC TAT CTA TTC

T7 Promoter — Met Gal4p Activation Domain, 21 AA before *the Bgl II site* in pACT1 polylinker

(2) 5' CTA CCA GAA TTC GGC ATG CCG GTA GAG GTG TGG TCA

In the *ADH1* Terminator

A second method that is useful if you have antibodies to your bait protein is to immunoprecipitate the protein out of yeast extracts prepared from cells containing both hybrid plasmids and then use antibodies to the activation domain of Gal4p to detect binding of the fusion protein. Alternatively, you can immunoprecipitate ^{35}S-labeled yeast extracts to detect their coprecipitation (Lee et al. 1995). A new band should appear in extracts from cells containing the library plasmid that is not present in extracts from cells containing control plasmids. Since the two-hybrid system can detect interactions that are below the affinity detectable by immunoprecipitation or protein affinity chromatography, a negative result in these assays is not definitive. Moreover, some in vivo protein interactions are mediated by a third protein provided by yeast cells or a certain in vivo modification may be required for the interaction. A lack of binding in in vitro reconstitution does not rule out the possibility that the proteins do interact in vivo. Direct coimmunoprecipitation from the original cells expressing your two proteins will provide the definitive evidence that they interact in vivo. However, as just noted, this may not be possible.

A third method is to switch the bait and prey in their respective plasmids, that is, take the library insert out of pACT and insert it in-frame into pAS2, and place the original pAS2 insert into pACT. The majority of false positives will not interact in this test. It should be noted that some true positives may not activate for structural reasons, so only a positive result can be trusted. We have placed the pAS2 polylinker into pACT creating pACT2 to facilitate this transfer.

Limitations and General Considerations

As with all methods, the yeast two-hybrid system has limitations in addition to the false positive problem discussed in the previous section. First, a frequently encountered problem is the toxic effect of expression of some hybrid genes in yeast. In this case, a reporter strain expressing your bait protein cannot be established. These toxic proteins may include proteins that interfere with essential cellular functions; for example, the expression of cell cycle regulators cyclins A, B, and F is toxic. You can alleviate this problem by using low copy vectors or inducible systems. However, in some cases, the expression of certain proteins even at low levels is still harmful to yeast cells. Cotransformation of bait and library plasmids simultaneously may be helpful, but relatively large amounts of transforming DNA are required. One procedure we have used successfully to circumvent this problem is to select for yeast mutants that can tolerate the toxic effect of your bait expression (Bai et al. 1996). Another way is to use part of your protein or certain mutants of your protein as baits. For example, for a toxic protein kinase, a kinase-defective mutant might relieve toxicity but still retain the ability to interact with its targets. In the case in which truncated forms of baits are used, care must be taken when considering what portion of the protein to delete, and interacting clones isolated need to be tested for binding in vitro with the intact protein as well as the deleted form.

Second, the current two-hybrid assay is based on transcription. This eliminates screens with proteins displaying transcriptional activation activity. In addition, the nuclear localization requirement for transcription may be suboptimal for membrane proteins and possibly some cytoplasmic and extracellular proteins. Third, yeast can not provide all protein modifications required for certain protein interactions in higher eukaryotes, for example, tyrosine phosphorylation. Fourth, the interaction observed in yeast may not reflect a physiological interaction. For example, two proteins specifically expressed in different tissues, subcellular compartments, or different stages of development may be capable of interacting with each other when coexpressed in yeast but not in the organism itself.

In summary, the yeast two-hybrid system is a powerful way to identify genes encoding proteins that interact with a known protein. In this chapter, we have described methods to carry out a two-hybrid screen using the vectors and strains with which we are most familiar. The main criteria for successfully using a two-hybrid screen is the nature of your bait and the quality of the two-hybrid libraries. Some newer versions of the reporter strain and vectors are particularly useful in improving the efficacy of the screen and the time spent in eliminating false positives. Future generation plasmid constructs including tightly regulated expression systems and features that facilitate biochemical characterization will provide more efficient and powerful tools to identify interacting proteins in the near future. Extended application of this system could also lead to the isolation of molecules that disrupt or enhance a chosen pair of proteins.

References

Allen, J.B., Walberg, M. W., Edwards, M.C., and Elledge, S.J. (1995). Finding prospective partners in the library: the two hybrid system and phage display find a match. Trends Biochem. Sci. 20:511–516.

Ausubel, F.M., Brent, R., Kingston, R.E., Moore, J., Seidman, G., Smith, J.A., and Struhl, K., eds. (1994). *Current Protocols in Molecular Biology*. New York, Wiley.

Bai, C., Sen, P., Hofmann, K., Ma, L., Goebl, M., Harper, J.W., and Elledge, S.J. (1996). SKPI connects cell cycle regulators to the ubiquitin proteolysis machinery through a novel motif, the F-box. Cell 86:263–274.

Bartel, P., Chien, C-T., Sternglanz, R. and Fields, S. (1993). Elimination of false positives that arise in using the two-hybrid system. BioTechniques 14: 920–924.

Breeden, L., and Nasmyth, K. (1985). Regulation of the yeast HO gene. *Cold Spring Harbor Symposia on Quantitative Biology* Vol. 50: 643–650.

Chien, C-T., Bartel, P.L., Sternglanz, R., and Fields, S. (1991). The two-hybrid system: a method to identify and clone genes for proteins that interact with a protein of interest. Proc. Natl. Acad. Sci. USA 88:9578–9582 .

Dalton, S., and Treisman, R. (1992). Characterization of SAP-1, a protein recruited by serum response factor to the c-fos serum response element. Cell 68: 597–612.

Durfee, T., Becherer, K., Chen, P., Yeh, S., Yang, Y., Kilburn, A., Lee, W., and Elledge, S.J. (1993). The retinoblastoma protein associates with the protein phosphatase type 1 catalytic subunit. Genes Dev. 7:555–569.

Fields, S., and Song, O. (1989). A novel genetic system to detect protein-protein interactions. Nature 340:245–246.

Fields, S. (1993). The two-hybrid system to detect protein-protein interactions. Methods 5:116–124.

Fields, S., and Sternglanz, R. (1994). The two-hybrid system: an assay for protein-protein interactions. Trends Genet. 10:286–289.

Finley, R.L. and Brent, R. (1994). Interaction mating reveals binary and ternary connections between Drosophila cell cycle regulators. Proc. Natl. Acad. Sci. USA 91:12980–12984.

Gietz, R.D. and Woods, R.A. (1994). High efficiency transformation with lithium acetate. In *Molecular Genetics of Yeast: A Practical Approach* (Johnston, J.A., ed.) Oxford, Oxford University Press, pp. 121–134.

Guthrie, C., and Fink, G.R. (1991). *Guide to Yeast Genetics and Molecular Biology.* Methods in Enzymology Vol. 194.

Gyuris, J., Golemis, E., Chertkov, H., and Brent, R. (1993). Cdi1, a human G1 and S phase protein phosphatase that associates with Cdk2. Cell 75:791–803.

Harper, J.W., Adami, G.,Wei, N., Keyomarsi, K., and Elledge, S.J. (1993). The p21 Cdk-interacting protein Cip1 is a potent inhibitor of G1 cyclin-dependent kinases. Cell 75:805–816.

Kohrer, K., and Domdey, H. (1991). Preparation of high molecular weight RNA. In Methods in Enzymology, Vol. 194:398–404.

Lee, T. H., Elledge, S.J., and Butel, J.S. (1995). Hepatitis B virus x protein interacts with a probable cellular DNA repair protein. J. Virology 69:1107–1114.

Li, B., and Fields, S. (1993). Identification of mutations in p53 that affect its binding to SV40 large T antigen by using the yeast two-hybrid system. FASEB J. 7: 957–963.

Matsuoka, S., Edwards, M., Bai, C., Parker, S., Zhang, P., Baldini, A., Harper, J.W., and Elledge, S.J. (1995) p57^{KIP2}, a structurally distinct member of the p21^{CIP1} Cdk-inhibitor family, is a candidate tumor suppressor gene. Genes Dev. 9:650–662.

Mulligan, J.T., and Elledge. S.J. (1994). The construction and use of cDNA libraries for genetic selections. In *Molecular Genetics of Yeast: A Practical Approach* (Johnston, J.A., ed.) Oxford, Oxford University Press, pp. 65–81.

Serrano, M., Hannon, G.J., and Beach, D. (1993). A new regulatory motif in cell-cycle control causing specific inhibition of cyclin D/CDK4. Nature 366: 704–707.

Toyoshima, H., and Hunter, T. (1994). p27, a novel inhibitor of G1 cyclin-Cdk protein kinase activity, is related to p21. Cell 78:67–74.

Vojtek, A.B., Hollenberg, S.M., and Cooper, J.A. (1993). Mammalian Ras interacts directly with the serine/threonine kinase Raf. Cell 74:205–214.

3

Searching for Interacting Proteins with the Two-Hybrid System II

Anne B. Vojtek
Jonathan A. Cooper
Stanley M. Hollenberg

In this implementation of the two-hybrid system, fusion proteins to the DNA-binding domain of the *Escherichia coli* LexA protein and to the activation domain of the herpes simplex virus VP16 protein are expressed on 2μ high copy number plasmids in the *Saccharomyces cerevisiae* L40 reporter strain (Hollenberg et al. 1995). The L40 strain contains two integrated reporters: the yeast *HIS3* gene and the bacterial *lacZ* gene. The first reporter places the *HIS3* gene under control of four LexA operators, whereas the second reporter places the *lacZ* gene under control of eight LexA operators. Association of the two fusion proteins activates transcription of both the *HIS3* gene and the *lacZ* gene. As a result, the yeast grow in the absence of histidine, and turn blue when ß-galactosidase activity is assayed using the chromogenic substrate 5-bromo-4-chloro-3-indolyl-ß-D-galactoside (X-Gal). The strengths of this version of the two-hybrid system are twofold: (1) the initial identification of interacting proteins relies on the *HIS3* selection, allowing the use of large, high complexity libraries; and (2) the presence of the *lacZ* reporter allows an immediate and independent reassay of interacting proteins identified by the *HIS3* selection. Furthermore, in this approach, we have used a random primed library with inserts of small size. Screening this type of library allows the recovery of two classes of protein interaction domains that may be missed when screening a conventional oligo-dT primed library: (1) interaction domains that are located at or near the N-terminus of a protein; and (2) interaction domains from proteins with inhibitory sequences that, when in the context of the native protein, bind to and sequester the interaction domain.

Materials

Strains

The genotype of the L40 reporter strain is *MAT**a** HIS3Δ200 trp1-901 leu2-3,112 ade2 LYS2::(lexAop)4-HIS3 URA3::(lexAop)8-lacZ GAL4*.

The genotype of AMR70, which is often used for testing specificity of interactions by mating, is *MAT**a** HIS3 lys2 trp1 leu2 URA3::(lexAop)8-lacZ GAL4*; this strain was constructed by Rolf Sternglanz (State University of New York, Stony Brook).

Media

Yeast are grown in supplemented YC minimal medium to maintain selection for plasmids introduced into the strain and for the integrated reporter constructs. Omitting tryptophan (W) from the medium maintains selection for the LexA-fusion (bait) plasmid, whereas omitting leucine (L) from the medium maintains selection for the library plasmid. Omitting uracil (U) and lysine (K) maintains selection for the integrated *lacZ* and *HIS3* reporter constructs in the L40 strain, respectively. And, omitting histidine (H) selects for an interaction event between the bait and the library plasmid. One liter of YC medium is composed of the following: 1.2 g of yeast nitrogen base, minus amino acids and ammonium sulfate (Difco); 5 g of ammonium sulfate; 10 g of succinic acid; 6 g of sodium hydroxide; 0.75 g of WHULK minus amino acid and pyrimidine mix, 2% (w/v) glucose (100 ml of a 20% sterile stock solution, added after autoclaving), 20 g of agar (Difco), and essential amino acids, which may include a combination of the following: 0.05 g of H; 0.1 g of W, U, L, or K. WHULK minus mix (enough for 10 liters) is composed of 1 g each of adenine sulfate, arginine, cysteine, and threonine and 0.5 g each of aspartic acid, isoleucine, methionine, phenylalanine, proline, serine, and tyrosine.

YPAD is a rich medium of the following composition per liter: 10 g *y*east extract, 20 g Bacto-*p*eptone, 0.1 g *a*denine, 2% (w/v) glucose (20% stock added after autoclaving), 20 g agar, as required.

YPA is YPAD without glucose.

Plasmids

The bait plasmid, pBTM116, contains the entire coding region of the *Escherichia coli* LexA protein, expressed from the yeast alcohol dehydroge-

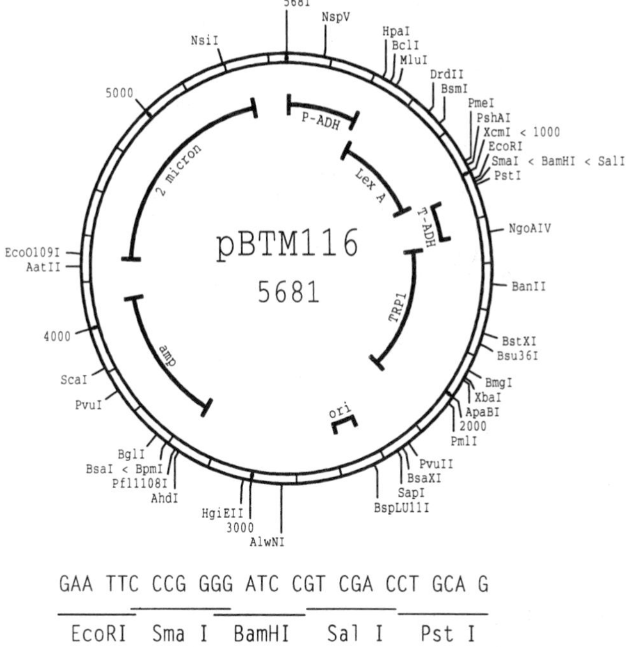

Fig. 3-1. Map and nucleotide sequence of the polylinker of the DNA-binding domain vector pBTM116. See text for vector description. Unique restriction sites are shown.

nase I (*ADH1*) promoter, followed by a polylinker for inserting cDNAs to generate in-frame fusions to LexA (Figure 3-1) (Bartel et al. 1993a). In addition, the bait plasmid contains the *TRP1* gene, which allows yeast containing this plasmid to grow in minimal medium lacking tryptophan; the 2µ origin of replication, for maintenance of the plasmid in yeast; a bacterial origin of replication; and, the ß-lactamase gene, which confers ampicillin resistance to *E. coli*. For Taq thermal cycle sequencing of cDNA inserts in pBTM116, use the following two primers: (1) sense primer, 5' CAGAGCTTCACCATTGAA 3', hybridizes 49 nucleotides upstream of the *Bam*HI polylinker cloning site; and (2) antisense primer, 5' GAAATTCGCCCGGAATT 3', hybridizes 19 nucleotides downstream of the *Bam*HI site.

The library plasmid, pVP16, contains the VP16 acidic activation domain, expressed from the yeast *ADH1* promoter, followed by a polylinker for inserting cDNAs to generate in-frame fusions to VP16 (Figure 3-2). Upstream of and in-frame to the VP16 coding region is the simian virus 40 (SV40) large T antigen nuclear localization sequence. In addition, the

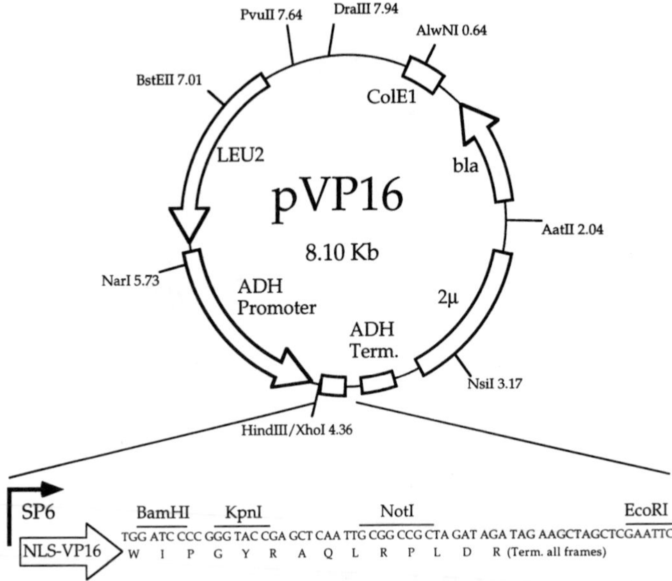

Fig. 3-2. Map and nucleotide sequence of the polylinker of the activation domain vector, pVP16. See text for a detailed description of the vector. Arrows indicate the direction of transcription of the ß-lactamase gene, the *LEU2* gene, and the *ADH1* promoter. Unique restriction sites are shown.

plasmid contains the *LEU2* gene, which allows yeast containing this plasmid to grow in the absence of leucine; the 2μ origin of replication; a bacterial origin of replication; the ß-lactamase gene; and an SP6 promoter, for in vitro synthesis of sense RNA (or of RNA/protein, using a coupled in vitro transcription and translation system, such as the Promega TNT system). Fusion proteins can be generated by inserting cDNAs at either the *Bam*HI or the *Not*I site. The reading frame of the *Bam*HI site in pBTM116 and pVP16 is the same; therefore, if *Bam*HI is used as the cloning site for the cDNAs of interest, then cDNAs can be readily transferred between bait and library vectors. For Taq thermal cycle sequencing of cDNA inserts in pVP16, use the following two primers: (1) sense primer, 5' GAGTTTGAGCAGATGTTTA 3', hybridizes 39 nucleotides upstream of the unique *Bam*HI polylinker cloning site; and, (2) antisense primer, 5' TGTAAAACGACGGCCAGT 3', hybridizes 54 nucleotides downstream of the *Bam*HI site.

pLexA-Lamin contains amino acids 66 to 230 of Lamin C inserted into the polylinker of pBTM116 (Bartel et al. 1993b). The LexA-Lamin-fusion protein is often a key component in strategies to eliminate false positives from among the isolates recovered from library screens.

Choosing a Library

The VP16 library used by Vojtek and coworkers to identify Ras effectors and by Hollenberg and coworkers to identify interacting partners for the *Drosophila* daughterless basic-helix–loop-helix protein was derived from mouse embryos at 9.5 and 10.5 days of development (Vojtek et al. 1993; Hollenberg et al. 1995). The cDNA for this library was synthesized using random primers at concentrations that resulted in the synthesis of products ranging from 350 to 700 nucleotides. The advantages of using a random primed, short insert size library are threefold: (1) random priming ensures that amino- and carboxy-terminal coding regions of proteins are equally represented in the library; (2) the short insert size enables those proteins that have both inhibitory and interacting domains to be recovered by separating the two domains; and (3) the short inserts can be rapidly sequenced, facilitating the rapid analysis of large numbers of interacting proteins. Possible disadvantages include: (1) use of a library with short inserts limits the partners recovered to those proteins with interacting domains that can be entirely contained within the insert size (approximately 233 amino acids or less for the 9.5-to-10.5-day mouse embryo library); (2) screens of additional libraries are required to obtain full-length clones for biological studies; and (3) the sequence of the interacting domain may not provide insights into function: longer clones may need to be isolated and sequenced before regions that indicate catalytic activity or familial relationships are identified.

Libraries using activation domains other than VP16 can also be used with the LexA reporter gene in the L40 reporter strain, as long as the auxotrophic marker on the library plasmid is the *LEU2* gene. For example, libraries using the Gal4p activation domain have been used successfully with the L40 reporter strain. The majority of the Gal4p activation domain libraries are oligo-dT primed and have large inserts (>1kb). The advantages of using these libraries include: (1) libraries from a variety of tissues and cell lines are available, and many libraries are commercially available; and (2) partners with large interaction domains (for example, greater than the 233 amino acid limitation imposed by the mouse embryo VP16 library) can be recovered. However, the clones recovered from screens using oligo-dT primed libraries often are biased towards those with interaction domains present at the carboxy-terminus of the protein. Three two-hybrid screens using H-Ras as bait have been described: two of these screens used oligo-dT primed libraries (Kikuchi et al. 1994; Hofer et al. 1994), while one screen used a random primed library (Vojtek et al. 1993 and unpublished data). All three screens recovered clones with carboxy-terminal interacting domains (either the Ral GDS or the Ral GDS-related protein), but only the screen that used the random primed VP16 library recovered clones that had amino-terminal interaction domains (c-Raf, A-Raf, and a novel phosphatidylinositol 3-kinase).

LexA does not contain a nuclear localization sequence and, therefore, does not preferentially partition to the nucleus (Silver et al. 1986). Therefore, when using libraries that also do not contain a nuclear localization sequence fused to the activation domain, some fusion proteins will remain in the cytoplasm. A derivative of pBTM116 with a nuclear localization sequence (NLS) fused to the carboxy-terminus of LexA, pBTM116-NLS, has been constructed (Vojtek and Cooper, unpublished data). The addition of the NLS may improve nuclear localization of some fusions; however, other fusions may interfere with the function of the NLS, since it is situated between LexA and the bait.

Insights into Bait Plasmids and Controls

The first decision that needs to be made when designing a bait plasmid is whether to fuse the full-length protein of interest to LexA or to fuse only a domain of this protein to LexA. Both types of fusions have been made and used successfully in library screens. However, higher backgrounds (transactivation of the reporter constructs by the bait plasmid alone) are often observed when domains of proteins are fused to LexA. The effect of background transactivation of the *HIS3* reporter construct can be reduced or eliminated by the inclusion in the medium of 3-amino-1,2,4-triazole (3-AT), an inhibitor of the dehydratase encoded by the *HIS3* gene (Jones and Fink 1982) (a protocol for use of 3-AT is outlined in the next section). Another possible disadavantage of using a fragment of a protein as a bait is that protein fragments may be more promiscuous in their interaction specificities. However, use of a protein fragment can be advantageous or essential if the native protein contains a domain capable of transcriptional activation, contains a region that is toxic to yeast, or contains an inhibitory domain, which sequesters the interaction domain of the bait. A second design decision is whether to add the *ADE2* gene to the bait plasmid. The presence of the *ADE2* gene will greatly reduce the tedium associated with segregation of the bait plasmid after positives have been obtained from the library screen, a necessary evil often required prior to analysis of the library isolates. pLex-A (constructed by S. Hollenberg) is a derivative of pBTM116 containing the *ADE2* gene. Once you have constructed the bait plasmid, introduce it into the L40 reporter strain and then test the resulting strain for activation of the reporter constructs.

1. Transform L40 with both the bait plasmid and the empty library vector (see Gietz et al. 1992 and Hill et al. 1991 for a small-scale transformation protocol). Plate the transformation mixture to YC-LW plates and incubate at 30°C for 2 to 3 days. At the same time, to make the strain needed for the library transformation, introduce the bait plasmid alone into L40 and plate the transformation mixture to YC-W plates. Note that expression of LexA alone is sufficient to detectably transactivate the reporter genes and, therefore, pBTM116 does not provide an appropriate negative control for

these experiments, although it can be used as a weak positive control. Fusions to LexA tend to decrease its ability to transactivate the reporter genes; therefore, an appropriate negative control for testing the bait plasmid is to introduce a plasmid expressing a nontransactivating LexA-fusion protein, such as pLexA-Lamin, together with pVP16 into L40. Introduce pLexA-Ras and pVP16-Raf into L40 for a positive control.

2. To determine if the bait plasmid transactivates the *HIS3* reporter gene, assay growth in the absence of histidine. Pick individual transformants from the plates in step 1 and streak for isolated colonies on YC-WHULK plates and to YC-LW plates. The expected results are as follows: (1) transformants containing pLexA-Lamin and pVP16 do not transactivate the *HIS3* reporter gene and, therefore, will not grow on YC-WHULK plates but will grow on YC-LW plates; (2) transformants containing positive control plasmids, such as pLexA-Ras and pVP16-Raf, or pLexA (pBTM116 without a cDNA insert) and pVP16, do transactivate the *HIS3* reporter gene, and yeast with these plasmid combinations grow on minimal medium lacking histidine (YC-WHULK plates); and (3) transformants containing the bait plasmid with pVP16 should not grow on YC-WHULK plates. Growth of the transformants with the bait plasmid implies that the fusion of your bait protein to LexA transactivates the *HIS3* reporter gene. In this case, a different bait plasmid will need to be constructed (for example, fuse a domain of the original bait to LexA) or 3-AT will need to be used to suppress activation (see next section).

Finally, not all fusion proteins will partition into the nucleus. Therefore, when possible, construct positive controls. If a partner of the bait of interest is known or even suspected, clone the partner into pVP16 for a positive control. If the bait of interest is cytoplasmic, consider adding a nuclear localization sequence, such as by cloning the bait into pBTM116-NLS.

Use of 3-AT to Reduce Nonspecific Transactivation of the *HIS3* Reporter Gene

To continue to work with a bait that transactivates the *HIS3* reporter gene in the absence of a partner, include 3-AT in the growth medium. Add 3-AT (1 M stock in water, filter sterilize, store at -20°C) to YC-WHULK plates (either spread in the appropriate amount onto the plates or add just prior to pouring the plates). The concentration of 3-AT required depends on the extent of transactivation of the *HIS3* reporter gene by the bait plasmid; initially, try including 0.5, 1, 2, 5, and 10 mM 3-AT in the medium. Determine the lowest possible concentration of 3-AT that will prevent growth of L40 containing the bait plasmid and pVP16 on YC-WHULK plates by streaking transformants for isolated colonies. Use of 3-AT at a higher than necessary concentration results in a reduction of plating efficiency.

Transformation of Yeast

The library transformation protocol that we have used successfully is a modification of published protocols (Gietz et al. 1992; Hill et al. 1991; Schiestl and Gietz 1989). Other transformation protocols can be used. The goal is to achieve an efficient transformation: one that yields primary transformants at three times the complexity of the library. [Screening through three times the library complexity ensures that 95% of the clones represented in the library have been examined (Clarke and Carbon 1976)].

1. Inoculate a 5 ml culture of L40 containing the bait plasmid in YC-WU; incubate overnight at 30°C. Minimal medium lacking W and U is used to maintain selection for the bait plasmid (W) and for the *HIS3* reporter gene (U).

2. Expand the culture from step 1 by inoculating 100 ml of YC-WU and incubating overnight at 30°C; make several dilutions of the overnight culture from step 1 to obtain a 100 ml culture that is at midlog phase (OD_{600} of 4 or less) after the overnight incubation.

3. Dilute the culture from step 2 in 1 liter of YPAD, prewarmed to 30°C, to a density of OD_{600} of 0.3. Incubate at 30°C with shaking for 2 to 4 hours to obtain yeast in logarithmic phase of growth. Note that in most situations the transformation protocol can be scaled down to 1/2 to 1/4 (that is, inoculate 250 to 500 ml YPAD in this step and scale all subsequent steps accordingly), and the scaled down protocol is sufficient to obtain enough primary transformants to represent three times the VP16 library complexity (5 to 10 million). Growth of a 1 liter YPAD culture at this step is required when the transformation efficiency of the L40 strain containing the bait plasmid is poor, either because the bait plasmid is toxic to the yeast or because a solution used in the transformation is not optimal (the likely candidate is the salmon sperm DNA).

4. Pellet the cells from step 3 by centrifugation at room temperature (1500 g, 5 minutes). Wash the pellet once in 500 ml 1× lithium acetate (0.1 M lithium acetate, pH 7.5), 0.5× TE (5 mM Tris, 0.5 mM EDTA, pH 7.5), and then resuspend in 20 ml of 1× lithium acetate, 0.5× TE. Transfer the cell suspension to a 1 liter glass flask.

5. Add a DNA *mixture* consisting of 500 µg of the library and 10 mg of denatured salmon sperm (prepared as described in Schiestl and Gietz 1989) to the cell suspension. (The addition of carrier DNA is not required if the library DNA is prepared by alkaline lysis and then extracted with phenol-chloroform, but NOT treated with RNase. The RNA functions as the carrier for the transformation.)

6. Add 140 ml of 1× lithium acetate, 40% polyethylene glycol (PEG) 3350, 1× TE and incubate 30 minutes at 30°C, with occasional mixing.

7. Add 17.6 ml of dimethyl sulfoxide and incubate at 42°C for 6 minutes, swirling the flask gently to mix periodically throughout the incubation. Immediately after the 6 minute incubation, dilute the cell suspension with 400 ml of YPA and rapidly cool to room temperature in an ice-water bath.

8. Pellet the cells at room temperature and wash with 500 ml of YPA. Resuspend the cells in 1 liter of prewarmed YPAD and incubate at 30°C for one hour on a shaking platform at a low speed.

9. To determine the primary transformation efficiency, pellet the cells from 1 ml of the 1 liter YPAD culture and resuspend in 1 ml of YC-WUL. Plate 10 and 1 µl aliquots ($1/10^5$ and $1/10^6$ of the total number) on YC-WUL plates. Incubate

the plates at 30°C for 2 to 3 days, until the colonies can be readily counted. This protocol gives an average of 100 million total transformants. Save the remainder of the 1 ml aliquot at 4°C to make additional dilutions for colony counts, if needed.

10. Pellet the rest of the cells, wash once in 500 ml YC-WLU, and resuspend in 1 liter of prewarmed YC-WLU. Incubate at 30°C for 4 to 6 hours or overnight to enable induction of *HIS3* expression. The shorter recovery period maintains the library representation better than the overnight growout, but is often less convenient than the overnight growout. Omitting W, L, and U from the growth medium maintains selection for the bait plasmid, the library plasmid, and the *HIS3* reporter gene, respectively.

11. Wash the cells twice with 500 ml YC-WHULK, then resuspend in 10 ml of YC-WHULK.

12. Plate aliquots of the cell suspension from step 11. His^+ colonies grow poorly at high cell densities. Therefore, to obtain plates with an appropriate cell density, plate 5, 10, 25, and 50 μl aliquots of the cell suspension to each of 10 YC-WHULK plates. Also, plate dilutions ($1/10^6$ and $1/10^7$ of the total) to YC-UWL plates to determine the number of doublings from the overnight growout. Using this number and the number of primary transformants from step 9, calculate the number of colonies that should be screened in order to cover three times the complexity of the library. For example, if the total number of transformants from step 9 is equal to the library complexity and if the cells underwent eight doublings in the overnight growout, then pick the His^+ colonies that grow from 117 μl of the cell suspension (1/256 of 10 mls × 3), or pick all the His^+ colonies from 12 plates with 10 μl of the cell suspension. Save the remaining cell suspension at 4°C for replating of additional dilutions, if necessary. Incubate the plates at 30°C for 2 to 3 days.

13. Transfer the His^+ colonies to two YC-WHULK plates, a master plate and a plate for the ß-galactosidase filter assay. Include a positive control for the ß-galactosidase assay on each plate, such as L40 containing pLexA-Ras and pVP16-Raf.

14. With a sterile toothpick, pick each His^+ $lacZ^+$ positive that will be studied further and regrow on a YC-WHULK plate; pick and regrow a total of two times. It is not necessary to streak for isolated colonies. Because multiple library plasmids may enter the cell during the transformation procedure, this step, by requiring multiple rounds of histidine prototrophy, selects for the library plasmid that is interacting with the bait, and allows the loss of any additional noninteracting library plasmids. Although this step is time consuming, it is not labor intensive, and can save much aggravation later on.

ß-Galactosidase Filter Assay

This protocol is adapted from Breeden and Nasmyth (1985).

1. Lay a dry nitrocellulose filter (BA-S85, 82 mm, Schleicher and Schuell) on the YC-WHULK plate; apply gentle pressure to the nitrocellulose filter to obtain close contact between the yeast and the filter.

2. Lift the filter from the plate and place it colony side up on an aluminum foil boat, floating on liquid nitrogen. Immerse the filter and boat into the liquid nitrogen for 5 seconds. If the filters crack when placed in liquid nitrogen or if you are using unsupported nitrocellulose (such as BA-85), wait 20 seconds prior

to immersing the filter and boat in liquid nitrogen. Let the filter thaw at room temperature, colony side up. Alternatively, instead of lysing the yeast in liquid nitrogen, place the filter in a plastic petri dish or large dish, cover, and incubate at -80°C for 10 minutes.

3. In the lid of a petri dish, place 1.5 ml of Z buffer [60 mM Na_2HPO_4, 40 mM NaH_2PO_4, 10 mM KCl, 1 mM $MgSO_4$ (pH 7.0)] and 10 to 30 μl of X-Gal (50 mg/ml in N,N-dimethylformamide), one Whatman number one filter circle (90mm), and then the nitrocellulose filter, again colony side up. Cover with the bottom of the petri dish and incubate at 30°C. Color development can take anywhere from 5 minutes to overnight. If incubating for over one hour, place the petri dishes in a plastic bag to prevent them from drying out.

Elimination of False Positives

The bane of the two-hybrid system is sorting the library isolates into two classes, clones of interest and false positives. A common strategy to eliminate false positives is to test the specificity of the interaction of the library isolate with fusion proteins to LexA, such as LexA-Lamin (Bartel et al. 1993). Clones of interest should interact with the LexA-fusion used in the screen (bait), but not with LexA-Lamin. Since the LexA-Lamin test does not eliminate all false positives, testing the interaction of the library isolate with one or two additional LexA-fusion proteins is recommended.

A second powerful method for the elimination of false positives is to screen the library isolates against a bait with a mutation that abolishes the interaction of interest. This approach was used to eliminate false positives in a search of the VP16 mouse embyro library for effectors of the small GT Pase H-Ras (Vojtek et al. 1993). In this screen, library isolates were initially screened against LexA-Lamin, and were subsequently rescreened with a Ras bait that contained mutations in the effector domain, a region of Ras absolutely required for Ras signal transduction. Three classes of interest, comprising a total of 13 clones out of 14 million library transformants screened, were identified by this approach. In a search of the VP16 mouse embryo library for Src interacting proteins, Howell and coworkers winnowed the library isolates by identifying those that required Src catalytic activity and an intact SH2 domain for the interaction (Howell, Gertler, and Cooper, 1997).

To apply these methods to eliminate false positives, you must separate the library isolate from the bait plasmid. Two strategies are outlined. The first is simply to isolate the library plasmid from the yeast and then to cotransform yeast with the library plasmid and pLexA-Lamin or other pLexA-fusion protein plasmid being used to test the specificity of the interaction. The second strategy is to remove the bait plasmid from each of the library isolates and then to introduce by mating the fusions being used to test specificity.

Isolation of Library Plasmid from Yeast

Protocols for isolating plasmids from yeast have been described in detail elsewhere (Rose, et al. 1990; Ward 1990). Once the DNA has been isolated from yeast, transform an aliquot of the yeast DNA mini-prep into a *recA E. coli* strain, such as DH5α or DB6507. DNA can be introduced into cells that have been made competent by treatment with calcium chloride or by electroporation. (Since the yeast DNA preps are a mixture of chromosomal and plasmid DNA and since chromosomal DNA can interfere with the efficiency of the *E. coli* transformation, if transformants are not recovered when using cells made competent with calcium chloride, try varying the amount of DNA used in the transformation.)

Consider using the *E. coli* strains HB101 or DB6507 if the DNA was prepared from a yeast strain with both library and bait plasmid. HB101 *(F- Δ(gpt-proA)62 leuB6 supE44 ara-14 galK2 lacYI Δ(mcrC-mrr) rpsL20 (Strr) xyl-5 mtl-1 recA13)* and DB6507 (HB101 *pyrF74::Tn5*) require leucine supplementation for growth, and the yeast *LEU2* gene is capable of complementing this defect. (In addition, both strains require proline and DB6507 requires uracil.) Therefore, by introducing the yeast DNA prep into this strain and plating the transformation mixture on minimal plates, *E. coli* containing the library plasmid can be selectively recovered. DB6507 is available from the American Tissue Culture Collection.

Plasmid Segregation to Remove the Bait

In Protocol A, the bait plasmid is segregated from the library isolates by growing the yeast under conditions that maintain a selection for the library plasmid, but not the bait plasmid (that is, in minimal medium lacking leucine but containing tryptophan, for the VP16 library screens). Use protocol A to segregate the bait plasmid if it does not contain the *ADE2* gene. Use Protocol B if the bait plasmid contains the *ADE2* gene. In this protocol, library isolates that have segregated the bait plasmid can be readily identified by simply observing the color after growing the library isolate in minimal medium with a low adenine concentration.

Protocol A

1. Inoculate 2 ml of YC-L with each library isolate. Incubate at 30°C for 2 days.
2. Dilute each culture and plate to YC-L plates. The goal is to obtain 100 to 200 colonies per plate and is usually achieved by plating 100 µl of a 1:10,000 dilution of each culture. Incubate the plates at 30°C for 2 days.
3. Transfer the colonies from each of the YC-L plates to a sterile velvet and then to a YC-LW plate and a YC-L plate, in this order. Incubate the plates at 30°C for 2 days.
4. From each pair of plates, identify a colony that has segregated the bait plasmid (pick colonies that grow on the YC-L plate but not the YC-LW plate).

Protocol B

Yeast mutant at the *ADE2* locus accumulate a red pigment as a result of the metabolic block in the adenine biosynthetic pathway (Jones and Fink 1982). Therefore, library isolates that have segregated the bait plasmid containing the *ADE2* gene accumulate the red pigment as a result of the metabolic block and are readily distinguished from the white ADE+ yeast that have not segregated the plasmid.

1. Grow each library isolate on YC-UL plates for 2 days at 30°C.
2. Streak each library isolate from step 1 to obtain single colonies on low adenine YC-UL plates (10 mg/liter adenine, 1% glucose). Incubate the plates at 30°C for 2 to 3 days. Pick a red colony for each library isolate.
3. Confirm that each red colony chosen in step 2 has lost the bait plasmid but retains the library plasmid by assaying growth on YC-ULW and YC-UL plates. If the bait plasmid was cured from the library isolate, then the yeast will grow on YC-UL plates but not YC-ULW plates.

Mating Assay

1. Transform a *MATα* yeast strain with the plasmids that will be used to test the specificity of the interaction, such as pLexA-Lamin. Also, introduce the original bait plasmid into this *MATα* strain. AMR70 is a *MATα* strain with an integrated copy of the *lacZ* reporter. It is not essential to use a mating strain with an integrated *lacZ* reporter; a generic yeast strain that is *MATα leu2 ura3 trp1* but lacks the integrated reporter is also a reasonable choice and often can be readily obtained from a yeast lab at your institution.
2. Inoculate 2 ml YC-L cultures with each of the L40 library isolates from which the bait plasmid has been eliminated and YC-W cultures of each of the *MATα* strains that will be used to test specificity of interaction. Incubate 1 to 2 days at 30°C.
3. Pipette 1 μl of each of the library isolates to a YPAD plate, then overlay with 1 μl of a tester strain. Include a positive control on each plate: mate each library isolate to the *MATα* strain containing the original bait plasmid. Incubate overnight at 30°C. (Alternatively, instead of growing yeast in liquid cultures, maintain the strains on plates (YC-L or YC-W) and use toothpicks to transfer yeast to YPAD plates for matings.)
4. Transfer the colonies from the YPAD plate to a sterile velvet and then to a YC-LW plate to select for diploids. Incubate 2 days at 30°C.
5. Transfer the diploids from step 4 to a sterile velvet and then to a YC-LW plate to test for activation of the *lacZ* reporter.

Characterization of Library Isolates

The library screen is a success! A partner for a protein of interest has been identified. What next? The next steps, which are often the most difficult, are to determine whether the interaction observed in yeast is direct and to place the interaction in a biological context.

Transactivation of the reporter constructs by the DNA-binding and activation domain fusion proteins indicates that a stable complex is formed between these proteins in the yeast nucleus. The complex may or may not contain additional components contributed by the yeast. To determine if the interaction is direct, the bait and partner can be synthesized as bacterial fusion proteins (for example, as fusions to glutathione S-transferase and maltose binding protein), which are readily purified, and then the interaction of the purified proteins can be assessed in vitro. If the two proteins interact in vitro, then the affinity of the interaction can be determined. Affinities should be 1 μM or better.

A protein-protein interaction identified in the two-hybrid system may not be biologically relevant. During a two-hybrid screen, proteins from distinct cellular compartments can be coexpressed, permitting interactions to take place that would normally be prevented by subcellular compartmentalization. Also, proteins from different cell types may be coexpressed, allowing interactions that never occur in an organism. And, finally, aberrant interactions may occur because the two-hybrid proteins adopt nonphysiological configurations, which could arise because the two-hybrid proteins are truncated, incorrectly modified by the yeast, or associated with yeast proteins. Therefore, it is essential to place a two-hybrid interaction in a biological context.

The approach to establish the biological relevance of an interaction will vary, but could include localization and coimmunoprecipitation studies. Mutant analysis is also a powerful approach for analyzing an interaction. When the bait and/or library partner has been well-characterized previously, point mutations can be introduced into critical residues that are known to affect the biological process of interest, and then the effects of the mutations on the interactions determined in the two-hybrid system (see Vojtek et al. 1993, for an illustration of this method). The effect of deleting domains of interest, such as SH2 or SH3 domains, from baits or partners can also be readily assessed. Alternatively, if mutations affecting the biological process are not known, then the two-hybrid system can be used to identify such mutations. Libraries of mutant baits or partners in two-hybrid vectors can be made by PCR mutagenesis and screened to identify white His$^-$ clones.

Summary

The two-hybrid system is a powerful approach to identify interacting proteins. In the two-hybrid system described here, the initial identification of interacting proteins relies on a *HIS3* selection, allowing the use of large, high-complexity libraries. A second reporter, the *lacZ* gene, enables the immediate and independent reassay of the positives identified from the initial selection. Finally, the use of a random primed, short insert size library enables the recovery of certain classes of library positives that are often missed when screening an oligo-dT primed library.

References

Bartel, P., Chien, C., Sternglanz, R., and Fields, S. (1993a). Using the two-hybrid system to detect protein-protein interactions in *Cellular Interactions in Development: A Practical Approach* (Hartley, D. A., ed.). Oxford: Oxford University Press, pp. 153–179.

Bartel, P., Chien, C., Sternglanz, R., and Fields, S. (1993b). Elimination of false positives that arise in using the two-hybrid system. BioTechniques 14:920–24.

Breeden, L. and Nasmyth, K. (1985). Regulation of the yeast HO gene. *Cold Spring Harbor Symp. Quant. Biol.* 50: 643–50.

Clarke, L. and Carbon, J. (1976). A colony band containing synthetic Col E1 hybrid plasmids representative of the entire E. coli genome. Cell 9:91–99.

Gietz, D., St. Jean, A., Woods, R. A., and Schiestl, R. H. (1992). Improved method for high efficiency transformation of intact yeast cells. Nucleic Acids Res. 20: 1425.

Hill, J., Donald, K. A., and Griffins, D. E. (1991). DMSO-enhanced whole cell yeast transformation. Nucleic Acids Res. 19:5791.

Hofer, F., Fields, S., Schneider, C., and Martin, G. S. (1994). Activated Ras interacts with the Ral guanine nucleotide dissociation stimulator. *Proc. Natl. Acad. Sci.* USA 91:11089–11093.

Hollenberg, S. M., Sternglanz, R., Cheng, P. F., and Weintraub, H. (1995). Identification of a new family of tissue-specific basic helix-loop-helix proteins with a two-hybrid system. Mol. Cell. Biol. 15:3813–3822.

Howell, B.W., Gertler, F.B., and Cooper, J.A. (1997). Mouse disabled (mDabl): a Src-binding protein implicated in neuronal development. EMBO J. 16:121–132.

Jones, E. W. and Fink, G. R. (1982) in *The Molecular Biology of the Yeast Saccharomyces cerevisiae* (Strathern, J. N., Jones, E. W., and Broach, J. R., eds.), New York, Cold Spring Harbor Laboratory Press. pp. 181–299.

Kikuchi, A., Demo, S., Ye, Z. H., Chen, Y.-W., and Williams, L. T. (1994). Ral GDS family members interact with the effector loop of ras p21. Mol. Cell. Biol. 14: 7483–7491.

Rose, M. D., Winston, F., and Hieter, P. (1990). *Laboratory Course Manual for Methods in Yeast Genetics.* New York, Cold Spring Harbor Laboratory Press. pp.128–132.

Schiestl, R. H. and Gietz, R. D. (1989). High efficiency transformation of intact cells using single stranded nucleic acids as a carrier. Curr. Genet. 16:339–346.

Silver , P., Brent, R., and Ptashne, M. (1986). DNA binding is not sufficient for nuclear localization of regulatory proteins in *Saccharomyces cerevisiae*. Mol. Cell. Biol. 6:4763–4766.

Vojtek, A. B., Hollenberg, S. M., and Cooper, J. A. (1993). Mammalian Ras interacts directly with the serine/threonine kinase Raf. Cell 74:205–214.

Ward, A. C. (1990). Single-step purification of shuttle vectors from yeast for high frequency back-transformation into *E. coli*. Nucleic Acids Res. 18:5319.

4

Searching for Interacting Proteins with the Two-Hybrid System III

Erica A. Golemis
Roger Brent

The interaction trap (Gyuris et al. 1993) is a LexA-based version of the two-hybrid system. In a continuation of work investigating the modular nature of transcription activators (Brent and Ptashne 1985), we had been studying nuclear-localized oncoproteins such as Fos and Myc that activated transcription when brought to DNA in yeast (Lech et al. 1988). We expressed mammalian libraries in yeast in attempts to isolate cDNAs encoding proteins that increased or decreased Fos- or Myc-dependent transcriptional activation function, reasoning that proteins that modulated the transcription phenotype of these oncoproteins might modulate oncogenic transformation in the cell types from which they came. The proposal for the two-hybrid system by Fields and Song (1989) provided a direct means of detecting the type of protein interactions we sought. Accordingly, we also constructed cDNA expression vectors that directed the synthesis of library-encoded proteins that contained transcription activation moieties.

The interaction trap selection system differs in a number of respects from other two-hybrid systems. First, the nutritional markers used to select the system components are different. Fusions to a DNA-binding domain provided by LexA are expressed from a *HIS3*-containing plasmid (the "bait" plasmid), while fusions to an activation domain-containing plasmid are expressed from a *TRP1*-containing library plasmid. More significantly, while a *LexAoperator-lacZ* reporter is *URA3* based, the auxotrophic selection uses a low background *LexAop-LEU2* fusion. Because of this difference, the investigator does not vary the sensitivity of the reporter genes by the use of inhibitors, such as 3-amino-1,2,4-triazole, to suppress the *HIS3* phenotype: rather, the investigator selects the *lacZ* and *LEU2* reporters from a series that bears different numbers and kinds of LexA operators upstream of the *lacZ* or *LEU2* reporter genes. We have shown that these reporters have different thresholds for response to LexA-dependent transcriptional activation (Estojak et al. 1995; Golemis and Brent 1992). Baits that activate can often be used with the less sensitive reporters.

Second, in the interaction trap, the activation domain fusion plasmid is expressed inducibly under the control of the *GAL1* promoter. This regulated expression allows rapid elimination of background colonies due to strain mutations, and, in addition, facilitates isolation from libraries of cDNAs whose continuous expression is disfavored or toxic in yeast. Further, the activation domain fusion proteins typically carry a nuclear localization sequence, while the bait proteins do not—a property intended to produce an excess of interacting protein over DNA-bound bait in the yeast nucleus and, thus, maximize chances of detecting an interaction.

Third, in the interaction trap, the activation domain is of only moderate strength [contributed by the B42 acid blob (Ma and Ptashne 1987)], in contrast to stronger activation domains such as that of Gal4p or VP16. The weak domain is used in order to circumvent possible toxic effects due to squelching (Gill and Ptashne 1988), and thus to broaden the spectrum of library-encoded proteins that might be detected.

Fourth, through the collaborative and generous efforts of a number of early users of the system, a number of variations on the basic fusion plasmids have been created and made publicly available. Some of these described here permit the interaction trap to be used for proteins with constraints on their expression and function that make the basic reagents inappropriate.

Fifth, the initial series of controls used to establish the suitability of particular bait fusions for use in interaction trap screens is extensive and includes test constructs with different reporter strains and plasmids to establish an activation profile, as well as an assay of the ability of the construction to express a full-length protein that binds DNA (Golemis et al. 1996). These additional steps are included to circumvent the majority of the problems that might result later in the procedure from use of flawed bait proteins.

This chapter describes how to use the components of the interaction trap to screen an interaction library for proteins that interact with a given bait. The components of the interaction trap have also been successfully used for other applications. These include the determination of residues involved in individual protein-protein interactions in attempts to select inhibitors of protein interactions and in the selection of peptide aptamers, which are designed to mimic the function of antibody molecules (Colas et. al. 1996). Finally, interaction trap components have been used to perform interaction mating (Finley and Brent 1994; chapter 12) to establish extended networks of protein-protein interaction. This mating approach is complementary to the library screening procedure described here and is a powerful way to rapidly gain additional information about the function of novel proteins.

If all goes well in an interactor hunt, it takes approximately one week to perform yeast transformations, obtain colonies, and determine whether bait proteins are appropriate once the required constructions have been made. During a second week, library transformations are performed, transformants are replated to selective medium and positives are chosen. In the third week, plasmids are rescued from yeast and passaged through *E. coli*, fresh yeast are transformed, and the specificity of interactions is determined. Cumulatively, the experiments should be completed within a month.

Comprehensive list of Plasmids, Strains, and Libraries

Core Components

The standard parent plasmid for generating LexA fusions, pEG202 (Figure 4-1), is a derivative of 202 + PL (Ruden et al. 1991) that contains an expanded polylinker region. The strong alcohol dehydrogenase (*ADH1*) promoter is used to express LexA-fused proteins. The available cloning sites in pEG202 for making LexA-fused proteins include *Eco*RI, *Bam*HI, *Sal*I, *Nco*I, *Not*I, and *Xho*I; the reading frame is shown in the legend to Figure 4-1. Since the original presentation of this system, a number of groups have developed variants of this plasmid that address specialized research needs. All plasmids currently available for this system, as well as purposes for which they are suited, are listed in Table 4-1. Source information is included in the legend to this Table.

Interaction libraries used with the interaction trap are made in pJG4-5 or its derivatives (see Figure 4-2) (Gyuris et al. 1993). The pJG4-5 cDNA library expression cassette is under control of the *GAL1* promoter, such that library proteins are expressed in the presence of galactose (Gal) but not glucose (Glu). This conditional expression has the advantage that many false positives obtained in screens can be easily eliminated because they do not demonstrate a Gal-dependent phenotype. The expression cassette consists of an ATG to start translation; a nuclear localization signal to extend the interaction trap's range to include proteins that are normally found in the cytoplasm and to maximize the intranuclear protein concentration; an activation domain [acid blob (Ma and Ptashne 1987)]; the hemagglutinin epitope tag to permit rapid assessment of the size of encoded proteins (Kolodziej and Young 1991); *Eco*RI-*Xho*I sites designed to receive directionally synthesized cDNAs; and the alcohol dehydrogenase (*ADH1*) termination sequences to enhance the production of high levels of library protein. The plasmid also contains the *TRP1* marker and 2μ origin for propagation in yeast.

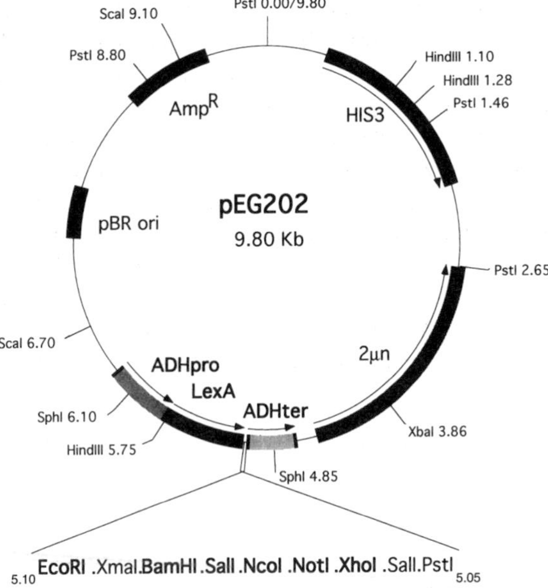

Fig. 4-1. pEG202 LexA-fusion plasmid. pEG202, a derivative of Lex202 + PL (Ruden et al. 1991), uses the strong alcohol dehydrogenase gene (*ADH1*) promoter (ADHpro) to express bait proteins as fusions to the DNA-binding protein LexA. A number of restriction sites immediately upstream of the *ADH1* terminator (ADHter) are available for insertion of coding sequences (those shown in bold type are unique, or otherwise suitable for fragment insertion). The reading frame for insertion is GAA TTC CCG GGG ATC CGT CGA CCA TGG CGG CCG CTC GAG TCG ACC TGC AGC. The sequence CGT CAG CAG AGC TTC ACC ATT G can be used to design a primer to confirm the correct reading frame for LexA fusions. The plasmid contains the *HIS3* selectable marker and the 2μ origin of replication to allow propagation in yeast, and the ampicillin resistance gene (AmpR) and the pBR origin (ori) of replication to allow propagation in *E. coli*. Endpoints of the *HIS3* and 2μ elements are drawn approximately with respect to the indicated restriction sites. Numbers indicate relative map positions.

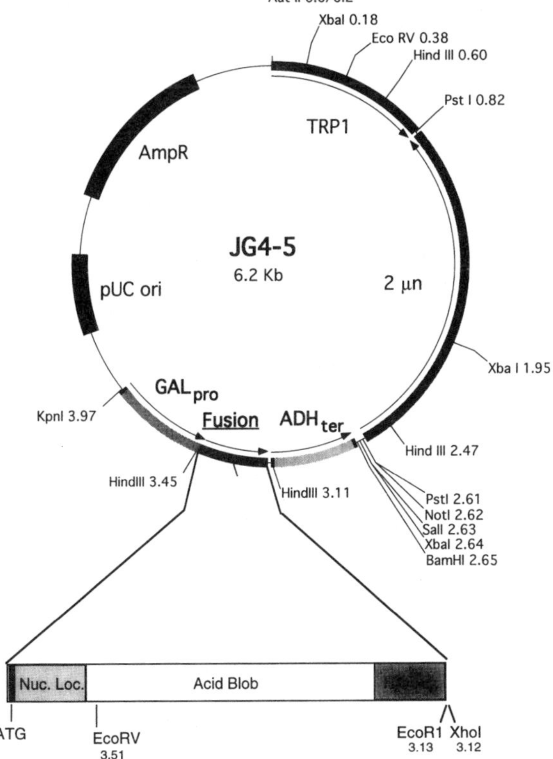

Fig. 4-2. pJG4-5 library plasmid. pJG4-5 (Gyuris et al. 1993) expresses cDNAs or other coding sequences inserted into the unique *Eco*RI and *Xho*I sites as translational fusion to a cassette consisting of the SV40 nuclear localization (nuc. loc.) sequence (PPKKKRKVA), the acid blob B42, and the hemagglutinin (HA) epitope tag (YPYDVPDYA). Expression of sequences is under the control of the *GAL1* inducible promoter. The sequence CTG AGT GGA GAT GCC TCC can be used as a primer to identify inserts or confirm the correct reading frame. The plasmid contains the *TRP1* selectable marker and the 2µ origin to allow propagation in yeast, and the ampicillin resistance (*AmpR*) gene and the pUC origin (PUC ori) to allow propagation in *E. coli*. Numbers indicate relative map positions.

Table 4-I Plasmids and strains required for the interaction trap

Plasmid	Markers	Characteristics
LexA-Fusion Plasmids		
pEG202[1]*	*HIS3*, 2μ *amp^R*	*ADH1* promoter expresses LexA followed by polylinker: basic plasmid used for cloning Bait.
pJK202[2]*	*HIS3*, 2μ *amp^R*	Like pEG202, but incorporates nuclear localization sequences between LexA and polylinker: used to enhance translocation of Bait to nucleus.
pNLexA[3]	*HIS3*, 2μ, *amp^R*	*ADH1* promoter expresses polylinker followed by LexA: for use with Baits where amino-terminal residues must remain unblocked.
pGilda[4]	*HIS3*, 2μ, *amp^R*	*GAL1* promoter expresses same LexA and polylinker cassette as pEG202: for use with Baits whose continuous presence is toxic to yeast.
pEE202I[5]	*HIS3*, *amp^R*	An integrating form of pEG202 that can be targeted into *HIS3* following digestion with *KpnI*: for use where physiological screen requires lower levels of Bait to be expressed.
Activation Domain-Fusion Plasmids		
pJG4-5[6]*	*TRP1*, 2μ, *amp^R*	*GAL1* promoter expresses nuclear localization domain, transcriptional activation domain, HA epitope tag, cloning sites: used to express cDNA libraries.
pJG4-5I[5]	*TRP1*, *amp^R*	An integrating form of pJG4-5 that can be targeted into *TRP1* by digestion with *Bsu*361 (NEB), to be used with pEE2021 to study interactions that occur physiologically at low protein concentrations.
LacZ Reporter Plasmids		
pSH18-34[7]*	*URA3*, 2μ, *amp^R*	8 LexA operators direct transcription of the *lacZ* gene: most sensitive indicator plasmid for transcriptional activation.
pJK103[2]*	*URA3*, 2μ, *amp^R*	2 LexA operators direct transcription of the *lacZ* gene: intermediate reporter for transcriptional activation.
pRB1840[8]*	*URA3*, 2μ, *amp^R*	1 LexA operator directs transcription of the *lacZ* gene: stringent reporter for transcriptional activation.
pJK101[2]*	*URA3*, 2μ *amp^R*	GAL1 upstream activating sequences followed by 2 LexA operators followed by *lacZ* gene: used in repression assay to assess Bait binding to operator sequences.
Positive and Negative Controls		
pRFHM1[9]*	*HIS3*, 2μ, *amp^R*	*ADH1* promoter expresses LexA fused to the homeodomain of bicoid to produce nonactivating fusion. Use as positive control for repression assay, negative control for activation and interaction assays.

Table 4-I *(Continued)*

Plasmid	Markers	Characteristics
pSH17-4[7]*	*HIS3*, 2μ, *amp*^{R}	*ADH1* promoter expresses LexA fused to Gal4p activation domain. Use as a positive control for transcriptional activation.
LEU2 Selection Strains		
EGY48[1]*	*MATα trp1, his3, ura3, 6ops-LEU2*	Basic strain used to select for interacting clones from a cDNA library: 6 LexA operators direct transcription from the LEU2 gene.
EGY191[1]	*MATα trp1, his3, ura3, 2ops-LEU2*	Like EGY48, but with 2 LexA operators rather than 6: a more stringent selection, and producing lower background with Baits with intrinsic ability to activate transcription.

Note: Interaction Trap reagents represent the work of many contributors: basic reagents (*) have been described in Gyuris et al. (1993) and were developed in the Brent laboratory, while others have been developed for specialized applications by the individuals noted. [1]E. Golemis, Fox Chase Cancer Center, Philadelphia; [2]J. Kamens, BASF, Worcester; [3]cumulative efforts of I. York, Dana-Farber Cancer Center, Boston and M. Sainz and S. Nottwehr, U. Oregon; [4]D. A. Shaywitz, MIT Center for Cancer Research, Cambridge; [5]R. Buckholz, Glaxo, Research Triangle Park; [6]J. Gyuris, Mitotix, Cambridge; [7]S. Hanes, Wadsworth Institute, Albany; [8]R. Brent, Dept. of Molecular Biology, MGH, Boston; [9]R. Finley, Wayne State University, Detroit. All plasmids and strains can be obtained by contacting the Brent laboratory at Massachusetts General Hospital, 617 726 5925 or brent@frodo.mgh.harvard.edu.

pRB1840, pJK103, and pSH18-34 comprise a series of *lacZ* reporters of different sensitivities to transcriptional activation that can be used to detect interactions of various affinities (Figure 4-3). These plasmids are LexA operator-containing derivatives of the plasmid LR1Δ1(West et al. 1984). In LR1Δ1, a minimal *GAL1* promoter lacking the upstream activating sequences (UAS_{GAL}) is located upstream of the bacterial *lacZ* gene. In pSH18-34, eight LexA operators have been cloned into an *Xho*I site located 167 bp upstream of the *lacZ* gene (Hanes and Brent, unpublished). pJK103 and pRB1840 contain two operators and one operator, respectively. Differential activation of LexA operator reporter constructs by a number of LexA-fused proteins (Golemis and Brent 1992), and by interacting pairs of activation domain-fused and LexA-fused proteins (Estojak et al. 1995) has been described.

EGY48 and EGY191 are *LexAoperator-LEU2* derivatives of the strain U457 (a gift of Rodney Rothstein) in which homologous recombination was used to replace the sequences upstream of the chromosomal *LEU2* gene with LexA operators derived from the colE1 gene (Ebina et al. 1983). EGY48 contains three such operators, and can bind up to 12 LexA monomers; EGY191 contains one such operator, and can bind up to four LexA monomers. Again, the differential sensitivity of these reporters allows the investigator to select reporters that can be used with baits with

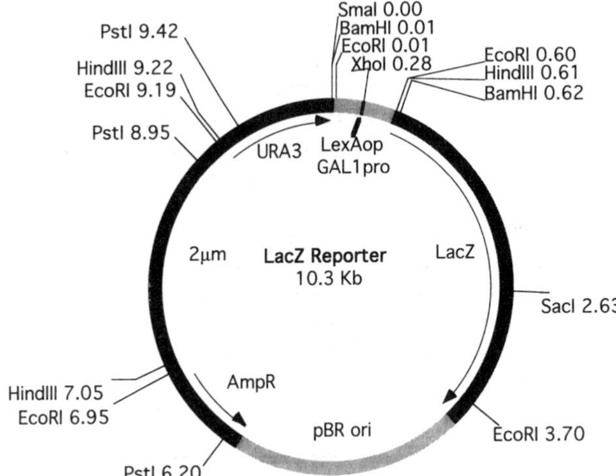

Fig. 4-3. *LacZ* reporter plasmid. pRB1840, pJK103, and pSH18-34 are all derivatives of LR1Δ1 (West et al. 1984) containing 8, 2, or 1 operator for LexA binding inserted into the unique *Xho*I site located in the minimal *GAL1* promoter (0.28 on map). The plasmid contains the *URA3* selectable marker and the 2 µ origin to allow propagation in yeast, and the ampicillin resistance (AmpR) gene and the pBR322 origin (pBR ori) to allow propagation in *E. coli*. Numbers indicate relative map positions.

weak intrinsic ability to activate transcription, and to exert some control over the affinity of the interactions being detected. Construction of these strains, as well as calibration of the sensitivity of the strains and *LexAoperator-lacZ* reporter series with interaction pairs of known affinity, is described in Estojak et al. (1995).

Controls

pJK101 is similar to pJK103, except it contains the *GAL1* UAS upstream of the two LexA operator sites. This plasmid is a derivative of Δ20B (West et al. 1984). It is used in the repression assay (see under heading of same name) (Brent and Ptashne 1984) to assess LexA-fusion binding to operator sites.

pSH17-4 encodes LexA fused to the activation domain of Gal4p. EGY48 cells bearing this plasmid will produce colonies on medium lacking leucine (Leu⁻), and yeast that additionally contain pSH18-34 will turn deep blue on plates containing X-Gal. This plasmid serves as a positive control for the activation of transcription.

pRFHM1 is a *HIS3*, 2μ plasmid that encodes LexA fused to the C-terminus of the *Drosophila* protein Bicoid. The plasmid has no ability to activate transcription, such that EGY48 cells that contain pRFHM1 and pSH18-34 do not grow on -Leu medium and remain white on plates containing X-Gal. LexA-Bicoid does bind DNA well; hence, it is also used as a positive control in the repression assay (see below). Finally, pRFHM1 is used as a control in specificity testing following a library screen, because it yields a transcriptional signal with library-encoded proteins that are unlikely to be genuine partners (R. Finley, unpublished observations).

Libraries

A number of libraries have been constructed in the vector pJG4-5. These include cDNAs derived from HeLa cells (constructed by J. Gyuris and R. Brent); serum starved WI-38 human fibroblasts (constructed by J. Gyuris, A. Mendelsohn, and R. Brent); human fetal brain (constructed by D. Krainc and R. Brent); CD4+ murine peripheral T cells (constructed by V. Prasad); CHO cells (constructed by V. Prasad); a series of *Drosophila* libraries derived from staged embryos, imaginal discs, and ovaries (constructed by R. Finley and R. Brent); and genomic DNA from *S. cerevisiae* (constructed by P. Watt and R. Brent). All can be obtained as described at the end of the chapter.

Preparation of Media and Other Required Solutions

Maintenance of Yeast

Most yeast strains can be grown on rich medium such as YEPD, and over short terms (<1 month) working stocks can be maintained as colonies on YEPD plates kept at 4°C. For long-term storage, yeast can be maintained by adding sterile glycerol to 15% volume of liquid cultures grown in YEPD, followed by freezing at -70°C.

The strains used in the following protocols have genetic deficiencies leading to their inability to grow on defined minimal medium lacking supplied tryptophan, histidine, uracil, and leucine. Leaving out one or more nutrients selects for yeast able to grow in its absence, for example, containing a plasmid that supplies the deficient gene. Thus, "dropout medium" lacking uracil (denoted -Ura in the following recipes) would select for the presence of plasmids with the *URA3* marker, etc.

Recipe, YEPD

YEPD plates (rich medium) : per liter, 10 g yeast extract, 20 g peptone, 20 g glucose, 20 g Difco Bactoagar. Autoclave ~18 minutes. Pour ~ 40 plates.

YEPD liquid: per liter, 10 g yeast extract, 20 g peptone, 20 g glucose. Autoclave ~15 minutes.

Recipe, CM-Plates and Liquid Media

General Directions for Defined Minimal Yeast Medium: All the minimal yeast media used in the following protocols, including liquid and plates, are based on the following three ingredients, which are sterilized by autoclaving 15 to 20 minutes:

per liter, 6.7 g *y*east *n*itrogen *b*ase (YNB) - amino acids (Difco 0919-15)
20 g glucose OR 20 g galactose + 10 g raffinose
2 g appropriate nutrient "dropout" mix (see last paragraph, this section)

For plates, 20 g Difco bacto-agar (Difco 0140-01) are also added. Approximately 35 to 40 plates should be obtained per liter.

A complete minimal nutrient mix includes the following:

Adenine	2.5 g
L-arginine	1.2 g
L-aspartic acid	6.0 g
L-glutamic acid	6.0 g
L-histidine	1.2 g
L-isoleucine	1.2 g
L-leucine	3.6 g
L-lysine	1.8 g
L-methionine	1.2 g
L-phenylalanine	3.0 g
L-serine	22.0 g
L-threonine	12.0 g
L-tryptophan	2.4 g
L-tyrosine	1.8 g
L-valine	9.0 g
uracil	1.2 g

These quantities of nutrients produce a quantity of dropout powder sufficient to make 40 liters of medium; it is advisable to scale down for most of the dropout combinations listed below. Premade dropout mixes are available from some commercial suppliers. Note: autoclaving too long can cause media for plates to become mushy, and not gel properly; this is to be avoided.

X-Gal Plates Start with the basic recipe for dropout plates, but mix basic ingredients in 900 mls of dH_20 rather than 1 liter.

Autoclave 20 minutes, and allow to cool to ~65°C or less. Then add 100 ml of sterile filtered 10× BU salts (1 liter 10× stock = 70 g $Na_2HPO_4{\cdot}7H_20$, 30 g NaH_2PO_4, pH adjusted to 7.0, filter sterilized), and 2 ml of a 40 mg/ml solution of X-Gal dissolved in dimethylformamide.

After plates have been poured, they should be stored at 4°C and protected from light. Note: pouring plates at temperatures between 50°C and 60°C rather than allowing them to cool further will reduce the tendency of bubbles to form in the solidifying agar.

Yeast Media Specifically Required for Interactor Hunts

Plates for growing yeast (100mm)

Defined minimal dropout plates, with glucose as a carbon source: Glu/CM-Ura; Glu/CM-Ura-His; Glu/CM-Ura-His-Trp; Glu/CM-Ura-His-Trp-Leu; Glu/CM-Ura X-Gal

Defined minimal dropout plates, with 2% galactose + 1% raffinose as a carbon source: Gal/Raff CM-Ura-His-Trp-Leu; Gal/Raff CM-Ura X-Gal

YEPD

Liquid medium for growing yeast

Defined minimal dropout media, with glucose as a carbon source:
Glu CM-Ura-His; Glu CM-Trp
YEPD

Plates for growing yeast library transformations (240 × 240 mm)

Defined minimal dropout plates, with glucose as a carbon source:
Glu/CM-Ura-His-Trp. Each plate requires ~200 to 250 mls medium.

Media for Bacterial Growth

At present, discrimination of the library plasmid from the other plasmids rescued from the yeast is generally accomplished by exploiting the ability of the yeast *TRP1* gene on the pJG4-5 vector to complement the *trpC* deficiency in the strain KC8. The second recipe provided describes how to make minimal plates appropriate for performing this selection.

Plates for Growing Bacteria (100mm)
LB supplemented with 50 μg/ml ampicillin

KC8 Plates for Selecting Library Plasmids

1. Autoclave 1 liter of dH_2O containing 15 g agar, 1 g $(NH_4)_2SO_4$, 4.5 g KH_2PO_4, 10.5 g K_2HPO_4, and 0.5 g sodium citrate·$2H_20$. Cool to 50°C.
2. Add 1ml sterile filtered 1M $MgSO_4.7H_2O$, 10 ml sterile filtered 20% glucose, and 5 ml each of 40 mg/ml sterile filtered stocks of L-histidine, K-leucine, and uracil. Add 1 ml of sterile filtered or autoclaved 1% thiamine HCl and 10 ml 10 mg/ml ampicillin. Pour. Note: sterile filtration should be done using a 0.45 or 0.2 micron filter.

Reagents to Prepare Carrier DNA for Transformations

High-quality salmon sperm DNA (for example, DNA Type III sodium salt from salmon testes, Sigma D1626, or equivalent grade from Boehringer Mannheim. Some investigators use Calf Thymus DNA D-8661 from Sigma.)
TE buffer, pH 7.5
TE saturated buffered phenol
1:1 (v/v) buffered phenol/chloroform
Chloroform
3 M sodium acetate, pH 5.2
100% (ice-cold) and 70% ethanol
Magnetic stirring apparatus, 4°C, and stir-bar
Sonicator with probe
High-speed centrifuge and appropriate tube
100°C and ice-water baths

Transformation Reagents (LioAC, LioAc-PEG, SSS DNA)

10 mM Tris-HCl pH 8.0, 1 mM EDTA, 0.1M lithium acetate, sterile-filtered through a 0.45 or 0.2 micron filter.
10 mM Tris-HCl pH 8.0, 1 mM EDTA, 0.1M lithium acetate, 40% PEG4000, sterile-filtered through a 0.45 or 0.2 micron filter.
Sterile DMSO
Sterile microscope slide and/or sterile glass balls, 4 mm, #3000, Thomas Scientific 5663L19
Sterile glycerol solution for freezing transformants (65% sterile glycerol, 0.1 M $MgSO_4$, 25 mM Tris-HCl pH 8.0)
Heat block or water bath set to 42°C.
Sterile flat-ended toothpicks (autoclaved on dry cycle for 30 to 40 minutes)

Reagents Required for Yeast Protein Lysates/Westerns

2× Laemmli Sample Buffer: 10% ß-mercaptoethanol, 6% SDS, 20% glycerol, 0.2 mg/ml bromphenol blue (10 mls can be prepared at a time, and is stable ~2 months at room temperature).
Antibody to LexA (polyclonal rabbit antiserum is available on request from the authors).

Reagents Required for Yeast Mini Preps for Recovery of Library Plasmids

STES lysis solution: (100 mM NaCl, 10 mM Tris-HCl pH 8.0, 1 mM EDTA, 0.1% SDS)
equilibrated phenol (pH ~7.0)
Chloroform
100% Ethanol
70% Ethanol
Acid washed sterile glass beads, 0.45 mm diameter

Characterization of the Bait

Before conducting an interactor hunt, you need to establish that your bait is suitable for screening purposes. Given the constraints of existing two-hybrid systems, some proteins are likely from first principles *not* to be suitable; for example, proteins with extensive transmembrane sequences are unlikely to translocate appropriately to the nucleus, or are likely to be improperly folded if they do so localize. Proteins known to be strong transcriptional activators will activate reporters to such a significant degree that it is impossible to score interactions over background. Others are unsuitable for unanticipated reasons; for example, a normally cytoplasmic protein kinase may turn out to have significant transcriptional activation potential when expressed as a LexA-fusion, making it necessary to truncate part of the protein to render it usable in the system. While a majority of proteins can be used successfully in two-hybrid approaches, it is necessary to avoid such obvious pitfalls.

The first group of five protocols below describes initial characterizations to establish parameters for successful screening. Protocol 1 describes a basic transformation of plasmid sets into yeast. Protocols 2 and 3 determine whether a bait activates transcription or not. Protocols 4 and 5 determine whether a bait binds LexA operator sites and whether it is expressed as a full-length protein. Finally, the summary of this section discusses recourses if any of the first protocols indicate that the bait is unsatisfactory for screening.

Protocol 1

1. Using standard subcloning techniques, insert the DNA encoding the protein of interest into the polylinker of pEG202 (Figure 4-1) or other LexA-fusion plasmid (Table 4-1) to make an in-frame protein fusion (pBait).
2. Perform six separate lithium acetate transformations of EGY48 using the following combinations of plasmids:
   a. Bait + pSH18-34 (test) + pJG4-5
   b. SH17-4 + pSH18-34 (positive control for activation) + pJG4-5
   c. RFHM1 + pSH18-34 (negative control for activation) + pJG4-5
   d. Bait + JK101
   e. RFHMI + JK101
   f. JK101

   A basic transformation procedure is as follows, and derives from protocols described in Ito et al. (1983):
3. The day before transformation, pick a single colony of EGY48 from a YEPD master plate into 5 mls of YEPD liquid medium. Grow with shaking overnight at 30°C.
4. The following morning, dilute the culture to a concentration equivalent to $OD_{600} = 0.15$ in a quantity of fresh YEPD, such that 10 mls are started for each intended transformation. Grow with shaking at 30°C until the culture has reached $OD_{600} = 0.50$ to 0.70 (usually about 4 to 6 hours).
5. Transfer the culture to 1 or more 50 ml conical (Falcon) tubes, and spin at

room temperature, 1500 × g, for 5 minutes to pellet the yeast. Pour off the supernatant.

6. Resuspend the yeast in an equivalent volume of sterile distilled water: repeat the spin conditions as in step 5.

7. Resuspend the yeast in TE/0.1M LiOAc solution, using 500 μl for each initial 100 mls (for example, 60 mls would reduce to 300 μl).

8. In an eppendorf tube, mix 2 to 5 μg of each DNA to be transformed (total volume of DNA less than 20 μl), 50 μl of yeast, and 350 μl of TE/0.1M LiOAc/PEG solution. Incubate 30 to 60 minutes at 30°C.

9. Place the tubes containing yeast and DNA into a 42°C heat block for 8 to 10 minutes, then remove them to room temperature.

10. Plate ~200 μl of each transformation mixture on Glu/CM-Ura-His-Trp (transformations a-c), Glu/CM-Ura-His (transformations d,e), or Glu/CM-Ura (transformation f) dropout plates. Incubate for two days at 30°C to select for yeast that contain both plasmids. Colonies obtained can be used simultaneously in tests for the activation of the *lacZ* (a-c, in Protocol 2) and *LEU2* (a-c, in Protocol 3) reporters, for repression (d-f, Protocol 4), and for bait expression (a-c, Protocol 5).

11. When colonies appear, make master plates containing ~8 to 10 independent transformants for each plasmid combination. Place the colonies in some kind of numbered and ordered array, so that it is possible to track individual colonies through subsequent tests. For reasons currently unknown, some fusion constructions are expressed only in a subset of colonies containing the plasmid; that is, in some cases, of 10 $Ura^+His^+Trp^+$ colonies, only 3 will be producing detectable LexA-Bait as determined by Western blot. Thus, it is prudent to perform library screens with a characterized individual colony. To streak, use the broad end of a sterile toothpick; touch this end to the colony, then lightly draw it for ~1 cm along a fresh plate.

Activation of the *lacZ* Reporter

Activation of transcription of a *LexAoperator-lacZ* reporter is one of the two tests used to detect positive interacting clones in a library screen. Hence, an important preliminary experiment is to determine the baseline activation of the *LexAop-lacZ* reporter by the bait and library vector.

Protocol 2

1. Prepare Glu/CM-Ura X-Gal and Gal/Raff CM-Ura X-Gal plates as described in the reagents section. Note, all X-Gal plates in these protocols drop out only uracil, even though in some cases yeast containing more than one plasmid are used on them. While yeast stocks should always be maintained on master plates with selection for all contained plasmids, in our experience, when patching from a master plate to X-Gal plates, plasmid loss is not a major problem even in the absence of selection, and the use of plates lacking only uracil allows one to assay sets of constructs on the same plate (to eliminate batch variation in X-Gal potency or plate preparation). Hence, we generally make plates either with complete minimal amino acid mix, or dropping out only uracil, to make the plates universally useful.

2. Streak transformations a-c from the master plate to a Gal/Raff CM-Ura X-Gal plate, and incubate it at 30°C. Streak at least 4 to 6 colonies from each transformation.

3. Examine the plate at intervals over the next 2 to 3 days. Strongly activating fusions and the pSH17-4 control should be visibly blue on the plate within 12 to 24 hours; moderate activators will be visibly blue after approximately 2 days; proteins that do not activate such as the pRFHMI negative control should still be essentially white after 4 days on the plate. Baits comparable to pRFHMI will be suitable for use in a screen; baits with some degree of activation may be (see Summary).

4. For baits that appreciably activate transcription under these conditions, there are several recourses. The first and simplest is to switch to the use of less sensitive *lacZ* reporter plasmids; use of JK103 and RB1840 may be sufficient to reduce background to manageable levels. If this switch fails to work, it is frequently possible to generate a truncated LexA-fusion that does not activate.

Activation of the *LEU2* Reporter

Actual selection in the interactor hunt is based on the ability of the bait protein and acid-fusion pair, but not the bait protein alone, to activate transcription of both the *LexAoperator-lacZ* gene and the *LexAoperator-LEU2* gene. The latter property allows growth on medium lacking leucine, and the test for the Leu requirement is thus the most important test of whether the bait protein is likely to have an unworkably high background. The pJG4-5 vector is included in the transformations scored because a small number of baits have been found to have weak background activation in the presence of the acid blob encoded as part of the activation domain fusion. The *LEU2* reporter in EGY48 is more sensitive than the pSH18-34 reporter for some baits (Estojak et al. 1995), so it is possible that a bait protein that gives little or no signal in a ß-galactosidase assay would nevertheless permit some level of growth on -Leu medium. If this growth occurs, there are several options for proceeding, the most immediate of which is to switch from EGY48 to EGY191, a less sensitive screening strain, and repeat the assay.

Protocol 3

1. Disperse a colony of EGY48 containing pBait and pSH18-34 reporter plasmids into 500 µl sterile water by vortexing in an eppendorf tube. Dilute 100 µl of the suspension into 1 ml sterile water. Make a series of 1/10 dilutions in sterile water to cover a 1000-fold concentration range.

2. Plate 100 µl from each tube (undiluted, 1/10, 1/100, and 1/1000) on Gal/Raff CM-Ura-His-Trp dropout plates and on Gal/Raff CM-Ura-His-Trp-Leu dropout plates. Incubate several days at 30°C. There will be a total of eight plates. Gal/Raff CM-Ura-His-Trp dropout plates should show a concentration range from 10 to 10,000 colonies and Gal/Raff CM-Ura-His-Trp-Leu dropout plates should have no colonies.

3. In parallel, from the master plate make patches of at least 4 to 6 colonies (the same colonies used for the *lacZ* test, Protocol 2) onto Gal/Raff CM-Ura-

His-Trp-Leu. While not as quantitative as the first assay, this assay will allow a determination of the activation of a large number of yeast cells, and represents a stringent test.

4. Inspect the growth of the yeast on the Gal/Raff CM-Ura-Trp-His and Gal/Raff CM-Ura-His-Trp-Leu plates over the next 4 to 5 days. If no growth is induced by the bait plasmid on the Leu plates in this period, the bait will probably be suitable. If extensive growth on plates lacking leucine is seen within 2 days (as should be seen with the positive control pSH17-4), the bait will almost certainly be unsuitable under any conditions, and the protein will have to be truncated.

If an intermediate phenotype is observed—for example, 1 to 3 positive colonies growing on Gal/Raff CM-Ura-His-Trp-Leu for every 1000 growing on Gal/Raff CM-Ura-His-Trp (step 2), or a limited amount of growth is seen for yeast patched thickly on Gal/Raff CM-Ura-His-Trp-Leu within 3 days (step 3), the pBait may be acceptable for screening, although it is possible that it will generate a high background. In that case, one option is to repeat the transformations and tests using a less sensitive *LexAoperator-LEU2* strain, such as EGY191.

Repression Assay

For LexA fusions that do not activate transcription, it is important to confirm that the LexA fusion protein is synthesized in yeast (some proteins are not, or are apparently clipped by intracellular proteases) and binds to the LexA operator sequences. This test is done by performing a "repression assay" (Brent and Ptashne 1984).

The principle behind a repression assay is that LexA bound to operator DNA situated between an enhancer element (UAS) and a transcriptional start site blocks the ability of the enhancer to activate transcription. Thus, the plasmid pJK101 contains the UAS for galactose (UAS_{GAL}), followed by LexA operators upstream of the transcriptional start site for the *lacZ* coding sequence. Yeast containing pJK101 have significant ß-galactosidase activity when grown on medium in which galactose is the sole carbon source because of binding of endogenous yeast Gal4p to the UAS_{GAL}. LexA-fused proteins that are made, enter the nucleus, and bind to the LexA operator sequences will block activation from the UAS_{GAL}, repressing ß-galactosidase activity to 1/3 to 1/20 the level of comparable yeast lacking a LexA-fusion. Note, on Glu/CM X-Gal medium, yeast containing JK101 should be white because UAS_{GAL} transcription is repressed.

Protocol 4

1. Transformations d-f from Protocol 1 will be used for this test.

2. Streak at least 4 to 6 colonies for each of these three transformations side by side on the same Glu/CM-Ura X-Gal and the same Gal/Raff CM-Ura X-Gal plate (that is, each plate should contain 12 to 18 colonies) and incubate the plates overnight at 30°C.

3. Examine the plates at ~12, ~24, and ~36 hours after streaking. All colonies

should be white on Glu/CM-Ura X-Gal, as the UAS_{GAL} should be repressed on this medium. On Gal/Raff CM-Ura X-Gal, yeast containing pRFHMI + pJK101 should be noticeably lighter in blue color than yeast containing pJK101. If your test pBait is comparable to the pRFHMI + pJK101 containing yeast, it is likely to be encoding a LexA-fusion that is capable of binding DNA.

4. Note: if this assay is run for more than 36 hours, the high basal *lacZ* activity of pJK101 will make differential activation of pJK101 impossible to see.

Also note: in cases where a pBait possesses intrinsic ability to activate transcription (as would be determined in Protocols 2 and 3), even though it may bind DNA, it will not repress. In such cases, the pBait + pJK101 combination will be comparable to or darker blue than pJK101 alone.

Finally, a small number of proteins do not score as positive in this assay, (suggesting that they occupy the operator less than 50% of the time) but occupy the operator sufficiently to work in library screens.

Synthesis of Full-Length Protein

If a bait protein neither activates nor represses, an immunoblot of a crude lysate using antibodies against LexA or the fusion domain should be performed to verify protein synthesis. Even if a bait protein represses, it is generally a good idea to verify that the LexA-fusion migrates appropriately on an SDS-PAGE gel, as occasionally some fusion proteins are cleaved by endogeneous yeast proteases. This clipping is most common with proteins greater than ~60 kD, and its effect is that a library screen performed with such a bait generally only identifies proteins that interact with regions of the bait amino-terminal (that is, LexA-proximal) to the clip site.

Protocol 5

1. From the Glu/CM-Ura-His-Trp master plate (Protocol 1), start a 2 ml culture in Glu/CM-Ura-His liquid medium for each bait being tested, and a 2 ml culture for a positive control for protein expression (for example, pRFHMI or pSH17-4). For each construct assayed, it is a good idea to grow colonies from at least two primary transformants, as levels of bait expression sometimes differ between colonies. Grow the cultures overnight at 30°C.
2. The following morning, from each overnight, restart a fresh 5 ml culture in Glu/CM-Ura-His at OD_{600} ~0.15. Grow the cultures again at 30°C.
3. When the culture has reached an OD_{600} of between 0.45 and 0.7 (approximately 4 to 6 hours), remove 1.5 mls to an eppendorf tube.
4. Spin the cells in a microfuge at 13,000 × *g* for 3 minutes. Inspection of the tube should reveal a pellet of ~1 to 3 µl volume. If this is not visible, spin the tube for another 3 minutes. When a pellet is visible, decant/aspirate the supernatant.
5. Working rapidly, add 50 µl of 2× Laemmli sample buffer to the tube, vortex, and place the tube on dry ice. Samples may be frozen at this stage at –70°C.
6. When you are ready to run a polyacrylamide gel (SDS-PAGE) prior to Western analysis, remove the samples directly to a boiling water bath or a PCR machine set to 100°C; boil them for 5 minutes.

7. Spin the tubes for 5 seconds in a microfuge to pellet the large cell debris.
8. Use ~20 to 50 µl to load each gel lane. Perform PAGE, blotting, and Western analysis by standard protocols. LexA-fusions can be visualized using an antibody to the fusion domain or an antibody to LexA.
9. Note that, for some LexA-fusion proteins, levels of the protein diminish as cells approach stationary phase. This decrease is caused both by the diminished activity of the *ADH1* promoter in late growth phases and the relative instability of particular fusion domains. Thus, it is not a good idea to let cultures become saturated in the hopes of getting a higher yield of protein to assay.
10. If a bait protein is made but does not repress, it may be inefficiently localized to the nucleus, resulting in poor results in a two-hybrid screen. Although this phenomenon is rare (Breitwieser and Ephrussi, personal communication), when it does occur it can be circumvented by recloning into a LexA-fusion vector that contains a nuclear localization motif, such as pJK202, or by modifying/truncating the fused domain to remove motifs that target it to other cellular compartments (e.g., myristoylation signals).

Summary

The preceding protocols suggest that the investigator begin with the most sensitive reporters followed by substitution with less sensitive reporters if the bait activates. You might wonder, why not start out with extremely stringent reporters and thus learn immediately whether the system is workable? In fact, some researchers routinely use a combination of pJK103 or pRB1840 with EGY191, and obtain biologically relevant partners. However, extensive comparison studies using interactors of known affinity with different *lacZ* and *LEU2* reporters (Estojak et al. 1995) have indicated that while the most sensitive reporters (pSH18-34) may in some cases be prone to background problems, the most stringent reporters (EGY191, pRB1840) can miss real interactions. Consistent with this idea, some baits have proven to isolate significant partner proteins from screens using EGY48, albeit amid high background, and these baits yielded no partners at all in parallel screens conducted with EGY191. In the end, the choice of reporters devolves to the preference of individual investigators, and results will be determined by the particular bait used; our bias is to cast a broad net in the early stages of a screen, and hence to use more sensitive reporters whenever practicable.

It happens very rarely that a bait that appeared to be well behaved and negative for transcriptional activation through all characterization steps will suddenly develop a very high background of transcriptional activation following library transformation, with very large numbers of colonies rapidly growing on -Leu medium. The reason(s) for this result is currently obscure, and no means of addressing this problem has as yet been found; such baits are generally inappropriate for use in screens.

Performing a Library Screen (Interactor Hunt)

Following characterization of a bait and establishment of optimal screening conditions, the next step is to perform a library transformation and screen for interactors. The following protocols describe how to produce carrier DNA for use in optimizing transformations (Protocol 6), how to perform a large scale transformation of yeast (Protocol 7), and how to isolate colonies with interacting proteins (Protocol 8). Subsequently, a procedure for reisolating positive cDNAs from yeast is described (Protocol 9), as well as a procedure for final confirmations of the specificity of the interactions (Protocol 10).

In contrast to *Escherichia coli*, the maximum efficiency of transformation that is routinely obtained for *S. cerevisiae* is ~10^4 to 10^5 per μg input DNA. It is extremely important to optimize transformation conditions before attempting an interactor hunt. Perform small-scale pilot transformations to ensure that this efficiency is attained to avoid having to use prohibitive quantities of valuable library DNA.

Protocol 6: Preparing High-Quality Carrier DNA

This protocol is derived from those described (Gietz et al. 1992; Schiestl and Gietz 1989); it generates high-quality sheared DNA for use as carrier in the transformation in Protocol 7.

1. Dissolve desiccated salmon sperm DNA in TE buffer at a concentration of 5 to 10 mg/ml by pipetting up and down in a 10-ml glass pipette. Place the DNA in a beaker with a stirbar and stir it overnight at 4°C to obtain a homogeneously viscous solution. It is important to use high-quality salmon sperm DNA. Sigma Type III sodium salt from salmon testes has worked well, as has a comparable grade from Boehringer Mannheim. Generally, it is convenient to prepare 20- to 40-ml batches at a time.
2. Shear the DNA by sonicating briefly using a large probe inserted into the beaker. The goal of this step is to generate sheared salmon sperm DNA (sss-DNA) with an average size of 7 kb, ranging from 2 to 15 kb. Oversonication (such that the average size is closer to 2 kb) drastically decreases the efficacy of the carrier in enhancing transformation. The original version of this protocol calls for two 30-sec pulses at three-quarter power, but optimal conditions vary between sonicators. The first time this protocol is performed, it is worthwhile to sonicate briefly, then test the size by fractionating a small aliquot alongside molecular weight markers on an agarose gel containing ethidium bromide, and sonicating further as needed.
3. Once DNA of the appropriate size range has been obtained, extract the sss-DNA solution by mixing it with an equal volume of TE-saturated buffered phenol in a 50-ml conical tube, shaking vigorously to mix, and centrifuging for 5 to 10 min at 3000 × g (or until a clear separation of phases is obtained). Transfer the upper phase containing the DNA to a fresh tube.
4. Repeat the extraction using 1:1 (v/v) buffered phenol/chloroform, then chloroform alone. Transfer the DNA to a tube suitable for high-speed centrifugation.

5. Precipitate the DNA by adding 1/10 vol of 3 M sodium acetate and 2.5 vol of ice-cold 100% ethanol. Mix by inversion, then pellet the DNA by spinning for 15 min at ~12,000 × g.

6. Wash the pellet with 70% ethanol. Briefly dry it either by air drying, or by covering one end of the tube with Parafilm with a few holes poked in and placing the tube under vacuum. Resuspend the DNA in sterile TE at 5 to 10 mg/ml. Do not overdry the pellet or it will be very difficult to resuspend.

7. Denature the DNA by placing the tube in a 100°C water bath for 20 min, then immediately transfer it to an ice-water bath.

8. Aliquot the DNA into microcentrifuge tubes and store them frozen at −20°C. Thaw a tube as needed. The DNA should be boiled again briefly (5 minutes) immediately before its addition to transformations.

9. Ideally, before using a new batch of sss-DNA in a large-scale library transformation, perform a small-scale transformation using suitable plasmids to determine the efficiency. Optimally, use of sss-DNA prepared in the manner described will yield transformation efficiencies of $>10^5$ colonies per μg input plasmid DNA.

Protocol 7: Transformation with an Interaction Library

1. When you are ready to screen a library, use the procedure outlined in Protocol 1 to transform yeast of the appropriate strain with pBait and the desired *lacZ* reporter plasmid. Plate the transformants on Glu/CM-His-Ura and incubate the plates at 30°C until colonies have been obtained, and make a master plate as before. If any heterogeneity of expression of the bait was observed in the first round of controls, it is a good idea to repeat the Western analysis to identify a colony expressing the Bait to be used in library transformation.

It is desirable to perform any final characterizations expeditiously; perhaps in part because of a gradual decline in plasmid copy number with time in storage, library screens work best when initial transformation of baits and reporters is done within 10 days of the subsequent library transformation.

It is important that one begins a library transformation with a strain that already contains both the reporter and the bait. Do not attempt to save time by cotransforming pBait, the *lacZ* reporter, and the library; this will result in an enormous decrease in the total number of transformants obtained, and thus constitute an inefficient use of the library.

The procedure detailed here describes a full transformation of 30 plates. If there is any doubt about the suitability of the bait, or if the investigator wishes to screen more than one library, it may be preferable to perform two or three smaller screens (10 to 15 plates each) to avoid wasting time and materials.

2. At least 2 to 3 days before a transformation is intended, pour the large transformation plates to be used; best results are obtained if these are allowed to dry out slightly for 1 to 2 days at room temperature before use. Note: it is easy to contaminate these plates; some researchers routinely flame them and expose them (uncovered, in tissue culture hoods) for 10 to 20 minutes of germicidal UV irradiation immediately after their setting.

3. Grow an ~20-ml culture of a chosen colony of EGY48 or EGY191 containing a *LexAop-lacZ* reporter plasmid and pBait in Glu/CM-Ura-His liquid dropout medium overnight at 30°C.

4. In the morning, dilute the culture into 300 ml Glu/CM-Ura-His liquid

dropout medium to ~2×10^6 cell/ml (OD_{600} ~0.10). Incubate at 30°C until ~1×10^7 cells/ml (OD_{600} ~0.50).

5. Centrifuge the yeast for 5 min at 1000 to 1500 × g in a low-speed centrifuge at room temperature to harvest the cells. Resuspend them in 30 ml sterile water and transfer them to a 50-ml conical tube.

6. Centrifuge the cells for 5 min at 1000 to 1500 × g. Decant the supernatant and resuspend the cells in 1.5 ml TE buffer/0.1 M lithium acetate.

7. Add 1 µg pJG4-5/library DNA and 50 µg high-quality sheared salmon sperm carrier DNA to each of 30 sterile 1.5-ml microcentrifuge tubes. Add 50 µl of the resuspended yeast solution from step 6. The total volume of library and salmon sperm DNA should be <20 µl, and preferably <10 µl.

8. Add 300 µl of sterile 40% PEG 4000/0.1 M lithium acetate/TE buffer pH 7.5 and invert the tubes to mix thoroughly. Incubate for 30 min at 30°C.

9. Add DMSO to 10% (~40 µl per tube) and invert the tubes to mix. Heat shock for 10 min in a 42°C heating block.

10. For 28 tubes: Plate the complete contents of one tube per 24 × 24-cm Glu/CM-Ura-His-Trp dropout plate and incubate the plates at 30°C. The easiest way to spread yeast rapidly and evenly on these large plates is to drop ~10 to 30 small sterile (autoclaved) glass balls (4 mm, from Thomas Scientific) onto the surface of the plates, close the lids, and agitate vigorously for ~2 minutes.

11. For the two remaining tubes: Plate 360 µl of each tube on 24 × 24-cm Glu/CM-Ura-His-Trp dropout plate. Use the remaining 40 µl from each tube to make a series of 1/10 dilutions in sterile water. Plate dilutions on 100-mm Glu/CM-Ura-His-Trp dropout plates. Incubate all plates at 30°C until colonies appear (2 to 3 days). The dilution series will give you an idea of the transformation efficiency and allow an accurate estimation of the number of transformants obtained.

Replica-plating does not work well in the selection process (Protocol 8) because so many cells are transferred to new plates that very high background growth inevitably occurs. The procedure described below creates a slurry in which cells derived from >10^6 primary transformants are homogeneously dispersed. A precalculated number of these cells from each primary transformant is then plated on selective medium.

12. Cool the thirty 24 × 24-cm plates containing transformants for several hours at 4°C to harden the agar.

13. Wearing gloves and using a sterile glass microscope slide, gently scrape yeast cells off the plate. Pool cells from the 30 plates into one or two sterile 50-ml conical tubes. This is the step where contamination is most likely to occur, so be careful.

14. Wash the cells by adding a volume of sterile water at least as large as that of the pellet, shaking the tube to resuspend yeast, and centrifuging for ~5 min at 1000 to 1500 × g at room temperature, and discarding the supernatant. Repeat the wash. After the second wash, the pellet volume should be ~25 ml cells from every 1.5×10^6 transformants.

15. Resuspend the pellet in 1 vol glycerol solution, mix well, and freeze the suspension in 1-ml aliquots at -70°C (stable at least 1 year). A typical transformation will yield ~30 to 50 mls of slurry.

Protocol 8: Selection of Putative Positive Colonies

1. Remove one tube of library transformation from the freezer. Dilute it in Gal/Raff CM-Ura-His-Trp until OD_{600} ~0.5, assumed to represent a concentration of 10^7 cells/ml. Grow the culture at 30°C with shaking for 4 to 6 hours to induce the library.

2. Plate 100 µl (10^6) of these cells on each of as many 100-mm Gal/Raff CM-Ura-His-Trp-Leu dropout plates as are necessary for full representation of the transformants. It is desirable that for each colony originally obtained in a transformation (determined in Protocol 7, step 11), 3 to 10 cells should be plated on selective medium. Hence, for a transformation in which 2×10^6 transformants had been obtained, you would wish to plate between 6×10^6 and 2×10^7 cells on 6 to 20 10 cm plates. Plating at higher density than 10^6 per plate can contribute to cross-feeding between yeast, resulting in spurious background growth.

3. Incubate the plates for 2 to 7 days at 30°C. Observe the plates each day. Generally, in a successful screen, interactors will cause colonies to arise 2 to 3 days after plating in EGY48, and slightly later in EGY191.

4. Carefully pick colonies arising to a Glu/CM-Ura-His-Trp master dropout plate. Incubate the plate at 30°C. A good strategy is to pick a master plate with colonies obtained on day 2, a second master plate (or set of plates) with colonies obtained on day 3, and a third with colonies obtained on day 4, and so forth. If a large number of colonies are obtained (>100), focus first on those arising during days 2 through 4. If no colonies appear within a week, those arising at later time points are likely to be artifactual.

Note: if contamination by airborne yeasts or mold has occurred at an earlier step (for example, during plate scraping), this is generally signaled by the growth of a very large number of colonies (>500/plate) within 24 to 48 hours after plating on selective medium.

5. To test for the Gal dependence of the Leu+ and *lacZ* phenotypes to confirm that they are attributable to expression of the library-encoded proteins, restreak or replica-plate potential positive colonies from the Glu/CM-Ura-His-Trp master plate to each of the following plates, and incubate them at 30°C:

a. Glu/CM-Ura X-Gal
b. Gal/Raff CM-Ura X-Gal
c. Glu/Raff CM-Ura-His-Trp-Leu
d. Gal/Raff CM-Ura-His-Trp-Leu

6. Examine plates a-d for the next several days. The X-Gal plates should be most informative at 1 to 2 days after streaking; results on the -Leu plates will be clear at 2 to 4 days after streaking. Colonies and the library plasmids that they contain are tentatively considered positive if they are blue on Gal/Raff CM-Ura X-gal but not blue, or only faintly blue, on Glu/CM-Ura X-gal plates, and if they grow on Gal/Raff CM-Ura-His-Trp-Leu but not Glu/CM-Ura-His-Trp-Leu plates.

7. The number of positives obtained will vary drastically from bait to bait. How they are processed subsequently will depend on the number initially obtained, the preference of the investigator, and the available manpower. If no positives are obtained using EGY48 as the reporter strain, it may be worth attempting to screen a library from a different tissue source. If a relatively small number (1 to 30) are obtained, the steps detailed in Protocol 9 can be followed essentially as written. However, sometimes searches will yield large numbers of

colonies (30 to 300 or more). In these cases, there are various options. Perhaps the most straightforward course is to "warehouse" the majority of the positives, and to work up the first 30 arising, following the observation that those growing fastest are frequently the strongest interactors. These can be checked for specificity, and restriction digests can be used to establish whether they are all independent cDNAs, or whether they represent multiple isolates of the same, or a small number, of cDNAs. If the former is true, it may be advisable to repeat the screen in a less sensitive strain (that is, switch from EGY48 to EGY191), as obtaining many different interactors can be a sign of a nonspecific background.

Protocol 9: Rapid Mini-Prep Reisolation of Plasmids

1. Starting from the Glu/CM-Ura-His-Trp master plate, pick colonies with the appropriate phenotype on selective plates into 5 ml of Glu/CM-Trp, and grow the cultures overnight at 30°C. Omitting the -His-Ura selection in this situation encourages loss of non-library plasmids, and facilitates isolation of the desired library plasmid.
2. Pellet 1 ml of the culture in a microfuge at 13,000 × g for 1 minute at room temperature.
3. Resuspend the pellet in 200 μl of lysis solution STES, and add ~100 μl of 0.45 mm diameter sterile glass beads. Vortex vigorously for 1 minute.
4. Add 200 μl equilibrated phenol, and vortex vigorously for another minute.
5. Spin the tube in a microfuge at 13,000 × g for 2 minutes at room temperature, and transfer the aqueous phase to a new microfuge tube.
6. Add 200 μl phenol and 100 μl chloroform, and vortex 30 seconds; spin as described in step 5, and transfer aqueous phase to a fresh tube.
7. Add two volumes 100% ethanol (approximately 400 μl), mix by inversion, and chill at -20°C for 20 minutes. Then spin the tube in a microfuge at 13,000 × g for 15 minutes at 4°C.
8. Pour off the supernatant. Wash the pellet with chilled 70% ethanol, and spin again at 13,000 × g for 2 minutes at 4°C. Pour off the ethanol, and dry the pellet briefly in a speed vacuum. Resuspend the pellet in 5 to 10 μl TE (10 mM Tris-HCl pH 8.0, 1 mM EDTA).
9. Use the isolated DNA to transform by electroporation KC8 bacteria (*pyrF, leuB600, trpC-9830, hisB463*, constructed by K. Struhl), and plate the bacteria on LB plates containing 50 to 100 μg/ml ampicillin (LB/Amp). The use of electroporation is essential to obtain transformants with KC8, as the strain is difficult to transform by other methods.

Note: an alternative procedure for rescuing plasmids from yeast into *E. coli* is the "Double-Zap" protocol. In this procedure, transformed yeast and a strain such as KC8 are suspended in water in an electroporation cuvette, and the mixture is shocked twice, once at high voltage to get plasmid DNA out of the yeast, and then at low voltage to make some of the *E. coli* permeable to it. Transformants are selected by plating this mixture on LB/Amp plates as above. This procedure is detailed at http://xanadu.mgh.harvard.edu.

10. The yeast *TRP1* gene can complement the bacterial *trpC-9830* deficiency, allowing the library plasmid to be easily distinguished from the other two plasmids contained in the yeast. For each original yeast positive, pick ~10 colonies onto defined minimal bacterial plates (KCB plates) supplemented with uracil, histidine, and leucine but lacking tryptophan; those colonies that grow will con-

tain the library plasmid. Some investigators omit the initial plating on LB/Amp media, plating directly onto -Trp or -Trp/Amp bacterial plates; in our experience, we find the two-step procedure helpful, as the less stringent LB/Amp selection maximizes the likelihood of obtaining transformants.

11. Grow KC8 containing the library plasmid in LB/Amp, and do mini preps by standard methods. This procedure will yield enough DNA for >5 transformations of yeast.

12. Some investigators are tempted to immediately sequence DNAs obtained at this stage. However, note that at this point, it is still possible that none of the isolated clones will express *bona fide* interactors; thus, we urge completion of the following specificity tests before committing the effort to sequencing.

Because multiple 2μ plasmids with the same marker can be simultaneously tolerated in yeast, it sometimes happens that a single yeast cell will contain two or more different library plasmids, only one of which encodes an interacting protein. Although the frequency of multiple transformants varies in the hands of different investigators, this phemomenon does account for the disappearance of apparent positives when the wrong cDNA is picked. When you choose colonies to mini prep, it is generally useful to work up at least two individual bacterial transformants for each yeast positive. These minipreps can then be digested with *Eco*RI + *Xho*I to release the cDNA inserts, and the size of inserts can be determined on an agarose minigel, to confirm that both plasmids contain the same insert.

An additional benefit of analyzing insert size is that it may provide some indication as to whether repeated isolation of the same cDNA is occurring, generally a good indication concerning the biological relevance of the interactor. Because the methodology of the screen involves plating multiple cells to Gal/Raff CM-Ura-His-Trp-Leu medium for each primary transformant obtained, multiple reisolates of true positive cDNAs are frequently obtained. If a large number of specific positives are obtained, it is generally a good idea to attempt to sort them into classes; for example, digesting mini preps of positives with *Eco*RI + *Xho*I + *Hae*III will generate a "fingerprint" of sufficient resolution to determine whether multiple reisolates of a small number of clones, or single isolates of many different clones, have been obtained. The former situation is the single best indication available that the system is working well for a given bait.

Establishment of the Specificity of an Interaction

Much spurious background will have been removed by the previous series of controls. Other classes of "false positives" are eliminated by retransforming purified plasmids into "virgin" *LexAop-LEU2/LexAop-lacZ/*pBait-containing strains that have not been subjected to Leu selection and verifying that interaction-dependent phenotypes are still observed. False positives include mutations in the initial EGY48 yeast that favor growth on Gal medium, library-encoded cDNAs that interact with the LexA DNA-binding domain, and proteins that result in transcriptional activity with multiple biologically unrelated fusion domains.

Protocol 10

1. In separate transformations, use purified plasmids from Protocol 9, step 11, to transform yeast that already contain the following plasmids and are growing on Glu/CM -Ura -His plates:
   a. EGY48 containing pSH18-34 and pBait
   b. EGY48 containing pSH18-34 and pRFHM1
   c. (Optional) EGY48 containing pSH18-34 and a nonspecific bait
2. Plate each transformation mix on Glu/CM-Ura-His-Trp dropout plates and incubate the plates at 30°C until colonies grow.
3. Create a Glu/CM-Ura-His-Trp master dropout plate for each library plasmid being tested. Adjacently streak five or six independent colonies derived from each of the transformation plates. Incubate the plates overnight at 30°C.
4. Restreak or replica-plate from this master plate to the same series of test plates used in the actual screen:
   a. Glu/CM-Ura X-Gal
   b. Gal/Raff CM-Ura X-Gal
   c. Glu/CM-Ura-His-Trp-Leu
   d. Gal/Raff CM-Ura-His-Trp-Leu

   As before, true "positive" cDNAs should make cells blue on Gal/Raff CM X-Gal but not on Glu/CM X-Gal plates, and should make them grow on Gal/Raff CM-Leu but not Glu/CM-Leu dropout plates only if the cells contain LexA-bait. cDNAs that meet such criteria are ready to be sequenced (for the primer, see legend to Figure 4-3) or otherwise characterized. Those cDNAs that encode proteins that result in a positive phenotype with either RFHM-1 or another nonspecific bait should generally be discarded, unless the interaction is of much less intensity than that with the specific bait.
5. If appropriate, or if there is any doubt about the results, conduct additional specificity tests. The three test plasmids outlined (pBait, pRFHM1, and nonspecific bait) represent a minimal test series. If other LexA-bait proteins that are related to the bait protein used in the initial library screen are available, substantial amounts of information can be gathered by additional specificity tests. For example, if the initial bait protein was LexA fused to the leucine zipper of c-Fos, specificity screening of interactor-hunt positives against the leucine zippers of c-Jun or Gcn4p in addition to that of c-Fos might allow discrimination between proteins that are specific for c-Fos versus those that generically associate with leucine zippers. Alternatively, added information about the interacting protein can be gained by testing it against a series of Lex-fusions containing subdomains of the original bait; this allows rapid mapping of the region(s) involved in interaction.
6. Note: DNA prepared from KC8 is generally unsuitable for dideoxy or automated sequencing even after use of Quiagen columns and/or cesium gradients. Although some investigators have reported it is suitable for cycle sequencing with ^{33}P (Cohen and Mendelsohn, unpublished), the more cautious approach is to introduce library plasmids from the KC8 miniprep into a more amenable strain such as DH5α before sequencing is attempted.

Variables Affecting Screens, and Reagent Availability

To maximize the chances of an interactor hunt working well, there are a number of parameters to be taken into account. To reiterate, before attempting a screen, you should carefully test bait proteins for low or no intrinsic ability to activate transcription. Baits must be expressed at reasonably high levels, and be able to enter the yeast nucleus and to bind DNA (as confirmed by the repression assay). By extrapolation from these conditions, proteins that have extensive transmembrane domains or are normally excluded from the nucleus are not likely to be productively used in a library screen. Proteins that are moderate to strong activators will need to be truncated to remove activating domains before they can be used. Optimally, the integrity and expression levels of bait proteins should be confirmed by Western analysis, using antibody either to LexA or to the fused domain [see Golemis and Brent (1992) for a discussion of relevant variables for LexA-fused proteins].

It is extremely important to optimize transformation conditions before attempting a library screen. Under the conditions we use, the maximum efficiency of transformation for *S. cerevisiae* is about 10^4 to 10^5/μg input DNA; small scale pilot transformations should be performed to ensure that this efficiency is attained to avoid having to use prohibitive quantities of library DNA. As for any screen, it is a good idea to obtain or construct a library from a tissue source in which the bait protein is known to be biologically relevant. Bendixen et al. (1994) and Finley and Brent (1994) have demonstrated the feasibility of using mating approaches, which should allow the library to be transformed once into a haploid strain of yeast. These transformants can be frozen in aliquots and used subsequently in screens with many different baits by mating this library of transformants against transformants carrying the baits. Investigators intending to do multiple interaction screens may find it worthwhile to investigate this option.

In hunts conducted to date, from zero to hundreds of positive colonies have been obtained for a given bait. Of these initial positives, from zero to practically all isolated plasmids have passed the final specificity test. If no initial positives are obtained, the tissue source for the library may not have been appropriate, and a different library may produce better results. However, there are some proteins that do not identify positives even from libraries in which cognate partners are thought to be expressed, suggesting that some proteins simply may not be amenable to a two-hybrid approach. One possible explanation for this is the fact that there are a number of cases on record in which two proteins known by other means (e.g., coimmunoprecipitation) to interact, can only be detected well in a two-hybrid assay with one component fused to the DNA-binding domain and the other fused to the activation domain [Estojak et al. (1995) "directionality"].

Conversely, some library-encoded proteins are known to be isolated repeatedly using a series of unrelated baits, and these proteins demonstrate at least some specificity. One of these, heat shock protein 70, might be explained by positing that it assists the folding of some LexA-fused bait proteins, or alternatively, that these bait proteins are not normally folded. Similarly, components of the ubiquitin-mediated degradation pathway are often isolated, consistent with the idea that the bait proteins may be subject to proteolysis. Other frequent isolates include certain ribosomal proteins containing extensive hydrophobic patches, or components of the cellular oxidative phosphorylation system. It is important to consider whether the protein isolated has plausible physiological relevance to its bait. It is also worth noting that although eukaryotic proteins that interact with LexA exist, in practice such proteins are isolated only very rarely.

When examining interactions using the *LEU2* and *lacZ* reporters in parallel, you will sometimes find that a combination of proteins will strongly activate one of the reporters, but only very slightly activate the other. There are many possible explanations for this result, including differences between the promoters, the fact that one reporter is integrated chromosomally and the other plasmid-borne, and so forth. In general, if a library plasmid activates both reporters to disparate degrees, but is specific for the bait used, it should not be arbitrarily discarded. The fact that different reporters can give different rank order of interaction affinity (Estojak et al. 1995) should lead one to regard quantitative transcriptional activity from the two-hybrid system as providing only a crude estimate of the affinity of two proteins.

How biologically relevant are protein interactions detected using this approach? Based on available data concerning the usual intracellular concentration of baits and library-encoded proteins (Golemis and Brent 1992), calibration experiments examining activation of different reporters by pairs of LexA- and activation domain-fused pairs of proteins of known in vitro interaction affinity (Estojak et al. 1995), and information from a large number of investigators regarding the outcome of their screens, it seems that for many proteins, the bulk of the interacting proteins isolated that pass specificity tests and that are isolated as multiple independent clones from a library will be significant. This is particularly true if a screen is run under stringent conditions (i.e., using less sensitive *lacZ* and *LEU2* reporters).

Finally, the fact that two proteins interact in the two-hybrid system is not rigorous proof that they associate directly or strongly in their native environment. It should be noted that a number of proteins are conserved extremely well between higher and lower eukaryotes; thus, an interaction between bait 1 and activation-domain-fused protein 2 might also include as mediating factors yeast proteins 3, 4, and 5. As with all assay systems, skepticism is a virtue.

Basic reagents for performing interaction traps can be obtained by contacting the Brent lab at Department of Molecular Biology, Massachusetts General Hospital, 40 Blossom Street, Boston, MA 02114; telephone (617) 726-5925; fax (617) 726-6893; email brent@frodo.mgh.harvard.edu. The majority of libraries for screening can be obtained from the Brent lab, with the exception of the CHO cell and murine T cell libraries, which can be obtained from Dr. Vinyaka Prasad at Albert Einstein Medical Center, New York; telephone (718) 430-2517; fax (718) 430-8711. Antibody to LexA can be obtained from the Brent lab or from Erica Golemis at W450, Fox Chase Cancer Center, 7701 Burholme Ave, Philadelphia PA 19111; telephone (215) 728-2860; fax (215) 728-3616; email EA_Golemis@fccc.edu. Further information on all topics mentioned here is available at the Brent lab/ protein interactions site, http://xanadu.mgh.harvard.edu.

Future Developments

For the library screens described in this chapter, the interaction trap is a mature technology. However, some developments promise to ease the task of library screening. The development of two-bait systems (Xu et al., unpublished) will aid the isolation of proteins that discriminate between closely related baits, such as those derived from wild-type and disease state gene alleles. The use of more sensitive reporter genes will allow a wider variety of partner proteins, including many biologically significant substrates for enzymes, to be detected. The use of mating methods (Bendixen et al. 1994; Finley and Brent 1994) may eventually obviate the need to transform the selection strain with a library, and will certainly continue to simplify the problem of distinguishing potentially signficant interactors from other positives.

For other applications of two-hybrid systems, many of which are described in this volume, the incremental improvements in technology will be equally valuable. Use of reporter genes of decreased sensitivity has already allowed the selection of peptide aptamers (Colas et al. unpublished data) that bind their targets more tightly. The advent of counterselectable reporter genes (chapter 7) may allow direct selection of aptamers, peptides, or small molecules that directly disrupt individual interactions. The use of fluorescent protein reporters (Colas, et al. unpublished data) will faciliate automated scoring and quantitation of interactions in arrays, and thus the charting of genetic regulatory networks. It is likely that these and other auxiliary two-hybrid strategies will provide a means of addressing the flood of data emerging from the Human Genome Project, and help shape the complex biology of the 21st Century.

ACKNOWLEDGMENTS Interaction trap technology as it currently exists is based on the cumulative efforts and generous contributions of a large number of individuals. Jeno Gyuris deserves much of the credit for the early work; he built the plasmid backbone and constructed the first libraries in pJG4-5. Russ Finley, Paul Watt, Vinyayka Prasad, and Dmitri Krainc constructed libraries as described in the text. Plasmids and strains currently used in this screening system were developed, in addition to the authors, by Joanne Kamens (pJK202, pJK101, pJK103), Steve Hanes (pSH18-34, pSH17-4), Russ Finley (pRFHMI), David Shaywitz (pGilda), Richard

Buckholz (pEE202I, pJG4-5I), and through the cumulative efforts of Manuel Sainz, Steve Nothwehr, and Ian York (pNLexA). Finally, many thanks are due to the many investigators who shared their results and comments with the authors, thereby contributing greatly to our understanding of the strengths and weaknesses of the system.

We are grateful to Current Protocols in Molecular Biology (John Wiley and Sons, New York) for permission to use Figures 1 through 3, and Table 1.

E.G. is supported in part by grants from the National Cancer Institute and the American Cancer Society. R.B. is supported by the Pew Scholars Program, Hoechst AG, and an ACS Faculty Research Award.

References

Bendixen, C., Gangloff, S. and Rothstein, R. (1994). A yeast mating-selection scheme for detection of protein-protein interactions. Nuc. Acids Res. 22:1778–1779.

Brent, R., and Ptashne, M. (1984). A bacterial repressor protein or a yeast transcriptional terminator can block upstream activation of a yeast gene. Nature 312:612–615.

Brent, R., and Ptashne, M. (1985). A eukaryotic transcriptional activator bearing the DNA specificity of a prokaryotic repressor. *Cell* 43:729–736.

Colas, P., Cohen, B., Jessen, T., Grishina, I., McCoy, J. and Brent, R. (1996). Genetic selection of peptide aptamers that recognize and inhibit cyclin-dependent kinase 2. Nature 380:548–550.

Ebina, Y., Takahara, Y., Kishi, F., Nakazawa, A., and Brent, R. (1983). LexA is a repressor of the colicin E1 gene. J. Biol. Chem. 258:13258–13261.

Estojak, J., Brent, R., and Golemis, E. A. (1995). Correlation of two-hybrid affinity data with in vitro measurements. Mol. Cell. Biol. 15:5820–5829.

Fields, S., and Song, O. (1989). A novel genetic system to detect protein-protein interaction. Nature 340:245–246.

Finley, R., and Brent, R. (1994). Interaction mating reveals binary and ternary connections between *Drosophila* cell cycle regulators. Proc. Nat. Acad. Sci. USA 91:12980–12984.

Gietz, D., St. Jean, A., Woods, R. A., and Schiestl, R. H. (1992). Improved method for high efficiency transformation of intact yeast cells. Nucl. Acids Res. 20:1425.

Gill, G. and Ptashne, M. (1988). Negative effect of the transcriptional activator GAL4. Nature 334:721–724.

Golemis, E. A. and Brent. R. (1992). Fused protein domains inhibit DNA binding by LexA. Mol. Cell. Biol. 12:3006–3014.

Golemis, E. A., Gyuris, J., and Brent, R. (1996). Interaction trap/two-hybrid system to identify interacting proteins. In *Current Protocols in Molecular Biology*, Ausubel, F. M., Brent, R., Kingston, R., Moore, D., Seidman, J., Struhl, S. J., and K., eds. New York, John Wiley and Sons. Ch. 20.1.1–20.1.28.

Gyuris, J., Golemis, E. A., Chertkov, H., and Brent, R. (1993). Cdi1, a human G1 and S phase protein phosphatase that associates with Cdk2. Cell 75:791–803.

Ito, H., Fukuda, Y., Murata, K., and Kimura, A. (1983). Transformation of intact yeast cells treated with alkali cations. J. Bacter. 153:163–168.

Kolodziej, P. A. and Young, R. A. (1991). Epitope tagging and protein surveillance. Methods Enz.194:508–519.

Lech, K., Anderson, K., and Brent, R. (1988). DNA-bound *Fos* proteins activate transcription in yeast. Cell 52:179–184.

Ma, J. and Ptashne, M. (1987). A new class of yeast transcriptional activators. Cell 51:113–119.

Ruden, D. M., Ma, J., Li, Y., Wood, K., and Ptashne, M. (1991). Generating yeast transcriptional activators containing no yeast protein sequences. Nature 350.

Schiestl, R. H. and Gietz, R. D. (1989). High efficiency transformation of intact yeast cells using single stranded nucleic acids as a carrier. Curr. Genet. 16:339–346.

West, R. W. J., Yocum, R. R., and Ptashne, M. (1984). *Saccharomyces cerevisiae* GAL1-GAL10 divergent promoter region: location and function of the upstream activator sequence UAS_G. Mol. Cell. Biol. 4:2467–2478.

5

Constructing an Activation Domain-Fusion Library

Li Zhu
David Gunn
Sailaja Kuchibhatla

The power of the two-hybrid system is, to a very large extent, based on its use to screen activation domain-fusion (AD-fusion) libraries in yeast for novel genes encoding proteins that interact with a protein of interest. Library screening is discussed elsewhere in this volume (see chapters 2, 3, and 4). In this chapter, we discuss the general principles of library construction and provide a detailed protocol for constructing an AD-fusion library in a plasmid vector. Although most widely used for two-hybrid screens to identify interacting proteins, AD-fusion libraries can also be used in one-hybrid screens to identify proteins based on their ability to bind specific DNA sequences (Luo et al. 1996; see chapter 17).

AD-fusion libraries have several critical features. First, as fusion libraries, each clone encodes a fusion of the Gal4p (or other) AD to whatever polypeptide is encoded in-frame by the cloned insert. Second, AD-fusion libraries are expression libraries, that is, the fusion proteins must be expressed in yeast cells for the two-hybrid assay to work. Third, because only protein-coding sequences are of interest in a two-hybrid screen, most AD-fusion libraries are cDNA libraries. In general, AD-fusion libraries made from genomic DNA can only be used for organisms such as bacteria and yeast that do not contain an extensive number of introns. Fourth, AD-fusion libraries must be constructed in—or converted to—plasmid vectors to allow transformation, maintenance, and selection in yeast cells. Specialized phagemid vectors, discussed at the end of this chapter, have been designed to permit the construction of AD-fusion libraries in the λ form, but these require subsequent conversion to plasmid form for use in two-hybrid assays.

Factors to Be Considered Before Constructing a Library

Choice of AD Plasmid Vectors

Almost without exception, the libraries used for two-hybrid screening are constructed in AD vectors rather than in the DNA-binding domain (BD) vectors that are also part of any two-hybrid screen. This vector choice is based on the extremely low probability that an insert will encode a sequence-specific DNA-binding protein that recognizes the promoters used in the yeast reporter strains. Therefore, libraries built in AD vectors usually do not produce high levels of false positives due to autonomous activation by an AD-protein X fusion. In contrast, the chance of an insert encoding a transcriptional activation domain is relatively high (Ruden et al., 1991), leading to high levels of autonomously activating false positives for libraries built in BD vectors (Bartel et al. 1993). Furthermore, a library constructed in an AD vector can also be used for one-hybrid screens.

Most AD libraries are constructed by cloning the inserts 3' to the end of the AD coding sequence (i.e., by making a fusion to the carboxy terminus of the Gal4p or VP16 AD). This is generally more convenient than cloning inserts upstream of the AD coding sequence for two reasons: (1) stop codons in the inserts still allow expression of a fusion protein (instead of terminating before the AD); and (2) translation of the fusion protein in all recombinant clones is automatically initiated from the same initiation codon (i.e., the ATG of the AD).

AD vectors each contain the following functional elements [illustrated in the map of pGAD10 (Figure 5–1)]:

1. A well-characterized AD (e.g., amino acids [a.a.] 768–881 of the yeast Gal4 protein, a.a. 411–455 of the herpes simplex virus VP16 protein or the *Escherichia coli* B42 sequence) under the control of either a constitutive (e.g., the yeast alcohol dehydrogenase [*ADH1*]) promoter or an inducible (e.g., *GAL1*) yeast promoter sequence; followed by
2. A multiple cloning site (MCS); followed by
3. Stop codons in all three reading frames and a transcriptional termination signal (e.g., the *ADH1* terminator);
4. A yeast selectable marker, such as the *LEU2* gene, and a yeast origin of replication (usually the 2 μ origin);
5. A bacterial selectable marker gene, such as the ampicillin-resistance gene (*bla*), and a bacterial origin of replication (for example, col E1).

Besides these essential features, there are numerous other functional features that can be incorporated into AD vectors. For example, most AD vectors have a nuclear localization signal (NLS) sequence (e.g., the NLS

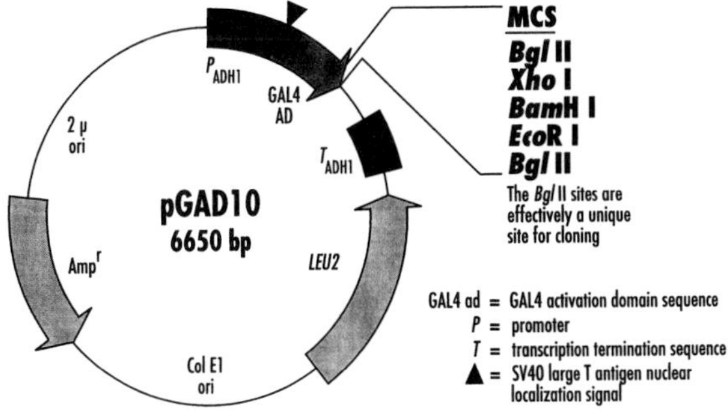

Fig. 5-1. pGAD10, a typical AD plasmid vector.

from the SV40 large T antigen; Kalderon et al. 1984) incorporated into the AD coding sequence to facilitate localization of the fusion protein to the yeast nucleus (Chien et al., 1991). In some cases, additional peptide sequences such as the hemagglutinin (HA) epitope are also incorporated into the AD coding sequence to allow detection of the fusion protein in yeast with antibodies to the epitope tag (Durfee et al. 1993). Finally, a strong bacterial promoter (such as that for T7 RNA polymerase) and an internal translation initiation codon can be integrated into the MCS region to allow in vitro transcription and translation of a tagged insert without the associated AD (pWITCH; Yavuzer and Goding 1995).

Several of the most commonly used AD plasmid vectors are described in Table 5-1, and several vector maps are provided (Figure 5-2). Some of the critical differences among various of the AD vectors are: (1) the choice of restriction sites in the MCS; (2) the selectable marker for yeast transformation; (3) the strength (and inducibility) of the promoter sequences that control the expression level of the AD-fusion protein in yeast; and (4) other functional features as discussed in the previous paragraph. It is convenient to use an AD vector in which the sites and reading frame of the MCS match the MCS of the BD vector. This will facilitate the exchange of inserts between AD and BD vectors once interacting proteins have been found. The yeast selectable marker on the AD vector must allow selection of the AD plasmids in the yeast host strain to be used for screening. A critical feature of the two-hybrid system is that the promoter used to drive expression of the reporter gene(s) in the yeast host strains must be recognized by the BD in the bait vector. However, any AD library can be used, as long as the vector can be selected in the yeast host strain. Thus, many AD-fusions libraries can be used for screening with a bait built in either a Gal4p or a LexA BD vector. Similarly, most AD libraries can also be used for one-hybrid screens, again provided that the selectable marker complements the auxotrophic requirement of the yeast host strain. With regard to promoter strength, it is not clear whether vectors that give higher expression levels are always better than low-level expression vectors. While high-expression

Table 5-1 Commonly Used Ad Plasmid Vectors

Vector	Size	Promoter	AD	Yeast selectable marker	Reference
pGAD10	6.6 kb	*ADH1* (fragment)	Gal4p	*LEU2*	Bartel et al. (1993)
pGAD GH	7.8 kb	*ADH1* (full-length)	Gal4p	*LEU2*	Hannon et al. (1993)
pGAD GL	6.9 kb	*ADH1* (fragment)	Gal4p	*LEU2*	Hannon et al. (1993)
pACT (lACT)	8.1 kb (42.6 kb)	*ADH1* (full-length)	Gal4p	*LEU2*	Dufree et al. (1993)
pACT2 (lACT2)	8.1 kb (42.6 kb)	*ADH1* (full-length)	Gal4p	*LEU2*	Harper et al. (1993)
pVP16	8.1 kb	*ADH1* (full-length)	VP16	*LEU2*	Hollenberg et al. (1995)
PJG4-5	6.2 kb	*GAL1* (inducible)	acid blob (*E. coli* B42)	*HIS3*	Gyruis et al. (1993)

vectors might be expected to improve sensitivity, there are reports of interacting clones being isolated from a library cloned in pGAD GL (a low-level expression vector), which could not be isolated from an equivalent library constructed in pGAD GH (a high-level expression vector) (G. Hannon, personal communication). The use of a low-level expression vector [or an inducible promoter in the AD vector (Gyrius et al. 1993)] may permit detection of interacting clones that would otherwise be toxic to the yeast cells. Finally, other functional features, such as nuclear localization signals, epitope tags, and embedded bacterial promoters, may improve sensitivity and facilitate troubleshooting and subsequent in vitro experiments. The use of phagemid AD vectors is discussed in "Alternative Protocols" at the end of this chapter.

Source of RNA

In building an AD-fusion cDNA library, primary consideration should be given to the source of RNA. Certain cell lines or tissues, such as HeLa cells or placenta, may be useful for general purposes. However, many researchers will want to search for interacting proteins that may be expressed only in a specific tissue or cell line, at a specific differentiation stage, or under special conditions such as mitogen stimulation. Whenever possible, libraries should be made from RNA derived from the appropriate tissue or cell line based on these expectations. For example, to construct a library to screen for interacting proteins involved in the early stages of apoptosis, we

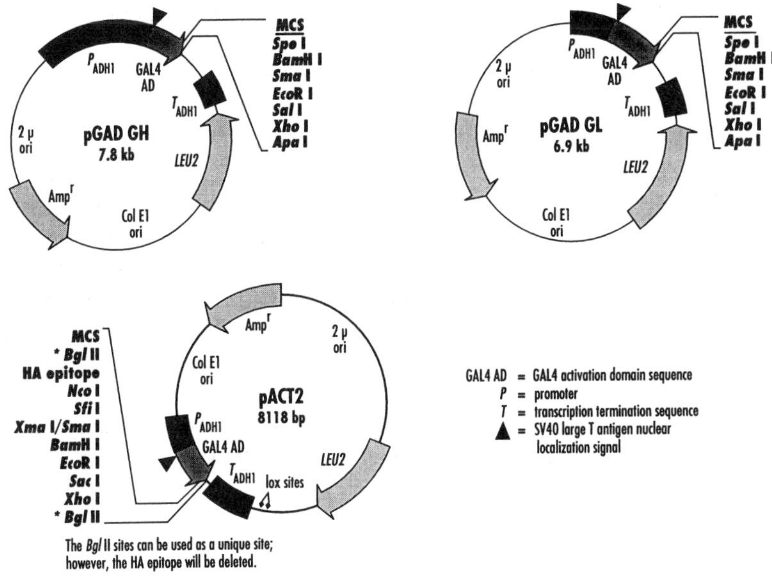

Fig. 5-2. Additional AD plasmid vector maps. For more information, see Table 5-1.

began with RNA isolated from the ovary glands of rats that had been treated with gonadotropin, which induces apoptosis in the ovarian follicle cells (Hseuh et al. 1994). However, it may not always be possible to get enough high-quality poly(A)+ RNA when the source material is limiting (10 μg of poly(A)+ RNA is recommended for the protocol described in this chapter).

Insert Length, Priming Method, and Directional versus Nondirectional Cloning

Unlike in the construction of other cDNA expression libraries, obtaining full-length cDNA is not absolutely essential for constructing a useful AD fusion library. Some investigators actually make the average insert length of their libraries quite short, based on the assumption that the minimum polypeptide length required for an interaction is probably in the range of 50 residues (Votjek et al. 1993; chapter 3). Shorter inserts also favor more efficient ligation of vector to insert. However, most researchers prefer longer inserts, based on the expectation that these will allow detection of multiple interactions or interactions that require complex protein folding or noncontiguous domains, while still permitting detection of interactions involving short contiguous domains.

The protocol described in this chapter combines these approaches by using two separate first-strand syntheses—one with random primers and one with oligo(dT). The random-primed reaction ensures that the library will contain clones encoding fusions to the amino-terminal and internal domains, while the oligo(dT)-primed clones will favor carboxy-terminal domains. The random-primed reaction also generates a wide size-range of cDNA, thus ensuring that both large and small inserts are included in the library. The oligo(dT)-primed reaction ensures that carboxy-terminal domains and full-length proteins will also be represented in the library. Insert length is further controlled by size fractionation of the adaptor-ligated double-stranded (ds) cDNA.

Some suggestions for directional cloning, which increases library complexity two-fold, are provided under "Alternative Protocols" at the end of this chapter.

Library Complexity

Library complexity is a critical determinant of whether or not screens of AD-fusion libraries will be successful. The complexity of a library is the number of independent clones present in the original, unamplified library. The more independent clones, the higher the complexity of the library and the better the chances are of finding even a very rare interacting protein. Generally, a library from any higher organism should contain at least one million independent clones. In order to maintain the initial complexity, libraries should be amplified by plating rather than in liquid culture. (Amplification is usually necessary to generate enough DNA for transformation into yeast and subsequent library screening.) Multiple rounds of library amplification should be avoided, since this can significantly reduce library complexity.

Other Tests of Library Quality

A second critical parameter for determining library quality is the percentage of recombinant clones. Before the original library is amplified, the percentage of recombinant clones should be determined either by PCR screening of randomly selected clones using primers that flank the cloning site or by digestion of mini-prep DNA with the restriction enzyme used for cloning. An example of insert screening by PCR is presented in Figure 5-3. (Blue-white screening, which is widely used to determine the percentage of recombinant clones with many library cloning vectors, is not possible with AD vectors.) One can also test AD-fusion libraries by checking (via hybridization) for normal representation of select housekeeping genes such as ß-actin.

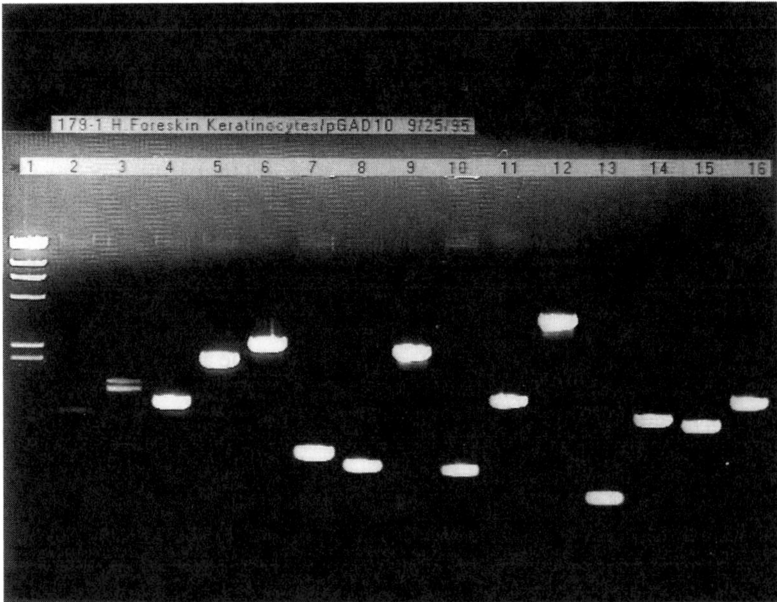

Fig. 5-3. Estimating the percentage of recombinant clones by insert-screeing. 15 individual colonies (lanes 2–16) were picked from the original plating of a human keratinocyte library cloned in pGAD10. PCR reactions were performed directly on each clone using primers that flank the cloning site. All 15 clones contained inserts, indicating that a very high percentage of the clones in this library contain inserts. Lane 1 is a DNA size marker.

A practical test of library quality is to screen the library with a bait that is known to interact with proteins that are expected to be expressed in the newly constructed library (L. Zhu, unpublished observations). For example, we have "prescreened" human brain libraries in this manner by using a bait construct that expresses the p53 protein to isolate clones that express 53BP2, a protein known to interact with wild-type p53 (Iwabuchi et al. 1994). If such screens fail to identify the expected interacting proteins, you may have trouble isolating proteins that interact with a novel bait protein. However, appropriate test proteins are not available for all libraries, and failure to isolate the expected clones is not an absolute indication that the library will yield no positives.

Protocol Overview

The protocol presented in this chapter describes the nondirectional cloning of a library primed with a combination of oligo(dT) and random primers into a plasmid vector. This procedure, outlined in Figures 5-4 and 5-5, has

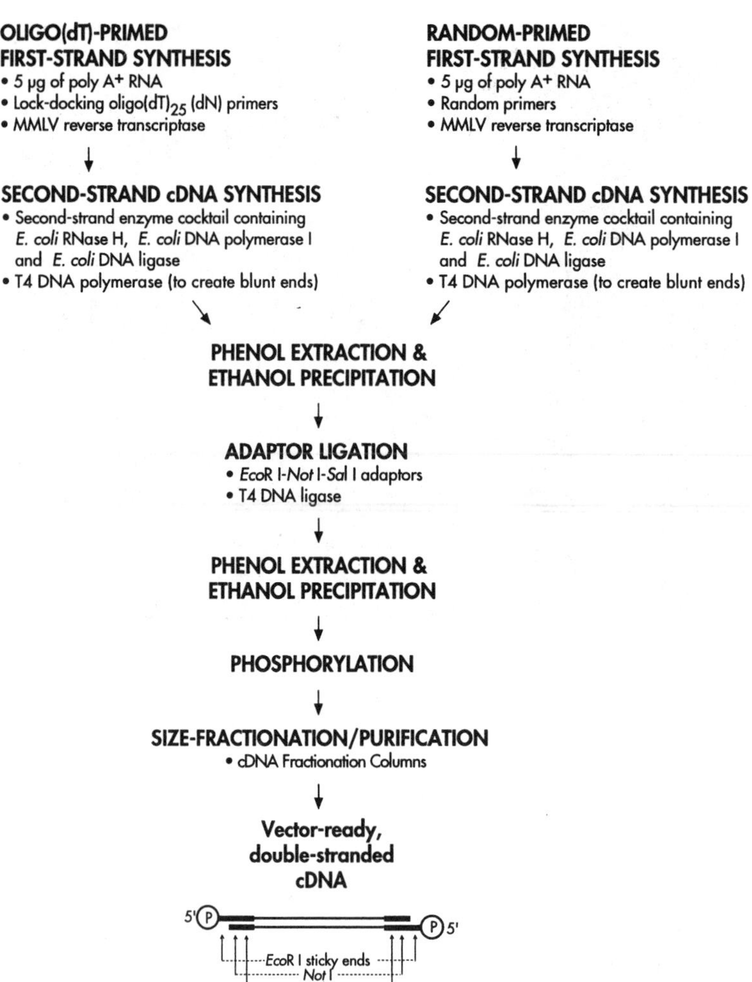

Fig. 5-4. Flowchart for AD-fusion library construction through size-fractionation of adaptor-ligated ds cDNA. The oligo(dT)- and random-primed cDNA samples are kept separate through completion of second-strand synthesis.

been used to prepare dozens of AD-fusion libraries at CLONTECH Laboratories. Options for directional cloning and phagemid vectors are discussed at the end of this chapter.

The protocol begins with two separate first-strand synthesis reactions: one with random primers and one with oligo(dT). In the oligo(dT)-primed reaction, a degenerate nucleotide can be incorporated onto the 3' end of the oligo(dT) primer to create a "lock-docking" oligo(dT) primer that positions the primer at the beginning of the poly-A tail and thus eliminates synthesis of lengthy poly(dT) (Borson et al. 1992). Both first-strand reactions are

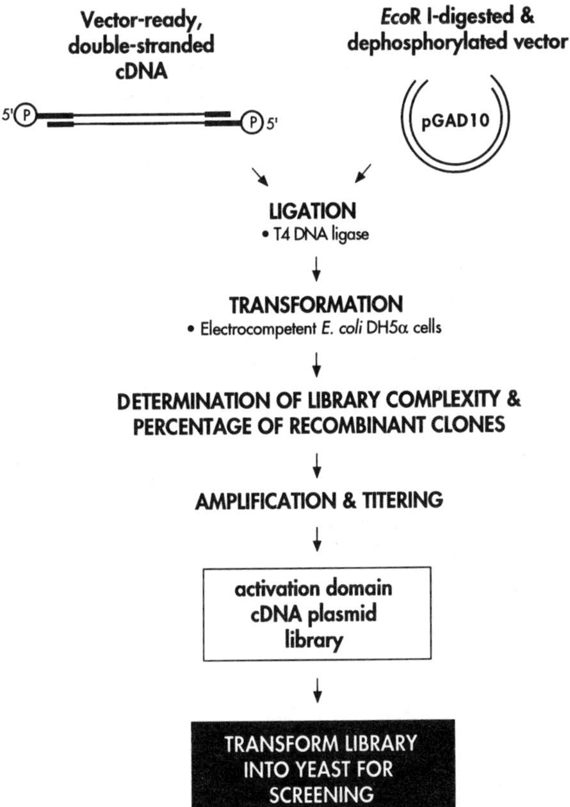

Fig. 5-5. Flowchart for final steps in construction of an AD-fusion library in a plasmid vector.

performed with MMLV reverse transcriptase. Second-strand synthesis is performed essentially according to the method of Gubler and Hoffmann (1983) using a "second-strand enzyme cocktail" in which the ratio of DNA polymerase I to RNase H has been optimized to increase the efficiency of second-strand synthesis and to minimize the formation of, and subsequent priming by, hairpin loops.

Following second-strand synthesis, the double-stranded (ds) cDNA is treated with T4 DNA polymerase to create blunt ends via the enzyme's 3'–5' exonuclease activity, and the two populations of cDNA (random- and oligo(dT)-primed) are pooled. The cDNA is ligated to an adaptor that has a pre-existing *Eco*RI "sticky end" (Figure 5-6). The use of an adaptor instead of a linker eliminates the need to methylate and then digest the cDNA with a restriction enzyme, and thus leaves internal restriction sites intact in the cloned cDNAs. The adaptor must be 5'-phosphorylated at the blunt end only to allow efficient ligation to the blunt-ended cDNA. Adaptors can be

EcoR I — *Not* I — *Sal* I

5' - AATTCGCGGCCGCGTCGAC
GCGCCGGCGCAGCTG -PO_4-5'

Fig. 5-6. Structure of an adaptor that can be used in the library construction protocol.

designed to include additional restriction sites that provide several options for excising cDNA inserts.

After adaptor ligation, the ds cDNA is phosphorylated and then purified and size-fractionated to remove unligated adaptors, unincorporated nucleotides, and most of the cDNA that is smaller than 300 bp. The ds cDNA is then ligated to a suitable AD vector that has been predigested with the appropriate enzyme and dephosphorylated. The resulting library is then transformed via electroporation into an *E. coli* host strain such as DH5α. Following amplification, titering, and DNA purification, the library can be transformed, along with the bait plasmid, into a suitable yeast host strain for a two-hybrid screen.

Protocols

A. Required Materials

MMLV reverse transcriptase (200 units/μl)
Conventional oligo$(dT)_{15-18}$ primers (15 μM) or "lock-docking" oligo$(dT)_{25}$(dN) primers (15 μM) (N = A, C, or G)
Random p$(dN)_6$ primers (15 μM; 0.1 μg/μl)
dNTP mix (20 mM each dNTP)
DTT (100 mM)
[α-^{32}P] dCTP (800 Ci/mmol)
5X First-strand buffer:
- 250 mM Tris-HCl (pH 8.3)
- 30 mM $MgCl_2$
- 375 mM KCl

Second-strand enzyme cocktail:

RNase H	0.25 units/μl
E. coli DNA polymerase I	6 units/μl
E. coli DNA ligase	1.2 units/μl

5X Second-strand buffer:
- 500 mM KCl
- 50 mM Ammonium sulfate
- 25 mM $MgCl_2$
- 0.75 mM ß-NAD
- 100 mM Tris-HCl (pH 7.5)
- 0.25 μg/ml Bovine serum albumin

T4 DNA polymerase (5 units/μl)

Sodium acetate (3 M; pH 4.8)
Ammonium acetate (4 M)
EDTA (0.2 M)
T4 DNA ligase (6 Weiss units/μl)
ATP (10 mM)
*Eco*RI/*Not*I/*Sal*I hemiphosphorylated adaptors (50 μM; see Figure 5-6)
10X ligation buffer (used for adaptor ligation, phosphorylation, and ligation of insert to vector)
 500 mM Tris-HCl (pH 7.8)
 100 mM $MgCl_2$
 100 mM DTT

This buffer is used in this protocol for adaptor ligation and phosphorylation, as well as for the ligation of the adaptor-ligated ds cDNA to the vector. However, we have obtained libraries of significantly higher complexity using any of several proprietary commercial buffers that give high-efficiency ligation of inserts to plasmid vectors. These buffers also permit the ligation time to be reduced to 30 minutes.

T4 polynucleotide kinase (30 units/ml)
Columns for size fractionation of ds cDNA. [These can be made and used as described on pp. 8.70–8.72 of Sambrook et al. (1989). Alternatively, there are several commercial sources of disposable columns for size-fractionation of cDNA.]
Glycogen (20 μg/μl)
Sterile H_2O, treated with diethylpyrocarbonate (DEPC) and autoclaved
Phenol:chloroform:isoamyl alcohol (25:24:1)
Chloroform:isoamyl alcohol (24:1)
95% Ethanol
70% Ethanol
STE buffer:
 30 mM NaCl
 10 mM Tris-HCl (pH 7.8)
 0.5 mM EDTA
pGAD10 AD vector (1 μg/μl; see Table 5-1 for alternative vectors)
*Eco*RI restriction enzyme and 5X buffer (or suitable alternative)
Calf intestinal alkaline phosphatase (CIAP; 1 U/μl)
10X CIAP buffer: Use the buffer supplied by the manufacturer with the enzyme
Electroporator (for example, Gene-Pulsar or *E. coli* Pulsar from BioRad)
Electrocompetent cells of a suitable *E. coli* host strain such as DH5α.

Electrocompetent cells should be prepared according to the protocol recommended by the manufacturer of your electroporator and resuspended at ~3×10^{10} cells/μl. Cells must have a transformation efficiency of 0.5–1.0×10^{10} cfu/μg as determined in a test ligation with an intact plasmid vector. Highly electrocompetent DH5α are available from several commercial sources. Chemically competent cells generally do not have high enough transformation efficiencies for constructing high-complexity plasmid libraries.

1 Liter LB

20 90-mm LB agar plates with appropriate antibiotic (typically 50 mg/ml ampicillin)

75 150-mm LB agar plates with appropriate antibiotic (typically 50 μg/ml ampicillin)

0.1-cm cuvettes for electroporation

(Optional) PCR primers for insert screening to determine the percentage of recombinant clones

B. General Considerations and Preparation of poly(A)+ RNA

1. Intact and pure poly(A)+ RNA is a prerequisite for synthesis of high-quality cDNA. To avoid contamination and degradation of RNA (and to minimize the presence of RNases), use the following precautions:
   a. Wear gloves to avoid RNase contamination from your hands.
   b. Treat H_2O with DEPC. Do not use DEPC to treat solutions containing Tris, nucleotides, or other amines.
   c. Rinse all glassware with 0.5 N NaOH, followed by DEPC-treated H_2O. Then bake the glassware at 160° to 180°C for 4 to 9 hours.
   d. Wipe pipettes with 70% ethanol or isopropanol before use with RNA.
2. AD-fusion cDNA expression libraries should be constructed from high-quality poly(A)+ RNA. Many procedures are available for the isolation of poly(A)+ RNA; for a comprehensive review, see Farrell (1993) or Sambrook et al. (1989). There are several commercial kits for the isolation of total and poly(A)+ RNA, as well as commercial sources of poly(A)+ RNA.

   The sequence complexity of the ds cDNA synthesized and, ultimately, of the cDNA library constructed, depends on the quality of the experimental RNA starting material. Therefore, before you use the RNA in a first-strand synthesis, we recommend that its integrity be checked by examining a sample on a denaturing formaldehyde/agarose gel.
3. Perform all reactions on ice, unless otherwise indicated.

C. First-Strand cDNA Synthesis

As just discussed, the following protocol includes separate reactions for random- and oligo(dT)-primed cDNA. The two separate pools of ds cDNA are combined at step D7. The use of a small amount of [α-^{32}P]dCTP in both first-strand synthesis reactions permits precise determination of which fractions of size-selected cDNA should be pooled at step G6, and facilitates troubleshooting of the cDNA synthesis reaction via examination of the cDNA by gel electrophoresis and autoradiography.

1. Combine the following in separate sterile microcentrifuge tubes:

   Tube 1: Random-primed cDNA synthesis

   5 μg poly(A)+ RNA sample

   5 μl Random $p(dN)_6$ primers

   Tube 2: Oligo(dT)-primed cDNA synthesis

   5 μg poly(A)+ RNA sample

   5 μl Conventional $oligo(dT)_{15-18}$ or lock-docking $oligo(dT)_{25}(dN)$ primers

2. Add sterile DEPC-treated H_2O to a final volume of 12.5 µl, mix the contents, and spin the tubes briefly.
3. Incubate the tubes in a 70°C water bath for 3 minutes.
4. Cool the tubes in an ice bath for 2 minutes. Spin the tubes briefly in a microcentrifuge.
5. Add the following to each reaction tube (each tube should already contain 12.5 ml of primed RNA):

   2.2 µl sterile DEPC-treated H_2O
   5.0 µl 5X first-strand buffer
   0.5 µl DTT
   1.3 µl dNTP
   1.0 µl [α-^{32}P]dCTP (800 Ci/mmol)
   ____
   22.5 µl total volume
6. Add 2.5 µl (500 units) of MMLV reverse transcriptase to each reaction tube.
7. Mix contents thoroughly, spin the tubes briefly, and incubate the tubes at 42°C for 1.5 hours.
8. Place the tubes on ice to terminate first-strand synthesis. Proceed directly to second-strand synthesis.

D. Second-Strand cDNA Synthesis

Keep tubes on ice while adding reagents.

1. Add the following to the separate reaction tubes containing the oligo(dT)- and random-primed first-strand cDNA (each tube already contains 25 µl):

   123.5 µl sterile DEPC-treated H_2O
   40 µl 5X second-strand buffer
   1.5 µl dNTP mix
   10 µl second-strand enzyme cocktail
   ____
   200 µl total volume

   Mix contents thoroughly and spin the tubes briefly.
2. Incubate the tubes at 16°C for 2 hours.
3. Add 3 µl (15 units) of T4 DNA polymerase.
4. Incubate the tubes at 16°C for 30 minutes.
5. Add 10 µl of 0.2 M EDTA to each tube to terminate second-strand synthesis.
6. Add 200 µl of phenol:chloroform:isoamyl alcohol to each tube, mix by continuous gentle inversion for 2 minutes, and spin the tubes in a microcentrifuge at 12,000 rpm for 2 minutes.
7. Remove the top aqueous layers from each tube (that is, the oligo(dT)- and random-primed ds cDNA) and combine both upper phases in a separate, clean microcentrifuge tube. Discard the interphases and lower phases.
8. Add 400 µl of chloroform:isoamyl alcohol to the combined aqueous layer, mix by continuous gentle inversion for 2 minutes, and spin the tube in a microcentrifuge at 12,000 rpm for 2 minutes.
9. Remove the top aqueous layer and place in a separate, clean 1.5 ml microcentrifuge tube. Discard the interphase and lower phase.
10. Divide the sample by removing half of the volume (~ 200 µl) and placing it in a second, clean microcentrifuge tube. (The pooled ds cDNA is divided

into two equal aliquots to maintain the optimal volumes and concentration for adaptor ligation and phosphorylation.)

11. Add 200 μl of 4 M ammonium acetate and 1 ml of cold 95% ethanol to each tube (1:1:5 ratio of sample, 4 M ammonium acetate, and ethanol). Mix by gentle rocking. (Ammonium acetate prevents dNTPs from coprecipitating with the cDNA.)
12. Place the tubes in a dry-ice/ethanol bath for 10 minutes, then spin the tubes in a microcentrifuge at 15,000 rpm for 20 minutes.
13. Carefully remove the supernatant with a pipette. Do not disturb the pellet. Save the supernatant until you are sure you have recovered your cDNA.
14. Dry the pellets under a vacuum for 1–2 minutes (or air dry for 15 minutes).
15. Check the radioactivity in the pellet with a hand-held counter. 100,000 cpm or higher is indicative of successful cDNA synthesis.

E. Adaptor Ligation

The following protocol should be performed with the two identical tubes prepared above. The pellet in each tube contains a mix of oligo(dT)-primed and random-primed ds cDNA.

1. Dissolve each pellet in 15 μl of sterile DEPC-treated H_2O.
2. Add the following to each tube:
   3 μl 10X ligation buffer
   3 μl 10 mM ATP
   7 μl *Eco*RI adaptors
   2 μl T4 DNA ligase (6 Weiss units/μl)
   30 μl total volume
   Mix the contents of each tube and spin briefly in a microcentrifuge.
3. Incubate at 16°C overnight.
4. Add 3 μl of 0.2 M EDTA to each tube to terminate the reaction.
5. Add 70 μl of sterile DEPC-treated H_2O to each tube.
6. Add 100 μl of phenol:chloroform:isoamyl alcohol to each tube and mix by continuous gentle inversion for 2 minutes, then spin the tubes in a microcentrifuge at 12,000 rpm for 2 minutes.
7. Combine the two samples by transferring both top (aqueous) layers to the same clean 1.5-ml microcentrifuge tube. Discard the interphases and lower phases.
8. Add 200 μl of chloroform:isoamyl alcohol and mix by continuous gentle inversion for 2 minutes, then spin the tube in a microcentrifuge at 12,000 rpm for 2 minutes.
9. Transfer the top (aqueous) layer to a clean 1.5-ml microcentrifuge tube. Discard the interphase and lower phase.
10. Add 20 μl of 3 M sodium acetate and 500 μl of cold 95% ethanol, mix by gentle rocking and place the tube in a dry-ice/ethanol bath for 1 hour.
11. Spin the tube in a microcentrifuge at 15,000 rpm for 20 minutes.
12. Carefully remove the supernatant with a pipette, taking care not to disturb the pellet. Save the supernatant until you are sure you have recovered your cDNA.
13. Dry the pellet under a vacuum for 1 to 2 minutes (or air dry for 15 minutes).

14. Confirm recovery of the ds cDNA by checking the pellet with a hand-held counter.

F. Phosphorylation of Adaptor-Ligated ds cDNA

1. Resuspend the pellet in 14 µl of sterile DEPC-treated H_2O.
2. Add the following to the tube:
   2 µl 10X ligation buffer
   2 µl 10 mM ATP
   2 µl T4 polynucleotide kinase
   20 µl total volume
   (The ligation buffer is also suitable for phosphorylation.)
3. Incubate at 37°C for 30 minutes.
4. Add 2 µl of 0.2 M EDTA to terminate the reaction.
5. Incubate at 70°C for 15 minutes.
6. Incubate in an ice bath for at least 2 minutes.

G. Size Fractionation of Adaptor-Ligated ds cDNA

It is critical to size-fractionate the cDNA prior to ligation to vector. Besides removing most of the cDNA fragments smaller than ~300 bp, this also removes the unincorporated nucleotides and unligated adaptors. The [^{32}P]-dCTP incorporated during synthesis will help you determine which fractions to take for ligation. Detailed protocols for size fractionation are described in Sambrook et al. (1989; pp. 8.70–8.72). Alternatively, there are several commercially available columns for the size-fractionation of ds cDNA. These should be used according to the manufacturer's protocol. Regardless of which protocol is used, all of the fractions selected should be pooled prior to ethanol precipitation and resuspended in a minimal volume of TE buffer as follows.

Ethanol precipitation of pooled fractions:

1. Add glycogen to a final concentration of 0.2 µg/µl to coprecipitate cDNA. This will make small cDNA pellets visible.
2. Add 1/10 volume of 3 M sodium acetate (pH 4.8). Then add 2.5 volumes (2.5 X the volume following addition of sodium acetate) of 95% ethanol. Mix by gently rocking the tube back and forth.
3. Place the tube in a dry-ice/ethanol bath for at least 1.5 hours to precipitate the DNA. Alternatively, overnight incubation at –70°C may produce better recovery.
4. Spin the tube in a microcentrifuge at 15,000 rpm at room temperature for 20 minutes.
5. Pour off the ethanol layer and dry the pellet.
6. Resuspend the pellets in a total of 15 µl of TE buffer.
   Store at –20°C until ready for use.

H. Preparation of Linearized, Dephosphorylated Vector DNA

The following protocol describes the preparation of *Eco*RI-digested, dephosphorylated pGAD10. The protocol can be modified accordingly for different vectors or restriction enzymes.

1. Combine the following reagents in a 1.5-ml microcentrifuge tube.

pGAD10 (1 μg/μl)	25 μl
5X *Eco*RI Buffer	30 μl
*Eco*RI (10 U/μl)	12.5 μl
H_2O	82.5 μl
Total	150 μl

2. Digest for 2.5 hours at 37°C. Do not reduce the incubation time.
3. Add the following reagents to tube:

10X CIAP Buffer	20 μl
CIAP (1 U/μl)	3 μl
H_2O	82.5 μl
Total	400 μl

4. Incubate 40 minutes at 37°C.
5. Extract twice with an equal volume of phenol:chloroform:isoamyl alcohol (25:24:1) and once with chloroform:isoamyl alcohol (24:1).
6. Ethanol precipitate and resuspend the DNA in TE at a final concentration of 0.15 μg/μl.
7. Prior to attempting to ligate your ds cDNA to the vector, it is wise to perform a control ligation (plus/minus ligase) and subsequent transformation with the linearized, dephosphorylated vector and with intact vector. If you observe a significant number of colonies with the linearized vector in either the plus ligase or no ligase condition, repeat the digestion and/or dephosphorylation reaction.

I. Ligation of cDNA to Linearized, Dephosphorylated AD Plasmid Vector

Three separate ligations are performed to determine the optimal ratio of cDNA to vector. It is also wise to perform a vector-only control ligation. The products of the three ligation reactions will be separately transformed into DH5α host cells in the next section, and the resulting transformants will be pooled at step J16 to form the original, unamplified library. In some cases, it may be necessary to subsequently perform a fourth, scaled-up ligation using the remaining cDNA in order to get a library of the desired complexity. cDNA libraries should be made as large as practically possible. Typically, at least 10^6 independent cDNA clones are required to ensure that low-abundance transcripts are represented in the library.

Before beginning, test your electrocompetent DH5α cells to be sure that they yield at least $0.5–1.0 \times 10^{10}$ cfu/μg as determined in a test ligation with an intact plasmid vector. You will also need several 90-mm LB agar plates containing 50 μg/ml of ampicillin.

1. Combine the following components in three 0.5-ml microcentrifuge tubes:

	Tube a	Tube b	Tube c
Vector (0.15 μg/μl)	1.0 μl	1.0 μl	1.0 μl
cDNA	0.5 μl	1.0 μl	1.5 μl
10X ligation buffer	0.5 μl	0.5 μl	0.5 μl
10 mM ATP	0.5 μl	0.5 μl	0.5 ml

T4 DNA ligase (6 Weiss u/μl)	0.5 μl	0.5 μl	0.5 μl
Sterile H_2O	2.0 μl	1.5 μl	1.0 μl
Total	5.0 μl	5.0 μl	5.0 μl

Store the unused CDNA at 4°C for later use.

The conditions described above will give libraries of adequate complexity. However, as noted in Section A, we have obtained libraries of significantly higher complexity using any of several proprietary commercial buffers that give high-efficiency ligation of inserts to plasmid vectors. These buffers should be used according to the manufacturer's recommendations.

2. Incubate at 16°C for 12 to 16 hours.
3. Add 80 μl of sterile DEPC-treated H_2O to each of the above mixtures. Add 1.5 μl of glycogen (20 μg/μl). Mix well with a pipette tip. Add 280 μl of ice-cold 95% ethanol. Mix by gently rocking the tube back and forth.
4. Place the tube at –70°C or in a dry-ice/ethanol bath for 1 to 2 hours. (Longer incubation times may increase the yield of DNA after ethanol precipitation. If transformations are to be performed at a later date, leave the ligation mixture at –70°C.)
5. Spin in a microcentrifuge at 15,000 rpm for 20 minutes at room temperature.
6. Carefully remove the ethanol layer without disturbing the pellet.
7. Air dry the pellet.
8. Resuspend each pellet (a, b, and c) in 5 μl of sterile DEPC-treated H_2O.

J. Transformation of Recombinant Plasmids into *E. coli*

In addition to the experimental transformations described below, we recommend that you perform a negative control transformation with no DNA and a positive control transformation with 10 to 100 ng of an intact plasmid such as pBR322 or pUC19.

1. Thaw electrocompetent DH5α cells on ice. Use the cells promptly after thawing to obtain maximum efficiency in electroporation.
2. Add 800 μl of LB broth to 14-ml polypropylene tubes labeled a, b, and c and to positive and negative control tubes.
3. Add 40–50 μl (1.5×10^9 cells) of thawed electrocompetent cells to each ligation reaction tube (a, b, and c from step I1) and to positive and negative controls. Mix thoroughly with a pipette tip.
4. Transfer the mixture to a chilled 0.1-cm cuvette.
5. Electroporate using the *E. coli* Pulsar (BioRad). Transform the DNA by pulsing the mixture at 1.8 kv for 0.4 milliseconds.
6. Immediately add 150 μl of LB broth to the cuvette and transfer the entire volume to the prelabeled polypropylene tubes containing 800 μl LB broth (prepared in step J2).
7. [Optional] Wash the cuvette with 50 μl of LB and transfer the washes to the respective polypropylene tubes.
8. Incubate with shaking (225 rpm) for 1 hour at 37°C.
9. During the incubation, label three 1.5-ml polypropylene tubes a, b, and c. Label tubes for positive and negative controls. Add 40 μl of LB broth to each of these tubes.

10. At the end of the 1-hour incubation, remove 10 μl of each transformation mixture and add it to the appropriate tube containing 40–200 μl of LB broth. Mix gently by swirling.
11. Spread the aliquot on a prewarmed 90-mm LB agar plate containing 50 μg/ml of ampicillin.
12. Allow the inoculum to soak into the plate for 10 minutes.
13. Invert the plates and incubate at 37°C overnight.
14. Store the remaining transformation mixture at 4°C.
15. The next day, examine the plates. Negative control plates should have no colonies, while plasmid positive control plates should have lots of colonies. Then examine the three plates that were plated with 1% of each transformation mixture. If all three plates are confluent, the original library (contained in the three transformation mixtures) contains about 10^6 independent clones. (A confluent plate contains several thousand colonies; since only 10 μl of each transformation mixture (of 1 ml total) was plated at step 11, the remaining transformation mixture contains 100 times more independent clones.)

    Often, one or two of the plates will have considerably fewer colonies than the optimal plate(s). If this is the case (that is, if you do not have three confluent plates), perform a fourth ligation using all of the remaining ds cDNA. Choose the ratio of cDNA to vector based on which of the original ligations a, b, and c (in step I1) produced the highest number of transformants, and scale up the volumes. Transform and plate this final ligation according to steps J1 through J13 above. If you still do not get three confluent or nearly confluent plates, you may wish to repeat the entire protocol starting with fresh RNA. (However, you can pool and amplify the clones that you do have at this point and combine them later with the second library.)

    To determine the approximate percentage of recombinant clones in each transformation at this stage, analyze the DNA from 15 independent clones in each transformation. If you cannot pick 15 isolated colonies from the periphery of each plate, prepare a fresh plate by streaking out 1 μl of the transformation. (Do not attempt to pick single colonies from the midst of a confluent plate.) You can screen for inserts by one of two methods: (1) by performing PCR directly on colonies or on mini-prep DNA using suitable primers, or (2) by restriction digestion of mini-prep DNA using the enzyme(s) used for cloning. We recommend pooling each transformation mixture in which at least 10 out of 15 clones contain inserts. (The percentage of inserts for the final amplified library can be determined at the end of section K).
16. Based on the above guidelines, pool the desired transformation mixtures to generate your original, unamplified cDNA library.
17. Store the original, unamplified library at 4°C until you are ready to proceed with amplification.

K. Amplification of Plasmid Libraries

Most AD fusion libraries must be amplified in order to obtain enough plasmid DNA for yeast transformations (and for long-term storage). Growing the transformants on solid media instead of liquid culture minimizes uneven amplification of the various clones. The cells are plated at a high density, so that the resulting colonies will be nearly confluent.

Before starting, prepare an adequate supply of 150-mm LB agar plates with 50 µg/ml of ampicillin. We suggest that you divide your transformation mixture among 75 plates. This allows for a library of up to 3×10^6 clones at a density of 40,000 colonies per plate; however, plating density depends on the size of the colonies and strains that produce larger colonies will have a somewhat lower plating density. For example, with 75 plates, a library of 1×10^6 clones will plate out at a density of 13,000 colonies per plate.

On the second day, you will need at least 600 ml of LB broth and a chilled, sterile polypropylene vessel large enough to hold the entire amplified library (600 ml for 75 plates).

1. Divide the total volume of your unamplified library by 75 to get the volume that should be spread onto each plate.
2. Spread the calculated volume of the transformation culture on a 150-mm, prewarmed LB agar plate containing 50 µg/ml of ampicillin. (If necessary, dilute the unamplified library with a small amount of LB to allow spreading over the entire surface area of each plate.)
3. Allow the inoculum to soak into the plate for 10 minutes.
4. Repeat steps 2 and 3 until you have plated the entire library.
5. Invert the plates and incubate at 37°C overnight.
6. Next morning, add 8 ml of LB broth to one plate. Scrape the colonies off the agar and pour the suspension into a large, sterile, prechilled polypropylene vessel. Keep the pooled library on ice while you scrape the remaining plates.
7. Repeat step 6 until you have scraped all the plates.
8. Remove the desired portion of the library for a large-scale plasmid prep (Sambrook et al. 1989).
9. Determine the titer of your library by following the protocol in section L.
10. Prepare the rest of the library for long-term storage by adding glycerol to a final concentration of 25%, then aliquoting it into desired volumes (typically 1 ml) in sterile tubes, and storing at –80°C. (Remember to adjust your titer for the glycerol added to frozen cultures.) Properly stored libraries with a titer of $>10^8$ cfu/ml should be stable for at least one year. Libraries with lower titers will be less stable (i.e., the titer of a library originally stored at 1×10^6 will decrease much faster than the titer of a library stored at 1×10^9). Stability may vary among different libraries.

    If desired, you can determine the percentage of recombinant clones in your final amplified library as described in step H15.

L. Titering Plasmid Libraries

Prior to freezing your library, you should determine the titer. To ensure a representative library, the titer should be at least 10-fold higher than the number of independent clones. In general, plasmid libraries should have a titer of at least 10^8 cfu/ml for long-term storage. Keep the following important points in mind when titering the library.

Diluted libraries are always less stable than undiluted libraries.

Once 10^{-3} and 10^{-6} dilutions of the library are made, use them within the next hour. Otherwise, drastic reductions in titer may occur.

A 2- to 5-fold range in titer calculations is reasonable, especially if more than one person is doing the titering.

Always use the recommended amount of antibiotic to ensure plasmid stability.

Use proper sterile techniques while aliquoting and handling libraries.

Design appropriate controls and include them during plasmid library growth to test for cross-contamination.

Use the following protocol to determine the titer of your library.

1. Prewarm LB agar + ampicillin (50 μg/ml) plates at 37°C.
2. Thaw an aliquot of the library and place on ice.
3. Remove 1 μl of the library, and add it to 1 ml of LB broth in a 1.5-ml microcentrifuge tube. Mix by gentle vortexing. This is the $1{:}10^3$ dilution (Dilution A).
4. Remove 1 μl from Dilution A, and add it to 1 ml of LB broth in a 1.5-ml microcentrifuge tube. Mix by gentle vortexing. This is the $1{:}10^6$ dilution (Dilution B). Note that the diluted library is not stable; therefore, plate it as soon as possible after dilution.
5. Add 1 μl from Dilution A to 50 μl of LB broth in a 1.5-ml microcentrifuge tube. Mix by gentle vortexing. Spread the entire mixture onto a prewarmed LB agar + ampicillin (50 μg/ml) plate.
6. Remove 50 μl and 100 μl aliquots from Dilution B and spread onto separate LB agar + ampicillin (50 μg/ml) plates.
7. Leave plates at room temperature for 15 to 20 minutes to allow the inoculum to soak into the agar.
8. Invert the plates and incubate at 37°C overnight.
9. Count the number of colonies to determine the titer (cfu/ml). Calculate the titer according to the following formulas:
   colony # from Dilution A $\times 10^3 \times 10^3$ = X cfu/ml
   (colony # from Dilution B/plating vol.) $\times 10^3 \times 10^3 \times 10^3$ = X cfu/ml

Alternative Protocols

AD Library Construction in Phagemid Vectors

As discussed above, AD-fusion libraries must exist in a plasmid form for transformation into yeast host strains. However, the libraries with the highest complexity have historically been made in phage vectors because of the higher efficiency of phage packaging (as compared to conventional methods of plasmid transformation). The development of lACT (Durfee et al. 1993; Harper et al. 1993) and other similar phagemid vectors enables researchers to circumvent the limitation of conventional phage vectors for two-hybrid applications. The l form of these vectors contains an embedded AD plasmid with all of the functional elements found in conventional AD plasmid vectors (Figure 5-7). Following construction of the l library, the library can be converted to a plasmid library using Cre-Lox recombination in

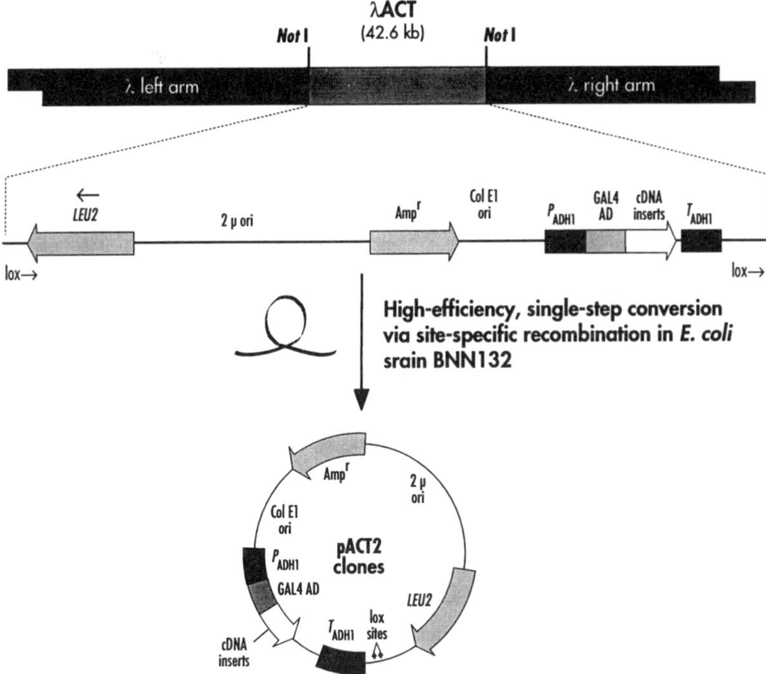

Fig. 5-7. Flowchart for conversion of λACT libraries to pACT plasmid libraries.

an appropriate host strain. Figure 5-8 provides an overview of the final steps in construction and characterization of phagemid libraries. However, most AD-fusion libraries are still constructed directly in plasmid libraries for two reasons: (1) the gain in library complexity may be lost in the conversion from phage to plasmid form; and (2) the development of highly competent bacterial cells, efficient bacterial electroporation, and optimized plasmid-insert ligation conditions now permits routine construction of plasmid libraries of adequate complexity for most applications.

Directional Cloning Strategies

Nondirectional cloning strategies, such as the one detailed in this chapter, are simpler than directional strategies; however, they create libraries in which half the inserts will be in the wrong orientation for expression. The nondirectional protocol described in this chapter can be adapted for directional cloning by using primers that have a restriction site incorporated into the 5'-end. The first-strand synthesis is then performed in the presence of methyl dCTP instead of dCTP (to prevent digestion of internal restriction sites) and the adaptor-ligated ds cDNA is digested with the appropriate restriction enzyme prior to size fractionation and vector ligation. However,

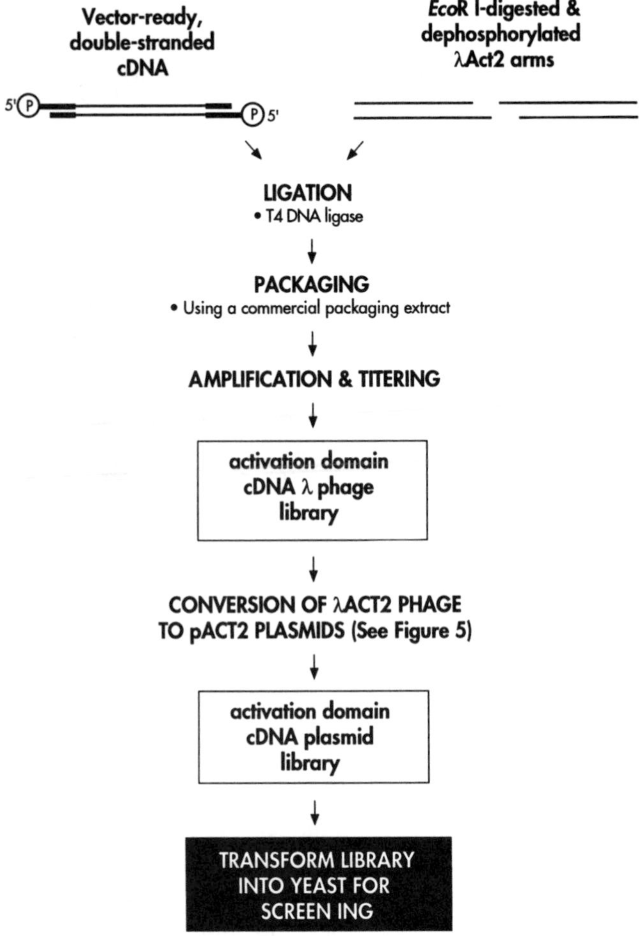

Fig. 5-8. Flowchart for final steps in construction of a phagemid AD library.

we have found it very difficult to generate high-quality primers that combine specific and random sequences. This problem, together with the extra steps, generally outweighs the advantages of directional cloning.

Genomic AD Libraries

As mentioned above, AD-fusion libraries are generally made from genomic DNA only when the genome of the organism being studied does not contain introns or extensive regions of noncoding DNA. Construction of genomic DNA libraries presents a different set of problems than cDNA libraries. While the source and amount of DNA is less likely to be a problem, the cloning strategy must ensure that any particular insert can be expressed in all three reading frames. (In the case of cDNA libraries, the random stopping of reverse transcriptase during first-strand synthesis ensures that virtually all of the original RNA sequences will be expressed in all three reading frames.) There are two common solutions to the expression problem for genomic DNA. Separate aliquots of genomic DNA that have been digested with a single restriction enzyme can be cloned into three related vectors which have the cloning site in each of the three reading frames (e.g., pGAD1, 2 and 3; Chien et al. 1991). Alternatively, the DNA can be mechanically reduced to the desired size by shearing or sonication. The DNA is then blunt-ended, ligated to adaptors, and size-fractionated prior to cloning as described above for plasmid vectors. Because it does not require ligation into three separate vectors, the shearing method is more widely used for the construction of genomic DNA libraries.

References

Bartel, P. L., Chien, C.-T., Sternglanz, R., and Fields, S. (1993). Using the two-hybrid system to detect protein-protein interactions. In *Cellular Interactions in Development: A Practical Approach*, D.A. Hartley, ed., Oxford, Oxford University Press. pp. 153–179.

Borson, N. D., Sato, W. L., and Drewes, L. R. (1992). A lock-docking oligo(dT) primer for 5' and 3' RACE PCR. PCR Methods and Applications 2:144–148.

Chien, C. T., Bartel, P. L., Sternglanz, R., and Fields, S. (1991). The two-hybrid system: A method to identify and clone genes for proteins that interact with a protein of interest. Proc. Nat. Acad. Sci. USA 88:9578–9582.

Durfee, T., Becherer, K., Chen, P-L., Yeh, S-H., Yang, Y., Kilburn, A. E., Lee, W-H., and Elledge, S. L., (1993). The retinoblastoma protein associates with the protein phosphatase type 1 catalytic subunit. Genes Dev. 7:555–569.

Farrell, Jr., R. E. (1993). *RNA Methodologies—A Lab Guide for Isolation and Characterization.* San Diego, California, Academic Press.

Gubler, U. and Hoffman, B. J. (1983). A simple and very efficient method for generating complimentary DNA libraries. Gene 25:263–269.

Gyruis, J., Golemis, E., Chertkov, H., and Brent, R. (1993). Cdi1, a human G1 and S phase protein phosphatase that associate with Cdk2. Cell 75:791–803.

Hannon, G. J., Demetrick, D., and Beach, D. (1993). Isolation of the Rb-related p130 through its interaction with CDK2 and cyclins. Genes Dev. 7:2378–2391.

Harper, J. W., Adami, G. R., Wei, N., Keyomarsi, K., and Elledge, S. J. (1993). The p21 Cdk-interacting protein Cip1 is a potent inhibitor of G1 cyclin-dependent kinases. Cell 75:805–816.

Hollenberg, S.M., Sternglanz, R., Cheng, P.F., and Weintraub, H. (1995). Identification of a new family of tissue-specific basic helix-loop-helix proteins with a two-hybrid system. Molec. Cell. Biol. 15:3813–3822.

Hsueh,, A. J. W., Billig, H., and Tsafriri (1994). Ovarian follicle atresia: a hormonally controlled apoptotic process. Endocrine Reviews 15 (6):707–724.

Iwabuchi, K., Bartel, P. L., Li, B., Marraccino, R. and Fields, S. (1994). Two cellular proteins that bind to wild-type but not mutant p53. Proc. Natl. Acad. Sci. USA 91:6098–6102.

Kalderon, D., Roberts, B. L., Richardson, W. D., and Smith, A. E. (1984). A short amino acid sequence able to specify nuclear localization. Cell 39:599–509.

Luo, Y., Vijaychander, S., Stile, J., and Zhu, L. (1996). Cloning and analysis of DNA-binding proteins by yeast one-hybrid and one-two-hybrid systems. BioTechniques. In Press.

Ruden, D. M., Ma, J., Li, Y., Wood, K., and Ptashne M. (1991). Generating yeast transcriptional activators containing no yeast protein sequences. Nature 350:250–252.

Sambrook, J., Fritsch, E. F., and Maniatis, T. (1989). *Molecular Cloning: A Laboratory Manual*, Second Edition. New York, Cold Spring Harbor Laboratory Press.

Vojtek, A., Hollenberg, S., and Cooper, J. (1993). Mammalian Ras interacts directly with the serine/threonine kinase Raf. Cell 74:205–214.

Yavuzer, U., and Goding, C. R. (1995). pWITCH: a versatile two-hybrid assay vector for the production of epitope/activation domain-tagged proteins both in vitro and in yeast. Gene 165:93–96.

Part II

ANALYZING INTERACTIONS WITH THE TWO-HYBRID SYSTEM

How can the two-hybrid system be used to determine the role of a particular interaction? In chapter 6, Amberg and Botstein describe their approach to analyzing actin mutants for their ability to interact with a variety of actin-binding proteins. This information is correlated to both the phenotype of the actin mutant and to the actin three-dimensional structure. Both Vidal (chapter 7) and White (chapter 8) describe related approaches to generate and analyze mutants that fail to interact with particular protein partners. White applies this approach to the study of Ras-binding proteins. The powerful approach developed by Vidal, the reverse two-hybrid screen, may also prove to be a useful strategy for the discovery of molecules that disrupt protein-protein interactions. These mutational approaches can simplify the generation of useful reagents, such as dominant negatives and compensating mutations, that may be valuable for other genetic analyses.

In chapter 9, Ozenberger and Young demonstrate that interactions between proteins that are normally extracellular can be detected by the two-hybrid system. Their experiments suggest that interactions often considered to be beyond the ability of the two-hybrid system to detect are, in fact, observable using this assay.

Investigators in the fields of signal transduction and cell cycle control were among the first to conduct two-hybrid searches. Because of the relatively long history of its use in these areas, the two-hybrid system has had an opportunity to make an impact on progress in these fields. In chapter 10, Sprague and Printen review the role of two-hybrid searches on our understanding of the yeast pheromone response pathway. Hannon reviews recent advances in our understanding of cell cycle control in chapter 11, highlighting contributions of the two-hybrid approach.

Finley and Brent discuss an approach to employing the two-hybrid system on a global scale in chapter 12. Their mating strategy allows the rapid identification of interactions involving members of a collection of DNA-binding and activation domain hybrids. Global approaches of this type should, in the future, allow the identification of all detectable interactions among the entire complement of an organism's proteins. This type of knowledge will add value to genome sequencing efforts by assigning protein partners to many of the novel proteins that are being identified.

6

Obtaining Structural Information about Protein Complexes with the Two-Hybrid System

David C. Amberg
David Botstein

The ability of the two-hybrid system to detect biologically relevant protein-protein interactions has been well established. When this system identifies a biologically relevant protein-protein interaction, a natural assumption is that the fundamental nature of this interaction is the same as when it occurs in its normal functional context. Indeed, the successful discoveries of a large number of protein interactions through use of the two-hybrid system should be taken as evidence supporting this assumption. Once this proposition is accepted, it follows that the two-hybrid system can be used not only to detect protein-protein interactions but also to study *how* proteins interact.

The most direct approach to obtaining information on the residues of a protein that are important in binding a ligand is to solve the structure of cocrystals. Even in these favorable cases, it has often been necessary to corroborate these structural studies with genetic assays, which confirm the amino acids that form critical stabilizing contacts by the identification of mutations that cause a failure to bind. Indeed, one can often use mutations in conjunction with in vitro binding assays to make inferences about the mechanisms of binding without the labor of solving cocrystals (for an example, see Honts et al. 1994). These genetic and biochemical methods, although less arduous than X-ray crystallography, require that all the mutant proteins be purified and that the interaction be strong enough to be detected under possibly suboptimal in vitro conditions. We have shown that this kind of information can be obtained by substituting the two-hybrid system for in vitro assay systems and obtain at least as much information about the structural requirements for binding with considerably less effort.

Our method of differential interactions requires that the interaction be amenable to two-hybrid analysis, that a three-dimensional structure exist of at least one of the protein ligands, and that an appropriate set of mutants exists in one or both of the binding partners. The advantages of this method are: the interactions are studied in vivo, weak interactions can be analyzed (apparent $K_d > 1 \mu M$; Phizicky and Fields 1995) as well as strong ones, and data can be obtained very rapidly with much less effort than more traditional methods.

In this chapter, we explain how the two-hybrid system can be used to describe interactions between yeast actin and several yeast actin ligands. Besides covering the mechanistic details of this analysis we will show examples of the data obtained using a large set of rationally designed mutations.

Construction of Suitable Mutations

Mutations can frequently affect the structure of a protein. How generalized these effects are will vary greatly from mutation to mutation in ways that may be difficult to predict. Since the goal is to use the mutations to try to delineate binding sites, one wants to use mutations that have only localized effects on the structure. It is, therefore, critical that the mutations be carefully designed to minimize structural disturbances. If the three-dimensional structure of the protein of interest is known, then rational decisions can be made in the design of these mutations. Two general philosophies might be used: alter charged amino acids on the surface of the protein in the hope of disrupting important ionic interactions; alternatively, alter surface exposed hydrophobic residues in the hope of disrupting hydrophobic interactions. The importance of hydrophobic interactions for protein-protein interactions is well established and has been made clear by work solving the structures of dimeric proteins (Green et al. 1995; Potts et al. 1995).

When the mutations that were employed in our actin analysis were constructed, the three-dimensional structure of actin was still unknown. The actin alleles were constructed by clustered charged-to-alanine scanning mutagenesis, a rational mutagenesis procedure designed to target surface exposed residues in proteins of unknown structure. In this method, the protein sequence is scanned in a window of five amino acids; when two or more charged amino acids appear in a window, they are altered to alanine (Bass et al. 1991; Bennett et al. 1991). This method yielded 36 mutant alleles of actin (Wertman et al. 1992). When the structure for actin was published, we discovered that the mutagenesis algorithm had worked well: all but one of the alleles altered surface exposed amino acids (Kabsch et al. 1990). More recently, many of these mutant proteins have been purified and assembled into filaments in vitro, indicating that their structure is not grossly damaged (Miller et al. 1995; Miller and Reisler, 1995). Clearly, these mutants were well suited for the two-hybrid analysis, or indeed any analysis of protein ligand binding.

It should be noted that charged-to-alanine scanning mutagenesis of yeast calmodulin did not have the same happy result. None of the calmodulin alleles constructed in this manner caused mutant phenotypes, indicating that in vivo the mutations had no effect on the important interactions calmodulin makes with other proteins. However, when phenylalanines (chosen on the basis of knowledge of calmodulin structure and interactions with peptide ligands; Ikura et al. 1992; Meador et al. 1992) were systematically changed to alanine, many mutants with interpretable phenotypes were obtained. These would be the most appropriate reagents for a structurally based two-hybrid analysis of calmodulin interactions (Ohya and Botstein 1994a, 1994b).

In summary, the best strategy would be to make mutations covering the surface of the protein of interest using knowledge of the protein's three-dimensional structure. Lacking such information, charged-to-alanine mutagenesis has been successful for a variety of proteins, including alpha-and beta-tubulin (Reijo et al. 1994; K. Richards and D. Botstein, personal communication), tissue plasminogen activator (Bennett et al. 1991), poliovirus (Diamond and Kirkegaard 1994) and yeast signal peptidase subunit Sec11p (E. Beasley and D. Botstein, personal communication).

Once one has decided which mutations to construct, a number of methods exist for the construction of these alleles. In the case of the actin alanine scan alleles, we chose to use the Kunkel method (Kunkel et al. 1987). This and other methods have been fully described elsewhere (for example, see Sambrook et al. 1989). Movement of the mutant alleles into the two-hybrid vector can done by standard subcloning methods; however, we have found a recombination strategy particularly convenient. When one transforms yeast with overlapping DNA fragments, one fragment carrying the bulk of the vector backbone and the other carrying the mutant allele, the recipient yeast cells will correctly reconstruct the plasmid by homologous recombination (Ma et al. 1987). In this way, passage of the new plasmid through *Escherichia coli* is entirely avoided.

Since the actin alleles were not constructed with the two-hybrid system in mind, we were forced to use a PCR strategy to move the alleles into the two-hybrid DNA-binding domain vector. To detect and avoid PCR-induced errors, each allele was amplified in three separate PCR reactions from which three independent clones were derived. All three isolates were then tested in the two-hybrid system. PCR-induced errors did occur frequently enough to present a problem: in 7 of the 36 alleles, one of the three independent isolates behaved differently from the other two. In all seven cases, the dissimilar constructions failed to interact with any actin ligands; no doubt, they contained PCR-induced mutations with a disastrous effect on the actin structure. This result allows us to re-emphasize the importance of the proper choice of mutant alleles: most uncritically obtained alleles will be unsuitable either because they have no effect on the structure or because the effect is not localized enough.

Details concerning the construction of the "bait" construct have been covered in (chapters 2, 3, and 4). Our approach with yeast actin was to make a Gal4p fusion to the N-terminus of actin and to insert a 4-residue alanine linker between the actin and Gal4p sequences to decrease the likelihood that the Gal4p sequences would block access of actin ligands to the actin surface. We feel that our concern was justified as the only activation domain fusion with actin found to dimerize with our bait actin fusion carried 24 additional amino acid residues between the actin and Gal4p sequences. It would probably be prudent in any new endeavor to investigate a variety of linker lengths.

Identification and Construction of Ligand-Gal4p Fusions

One may already know what ligand(s) the protein of interest interacts with. In this case, ligand sequences can be moved directly into the two-hybrid activation domain vectors and tested for their ability to interact with the bait Gal4p DNA-binding domain fusion prior to undertaking an extensive mutagenesis of the bait sequences. If the Gal4p domains interfere with the interactions of interest, a considerable amount of trying various linker lengths may be required.

If one has not already determined which proteins interact, one starts by screening a library of fusions to the Gal4p activation domain (see chapters 2, 3, and 4). This will not only identify potential ligand proteins, but will also provide them as activation domain fusions that are capable of interaction. In addition, use of the library may provide additional information such as the identity of the interaction domain within a putative ligand protein.

A library approach should provide a number of constructs with different fusion junctions all capable of interacting with the bait. When these constructs are used with the mutant forms of the bait, there should be strong agreement as to which mutations disrupt the bait-ligand interaction, regardless of the fusion junction. In our actin analysis, we generally found this to be the case.

Two-Hybrid Analysis with Mutants

Analysis of many mutant alleles and many putative ligand constructs in all pairwise combinations means that a very large number of comparisons must be made. The most direct and easy way to examine and compare interactions among a large number of mutants and a large number of ligands is to introduce the DNA-binding domain and activation domain plasmids into the same cells via mating of haploid yeast cells. The DNA-binding domain fusions bearing actin mutants (in triplicate) were transformed into haploid cells of one mating type while the constructs bearing ligand fusions to the Gal4p activation domain were transformed into yeast of the opposite mating type. The transformant strains were then mixed into small volumes (20 μl) of sterile water and pipetted into the wells of microtiter dishes and arranged so that transformants in one mating type were arranged in the columns of the microtiter dishes while those in the opposite mating type were placed in the rows of the microtiter dishes. This can all be done with a multichannel pipettor if one takes care to avoid cross-contamination between wells. The mixed cells can then be spotted (using either a multichannel pipettor or a multipronged implement called a "frog") onto a plate on which all the strains can grow and mate efficiently (such as YEPD; see chapters 2, 3, and 4); selection for plasmids appears generally not to be required. On the next day, the mated cells are replica-plated onto selective plates in which only diploids bearing both a DNA- binding domain plasmid and activation domain plasmid will be able to grow.

Once the diploids have grown into colonies, they can be assessed for activation of the two-hybrid reporters and, most importantly, the degree of activation can be compared among the whole set. We found that the β-galactosidase reporter was too variable for this type of analysis, although it is a useful and comforting confirmatory test. We favor using the His3p enzyme reporter. Enzyme levels are assessed by the ability of yeast to grow on minimal plates containing an inhibitor of the His3p enzyme, 3-amino-1,2,4-triazole (3-AT). We found that plate media containing 25, 50, or 100 mM 3-AT were adequate to assess the relative levels of activation resulting from interaction between the ligand-Gal4p activation domain fusion and the actin bait variants. The 3-AT selection is tightest when minimal media containing low adenine concentrations are used, presumably because adenine is a potential source of imidazole. In our experiments, we used strains Y187 and Y190 (Durfee et al. 1993) allowing the 3-AT selections to be done on SD medium containing 10 μg/ml adenine.

In our two-hybrid analysis, we examined combinations of 36 actin alleles (including wild type, an absolutely essential control on every plate) and 10 ligands for interaction, for a grand total of 360 assessments of protein interaction. We performed 36 (plus wild type) assays per plate and thus could examine all 360 combinations at three different 3-AT concentrations on a total of 30 plates.

Analysis of Differential Interaction Data

One can view a data set of differential interactions with a given ligand and a large number of mutant alleles from a variety of perspectives. One of these concerns the behavior of each mutant allele; another concerns which mutations affect the interaction with a given ligand; and the third concerns the spatial relationships among mutations that have common effects on interaction with a given ligand.

In our experiments with actin, mutant alleles could be placed into three groups based on how they behaved with the 10 ligands. The first class of alleles included those that encoded actin proteins that bound all the ligands as well as wild-type actin did. The second class comprised those that failed to bind any of the ligands. The third class contained those that bound some of the ligands as well as wild-type and other ligands poorly or not at all. It is this last class that is the most informative about protein-protein interactions.

Surprisingly, there was a correlation between the two-hybrid phenotype of the mutants and the severity of the growth defects of strains bearing the mutant alleles. Those of the first class, whose products bound all ligands well, were, with only one exception, alleles with no growth defect relative to yeast expressing wild-type actin. The surface characteristics and structure are presumably unperturbed in these mutant proteins. Alleles of the second class were, with one exception, recessive lethal alleles that presumably encode either extremely unstable proteins and/or proteins that are folded very differently than the wild-type protein.

Alleles of the third class that encode proteins that bind some ligands but not others, are less generalized in their effects on the surface and structure of actin than those mutations of the second class. It is from these mutations that we expected to extract structural information about the actin-ligand complexes. A more conservative view might be that only those mutations that affect a few interactions should be considered. In any case, one is tempted to weight more heavily the behavior of these alleles in developing a hypothesis about a ligand's binding site.

Figure 6-1 shows a bar graph of the mutations, in order of decreasing severity of their growth phenotypes, relative to how many interactions they disrupt in the two-hybrid system. There is a clear correlation between the severity of the two-hybrid defects and the severity of their growth defects. Interestingly, presumed dominant mutants affect few interactions, and, therefore, may be sequestering or otherwise inhibiting essential components of the actin cytoskeleton. Since actin self-assembles into filamentous actin (F-actin), it is possible that the dominant mutants might poison actin assembly.

We found that the data for a given ligand are generally internally consistent when modeled on representations of the known structure of actin. In some cases where information already exists about interactions between actin and a protein ligand, the data obtained by two-hybrid analysis are in agreement. Figure 6-2 shows a surface representation of the back of monomeric actin (G-actin). In dark gray (and outlined in black) are the mutations that affect the two-hybrid interaction between actin and Oye2p and in light gray are those mutations that have no effect. The data are internally consistent in that all the mutations are located very near each other in the structure of actin. The data are clearly informative: the region of actin

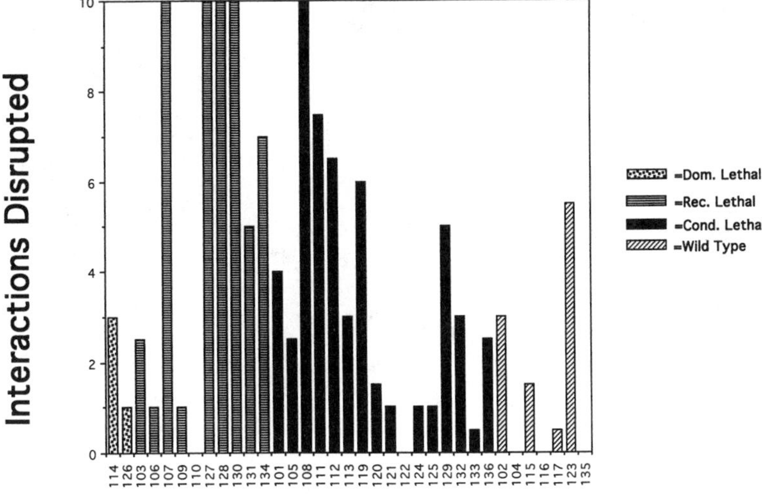

Fig. 6-1. Number of ligand interactions disrupted by the actin mutants. This bar graph displays the number of ligand interactions disrupted in the two-hybrid system by the actin alanine scan alleles. The growth phenotypes of the alleles are indicated and coded according to the accompanying key at right and the alleles have been grouped by decreasing severity of growth phenotypes moving from left to right.

identified is completely buried in F-actin and therefore Oye2p is expected to be an obligate G-actin binding protein.

The equipment used for modeling of our two-hybrid data was a Silicon Graphics Iris running GRASP (Nicholls et al. 1991) software. We have also used the Insight II (Biosym Technologies) software but we found GRASP to be more adept at rendering molecular surfaces. Complete information about all these programs are beyond the scope of this review and can be found elsewhere.

Prospects

We have found additional benefits to the construction of our two-hybrid reagents. Clearly, the analysis can be extended simply with any additional actin ligands we or others identify, because of our use of the mating of yeast to bring the interacting proteins together. Although we have yet to test this hypothesis, we expect that our yeast actin fusion will be capable of interacting with actin-binding proteins from organisms other than yeast. Many protein-protein interactions have been evolutionarily conserved and

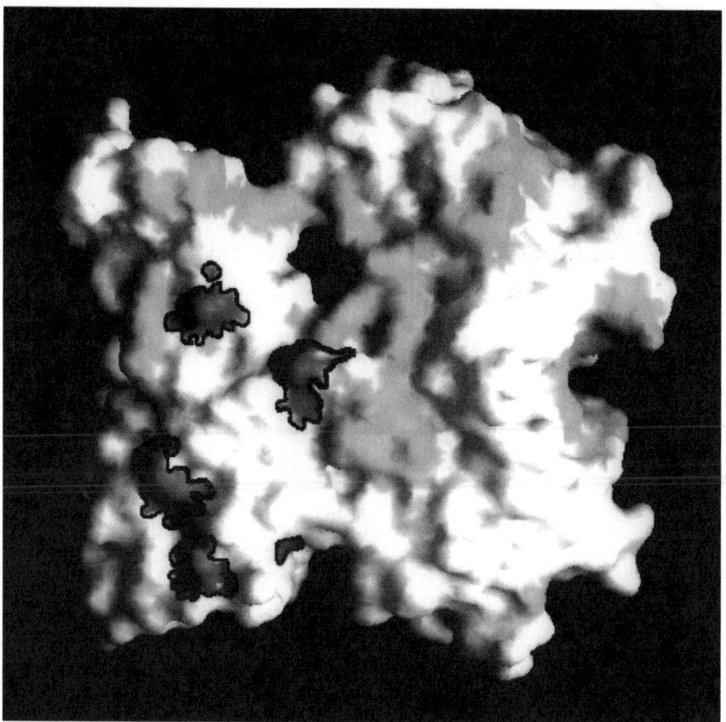

Fig. 6-2. Modeling of actin mutants that disrupt the actin-Oye2p interaction. Displayed is the calculated, solvent exposed surface of the back of G-actin. Mutants that disrupt the interaction between actin and Oye2p in the two-hybrid system are outlined in black and indicated in dark gray. Mutants that have no effect on the interaction are shown in light gray. This rendering was executed with GRASP software (Nicholls et al. 1991) on a Silicon Graphics Iris Computer.

we expect that in some cases the mechanics of these interactions are conserved as well.

We anticipate using the two-hybrid constructs to identify new mutations in either actin or the ligands that precisely affect particular interactions. Reintroduction of these mutants into yeast will then allow us to determine the functional significance of certain actin-ligand interactions. If the three-dimensional structure of the ligand is known, then the analysis affords the possibility of identifying the actin-binding site on the ligand. Such information will be very useful for modeling entire complexes and should lead to testable hypotheses concerning their structures.

Clearly, the two-hybrid system can be used for much more than merely detecting relevant protein-protein interactions and can be adapted to study the nature of protein-protein interactions as well.

References

Bass, S. H., Mulkerrin, M. G., and Wells, J. A. (1991). A systematic mutational analysis of hormone-binding determinants in the human growth hormone receptor. Proc. Natl. Acad. Sci. USA 88: 4498–4502.

Bennett, W. F., Paoni, N. F., Keyt, B. A., Botstein, D., Jones, A. J., Presta, L., Wurm, F. M., and Zoller, J. (1991). High resolution analysis of functional determinants in the human growth hormone receptor. J. Biol. Chem. 266: 5191–5201.

Diamond, S. E., and Kirkegaard, K. (1994). Clustered charged-to-alanine mutagenesis of poliovirus RNA-dependent RNA polymerase yields multiple temperature-sensitive mutants defective in RNA synthesis. J. Virol. 68: 863–876.

Durfee, T., Becherer, K., Chen, P. L., Yeh, S. H., Yang, Y., Kilburn, A. E., Lee, W. H., and Elledge, S. J. (1993). The retinoblastoma protein associates with the protein phosphatase type 1 catalytic subunit. Genes Dev. 7: 555–569.

Green, S. M., Gittis, A. G., Meeker, A. K., and Lattman, E. E. (1995). One-step evolution of a dimer from a monomeric protein. Nat. Struct. Biol. 2: 746–751.

Honts, J. E., Sandrock, T. S., Brower, S. M., O'Dell, J. L., and Adams, A. E. (1994). Actin mutants that show suppression with fimbrin mutations identify a likely fimbrin-binding site on actin. J. Cell Biol. 126: 413–422.

Ikura, M., Clore, G. M., Gronenborn, A. M., Zhu, G., Klee, C. B., and Bax, A. (1992). Solution structure of a calmodulin-target peptide complex by multidimensional NMR. Science 256: 632–638.

Kabsch, W., Mannherz, H. G., Suck, D., Pai, E. F., and Holmes, K. C. (1990). Atomic structure of the actin: DNase I complex. Nature 347: 37–44.

Kunkel, T. A., Roberts, J. D., and Zakour, R. A. (1987). Rapid and efficient site-specific mutagenesis without phenotypic selection. Methods Enzymol. 154: 367–382.

Ma, H., Kunes, S., Schatz, P. J., and Botstein, D. (1987). Plasmid construction by homologous recombination in yeast. Gene 58: 201–216.

Meador, W. E., Means, A. R., and Quiocho, F. A. (1992). Target enzyme recognition by calmodulin: 2.4 A structure of a calmodulin-peptide complex. Science 257: 1251–1255.

Miller, C. J., Cheung, P., White, P., and Reisler, E. (1995). Actin's view of actomyosin interface. Biophys. J. 68: 50–54.

Miller, C. J., and Reisler, E. (1995). Role of charged amino acid pairs in subdomain-1 of actin in interactions with myosin. Biochem. 34: 2694–2700.

Nicholls, A., Sharp, K. A., and Honig, B. (1991). Protein folding and association: insights from the interfacial and thermodynamic properties of hydrocarbons. Proteins 11: 281–296.

Ohya, Y., and Botstein, D. (1994a). Diverse essential functions revealed by complementing yeast calmodulin mutants. Science 263: 963–966.

Ohya, Y., and Botstein, D. (1994b). Structure-based systematic isolation of conditional-lethal mutations in the single yeast calmodulin gene. Genetics 138: 1041–1054.

Phizicky, E. M., and Fields, S. (1995). Protein-protein interactions: Methods for detection and analysis. Micro. Rev. 59: 94–123.

Potts, B. C., Smith, J., Akke, M., Macke, T. J., Okazaki, K. Hidaka, H. Case, D. A., and Chazin, W. J. (1995). The structure of calcyclin reveals a novel homodimeric fold for S100 Ca(2+)-binding proteins. Nat. Struct. Biol. 2: 790–796.

Reijo, R. A., Cooper, E. M., Beagle, G. J., and Huffaker, T. C. (1994). Systematic mutational analysis of the yeast beta-tubulin gene. Mol. Biol. Cell 5: 29–43.

Sambrook, J., Fritsch, E. F, and Maniatis, T. (1989). *Molecular Cloning: A Laboratory Manual*, 2nd ed. New York, Cold Spring Harbor Laboratory Press.

Wertman, K. F., Drubin, D. G., and Botstein, D. (1992). Systematic mutational analysis of the yeast ACT1 gene. Genetics 132: 337–350.

7

The Reverse Two-Hybrid System

Marc Vidal

Specific protein-protein interactions are critical to a wide range of biological processes, from the formation of cellular macrostructures and enzymatic complexes to regulatory processes such as signal transduction pathways. The formation of large cellular structures such as the cytoskeleton, the nuclear scaffold, and the mitotic spindle have long been thought to result from complex interactions between proteins. Relatively smaller structures such as nuclear pores, centrosomes, and kinetochores are beginning to be characterized and, in each case, protein-protein interactions seem to play a central role. Surprisingly, many enzymatic activities are also mediated by very large protein complexes. For example, a functional RNA polymerase II holoenzyme of approximately 1,200 kDa was purified that contains RNA polymerase II and a subset of general transcription factors and coactivators (Koleske and Young 1994). Similarly, a 20S protein complex was found to be required for specific cyclin degration activities (Tugendreich et al. 1995; Irniger et al. 1995; King et al. 1995).

Transmission of regulatory signals from the external environment to relevant locations in the cell was originally thought to consist of successive catalytic activities that could amplify a weak signal into a significant cellular response. However, more recent experiments suggest that in many signal transduction pathways, the catalytic activities involved, such as protein kinases, may bind strongly to their protein substrates. In addition, structural proteins required for signal transmission have been suggested to act as scaffolds bridging several proteins involved at consecutive steps in a signal transduction pathway (Choi et al. 1994; Printen and Sprague 1994; chapter 10). Thus, signal transduction pathways might be viewed as large protein structures along which a signal is being transmitted.

In recent years, new proteins belonging to particular macromolecular structures, or functioning in enzymatic complexes or in signal transduction pathways, have been identified on the basis of their interactions with previously known, functionally important proteins. Subsequent to the identification of interacting proteins, characterization of the structural basis for the interaction is important. Protein-protein interaction models are based on the identification of critical amino acids required for the interaction. Although models can be derived from X-ray or NMR structures, they need to be tested with single amino acid changes corresponding to the critical residues. In addition, prior knowledge of the position and nature of particularly deleterious amino acid changes can be informative for the elaboration of such structural models. More important is how the structure relates to the function of the complex. The isolation of mutant proteins modified in their ability to interact with an important partner but wild type for other functions is also critical for thorough functional analyses. Loss of function, dominant negative or conditional alleles affecting protein-protein interaction can be used to study the role of protein complexes. In each case, relatively rare single amino acid changes that dissociate protein-protein interactions under certain conditions need to be identified.

Alteration of protein-protein interactions is known to contribute to certain diseases. Hence, manipulation of protein-protein interactions that contribute to disease is a potential therapeutic strategy. Small molecules that dissociate or stabilize particular interactions could be critical therapeutic reagents.

The study of protein-protein interactions can thus be conceptually divided into three major parts: identification, characterization, and manipulation. Traditionally, the tools available to perform these experiments in multicellular organisms have been restricted to biochemical approaches. Despite obvious advantages, biochemical approaches can be time-consuming and genetic selection systems are often needed. For example, the molecular cloning of the genes encoding interacting proteins identified biochemically is often difficult and tedious. The yeast two-hybrid system overcomes that limitation and allows, in a single genetic selection, both the identification of a potential interacting protein and the isolation of its encoding gene.

The two-hybrid system does not allow genetic selection of events responsible for the dissocation of particular interactions and, hence, its use in the characterization and manipulation of protein-protein interactions has been limited. The "reverse two-hybrid system" described in this chapter facilitates such experiments. Yeast strains were generated in which expression of interacting hybrid proteins increases the expression of a counterselectable marker that is toxic under particular conditions (negative selection;

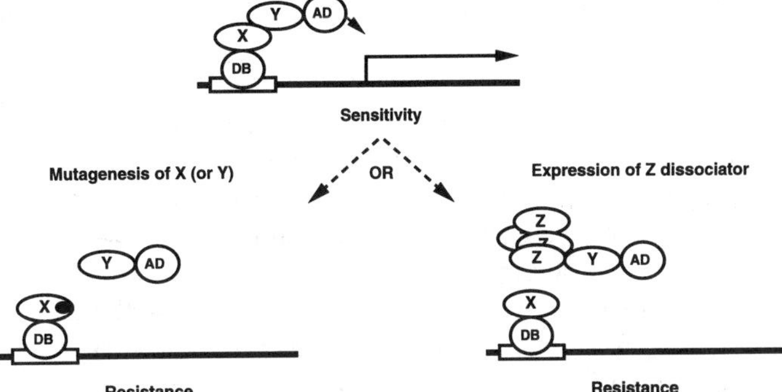

Fig. 7-1. The reverse two-hybrid. Expression of interacting proteins X and Y in fusion with a DNA-binding domain (DB) and an activation domain (AD), respectively, reconstitutes a transcription factor (Fields and Song, 1989). In the reverse two-hybrid system, this leads to activation of a reporter gene whose expression is lethal under particular growth conditions (sensitivity). A dissociator molecule (Z) or mutations in one of the interacting partners (black circle) affect the interaction resulting in decreased expression of the reporter gene and resistance under identical growth conditions (resistance). Modified from Vidal et al., Figures 1a and 1b, *Proc. Nat'l Acad. Sci.* 93: 10315–10320. Reprinted by permission of the National Academy of Sciences.

Vidal et al. 1996a). Under these conditions, dissociation of an interaction provides a selective advantage, thereby facilitating detection: a few growing yeast colonies in which hybrids fail to interact can be identified among millions of nongrowing colonies expressing interacting hybrid proteins (Figure 7-1). Hence, mutations that prevent an interaction can be selected from large libraries of randomly generated alleles (Vidal et al. 1996b). Similarly, molecules that dissociate or prevent an interaction could be selected from large libraries of peptides or compounds.

Principle of the Reverse Two-Hybrid System

The negative selection in the context of the two-hybrid system is based on the use of a toxic reporter gene whose expression is dependent upon the activity of a reconstituted transcription factor. The most widely used counterselectable marker in yeast genetics is *URA3*, which encodes orotidine-5'-phosphate decarboxylase, an enzyme required for the biosynthesis of uracil. Yeast cells that contain wild-type *URA3*, either on a plasmid or integrated in the genome, grow on media lacking uracil (Ura$^+$ phenotype). However, the *URA3*-encoded decarboxylase can also catalyze the conversion of a nontoxic analog, 5-fluoroorotic acid (FOA) into a toxic product, 5-fluorouracil.

Thus, in a large population of cells sensitive to FOA (Foa^S phenotype), a few cells with a mutant *URA3* gene can be conveniently selected on the basis of their resistance to FOA (Foa^R phenotype; Boeke et al. 1984).

In the reverse two-hybrid system, the *URA3* counterselectable marker is used in a different way. In this case, the expression level of *URA3* in response to a pair of interacting proteins (designated X and Y) is used as the basis for identifying changes in X-Y interaction. A *URA3* allele was designed whose expression responds tightly to a specific trancription factor and thus confers a Ura$^-$ Foa^R phenotype in the absence of the transcription factor (or of a pair of interacting hybrid proteins) as well as a Ura$^+$ Foa^S phenotype in its (their) presence. Thus, in a large population of yeast cells that express wild-type interacting hybrid proteins and exhibit a Foa^S phenotype, a few cells in which this interaction is affected can be conveniently selected on the basis of their Foa^R phenotype (Vidal et al. 1996a).

The *SPAL::URA3* Reporter Gene

Since the system described here makes use of Gal4p, Gal4p-responsive alleles of *URA3* were constructed. One difficulty in designing a reverse two-hybrid *URA3* reporter gene is that the basal levels of expression of most of the available Gal4p-inducible yeast promoters are high enough to confer a Ura$^+$ Foa^S phenotype in the absence of Gal4p. Thus, a promoter was designed that contains both a *cis*-acting Upstream Repressing Sequence (URS1) to maintain low levels of basal expression in cells lacking Gal4p, and a large number of Gal4p binding sites to allow Gal4p-dependent activation.

The URS1 element potently represses transcription of many unrelated genes (Buckingham et al. 1990; Luche et al. 1990; Park and Craig 1991; Bowdish and Mitchell 1993; Lopes et al. 1993; Vidal et al. 1995). The mechanism of repression by URS1 involves the binding of a protein complex (Luche et al. 1992; Luche et al. 1993; Strich et al. 1994; Anderson et al. 1995; Vidal et al. 1995) that includes the High-Mobility-Group-like protein Sin1p (Katcoff et al. 1993). One gene naturally containing such a URS element is *SPO13*, a gene involved in yeast sporulation (Buckingham et al. 1990). *SPO13* is repressed very tightly under conditions of mitotic growth and activated under sporulation conditions (Wang et al. 1987). A fusion between the *SPO13* promoter and the wild type *URA3* open reading frame confers a very tight Ura$^-$ Foa^R phenotype under normal growth conditions (R. Strich and R. E. Esposito, personal communication).

To obtain *URA3* expression upon reconstituting protein-protein interactions, derivatives of the *SPO13* promoter were constructed that contain various numbers of Gal4p binding sites (Giniger et al. 1985). By convention, *URA3* alleles under the control of the *SPO13* URS and the Gal4p binding sites are referred to as *SPALn::URA3*, where n represents the number of Gal4p binding sites. In order to ensure stability and reproducibility of expression levels, the different *SPALn::URA3* alleles were integrated

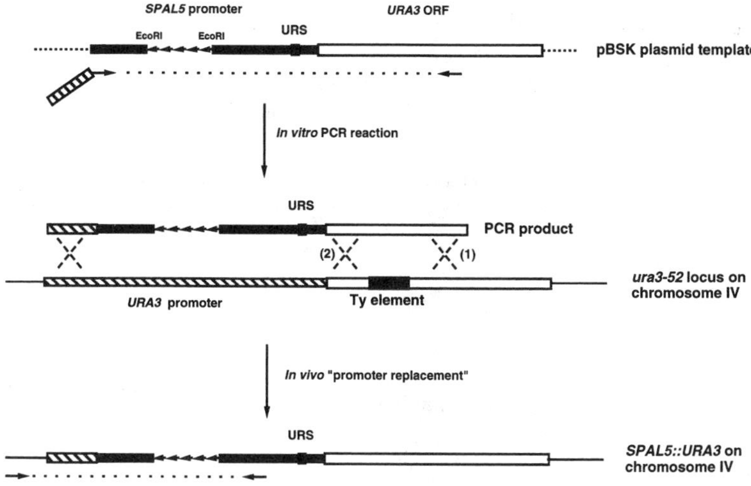

Fig. 7-2. GAL4-dependent *URA3* promoter replacement. The integration of a *SPAL5::URA3* construct is shown as an example. The pBSK derivative shown on top contains five Gal4p binding sites (small arrows) in the *Eco*RI sites of the *SPO13* promoter (gray box). The *SPO13* promoter contains an Upstream Repressing Sequence (black box). *SPAL::URA3* constructs can be amplified (dashed line and below) using two primers: (1) a 5' primer that anneals to the upstream *SPO13* promoter sequence (arrow) and also contains 40 nucleotides of upstream *URA3* sequence (shaded box), and (2) a 3' primer located downstream of the Ty integration site in *ura3-52* (arrow and see below). The *ura3-52* genomic locus on chromosome IV contains a disruptive Ty insertion of about 7 Kb (off-scale black box). The PCR product can integrate by double homologous recombination or gene conversion (dashed crosses) generating a stably integrated *SPAL5::URA3* on chromosome IV (recombination event "1"). This event can be verified by a genomic PCR reaction using the indicated primers (arrows).

into the yeast genome in place of the endogenous Ty1-disrupted *ura3-52* allele (Figure 7-2; Vidal et al. 1996a).

An alternative method to eliminate the Ura$^+$ phenotype conferred by *URA3* basal levels of expression uses media containing 6-azauracil, a pyrimidine biosynthetic inhibitor (Le Douarin et al. 1995). However, 6-azauracil is a relatively nonspecific inhibitor mediating some of its effects on subunits of RNA polymerase II (Nakanishi et al. 1995).

A Modified Version of the Two-Hybrid System

Different versions of the two-hybrid system have been published (Chien et al. 1991; Chevray and Nathans 1992; Dalton and Treisman 1992; Munder and Fürst 1992; Durfee et al. 1993; Gyuris et al. 1993; Vojtek et al. 1993; Le Douarin et al. 1995; chapters 2, 3, and 4), each of which has its own advantages in identifying novel interacting proteins. The purpose of the reverse two-hybrid system, however, is to select mutations or molecules that dissociate or prevent protein-protein interactions. To meet the requirements of such experiments, a modified version was designed by combining aspects of some previously described systems with the counterselectable marker described above.

Plasmids

Since the levels of expression of the DNA-binding domain hybrid protein (DB-X) and the activation domain hybrid protein (AD-Y) can dramatically influence the two-hybrid signal, mutations or molecules that dissociate or prevent interactions might not be detectable when the two hybrid proteins are grossly overexpressed. Thus, in the context of the reverse two-hybrid system, the expression levels of the two hybrid proteins should be maintained relatively low by the use of low-copy expression plasmids. The difference in expression levels obtained between high- and low-copy plasmids can be as much as twenty- to thirty-fold. Another way to control the expression levels is by promoter strength. For the conditions tested here, the constitutive and moderately active *ADH1* promoter was chosen.

The parent DNA-binding domain (pPC97) and activation domain (pPC86) plasmids used in the experiments presented here were previously described (Chevray and Nathans 1992) and contain the following elements (Figure 7-3):

- the pUC-based ori and ampicillin resistance (AMP^R) genes for maintenance and selection in bacteria,
- the *ARS4* sequence for autonomous replication in yeast,
- the *CEN6* centromeric sequence for maintainance at low-copy in yeast,
- the constitutive moderate strength promoter and the transcriptional terminator of *ADH1*,
- the *GAL4* sequence encoding DB (Gal4p 1-147) in pPC97 or the *GAL4* sequence encoding AD (Gal4p 768-881) in pPC86,
- the SV40 large T antigen sequence encoding a nuclear localization signal fused in frame to AD in pPC86 (the Gal4p DB naturally contains such a signal sequence),
- the *LEU2* (pPC97) or *TRP1* (pPC86) selectable marker for selection in yeast on synthetic complete media lacking leucine (Sc-L) or tryptophan (Sc-T), respectively.

Many genetic strategies described below rely upon selection methods to eliminate one of the two-hybrid plasmids from yeast cells (plasmid shuffling; Sikorski and Boeke 1991). In principle, the plasmid contains a marker such as *CYH2*^{S} that confers sensitivity to cycloheximide (CYH) and consequently, the cells in a population that have naturally lost the plasmid exhibit

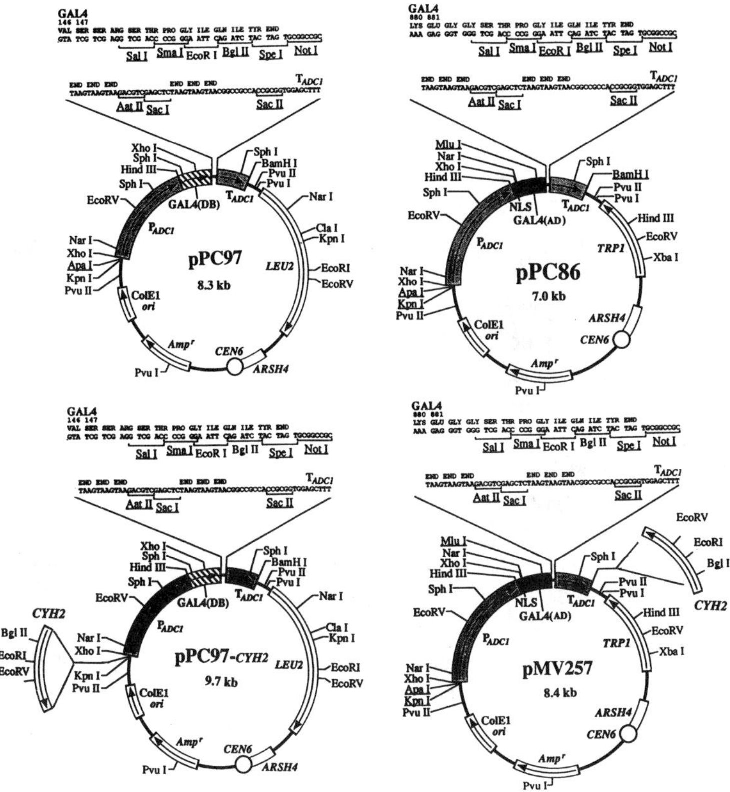

Fig. 7-3. Two-hybrid plasmids. Two plasmids (left) encode a Gal4p DNA-binding domain (DB) fusion and either lack (top) or contain a *CYH2* sensitivity marker (below). Two plasmids (right) encode a Gal4p activation domain (AD) fusion and either lack (top) or contain a *CYH2* sensitivity marker (below). Unique restriction sites in the polylinkers are underlined. The numbers under "GAL4" refer to Gal4p amino acids. Please note that addition of *CYH2* eliminates unique restriction sites in the polylinker site (*Eco*RI and *Bgl*II). Details of these plasmids can be found in the text or in a paper by Chevray and Nathans (Chevray and Nathans, 1992). Modified from Chevray and Nathans (1992), Proc. Nat'l Acad. Sci. 89: 5789–5793. Reprinted by permission of the authors.

a growth advantage on media containing the drug. Versions of plasmids pPC97 and pPC86 have been generated that contain the *CYH2*^{S} marker, plasmids pPC97-*CYH2* (*CYH2*^{S} is in the unique *Apa*I site) and pMV257 (*CYH2*^{S} is in the unique *Bam*HI site), respectively (Figure 7-3). In the experiments described below, only one of the two plasmids expressing the pair of interactors contains *CYH2*. The strains used here, including MaV103, contain a *cyh2*^{R} mutation (see subsection, "Yeast Strains"). In MaV103 cells containing a pair of pPC97-*CYH2* and pPC86 plasmids or a pair of pPC97 and pMV257 plasmids, the pPC97-*CYH2* or the pMV257 plasmid can be counterselected on media containing cycloheximide and leucine or tryptophan, respectively.

Reporter Genes

The negative selection against interaction provided by a counterselectable marker is important to detect dissociation. However, many experiments require, in the same yeast strains, the presence of reporter genes that allow positive selection for interaction, as well as a quantitative read-out. In addition, since the system should ideally be usable with many different protein-protein interactions exhibiting a wide range of affinities, both the negative and positive selections should be titratable.

Three reporter genes, *URA3*, *HIS3*, and *lacZ*, are used in the system described here (Figure 7-4). The read-outs of these reporters are qualitatively different and different promoter sequences regulate expression of each. *URA3* is a dual reporter gene that allows both positive and negative selection. Elevated levels of *URA3* expression in yeast cells provide a selective advantage on media lacking uracil and confer toxic effects on media containing FOA. *HIS3* allows positive selection. *HIS3* encodes imidazole glycerol phosphate dehydratase, an enzyme involved in histidine biosynthesis. This enzyme can be specifically inhibited in a dose-dependent manner by 3-aminotriazole (3-AT) (Kishore and Shah 1988). Higher expression levels of *HIS3* are required to reach the growth threshold when 3-AT is added to the media, and this phenomenon is dose-dependent in a given range of drug concentrations (Durfee et al. 1993). *LacZ* allows a quantitative estimate of the level of expression of the two-hybrid-dependent reporter genes. *LacZ* encodes the enzyme β-galactosidase (β-Gal) that converts X-Gal into a blue product. The intensity of the blue color reflects *lacZ* expression levels.

The *URA3* reporter gene is fused to the *SPO13* promoter containing Gal4p binding sites (*SPAL::URA3*; Vidal et al. 1996a). The *HIS3* reporter gene is fused to its own promoter lacking endogeneous UAS sequences and containing a 125bp GAL_{UAS} sequence (*GAL1::HIS3*; Durfee et al. 1993). The *lacZ* reporter gene is fused to the *GAL1* full-length promoter (*GAL1::lacZ*; Fields and Song 1989). The three reporter genes are stably integrated in single copies at different loci in the strains described in the following section.

Yeast Strains

The strains used in this system contain:

- a set of nonreverting auxotrophic mutations: *leu2* and *trp1* to allow selection of the two plasmids expressing the hybrid proteins, and *his3* for growth dependence upon *GAL1::HIS3* expression on media containing 3-AT.

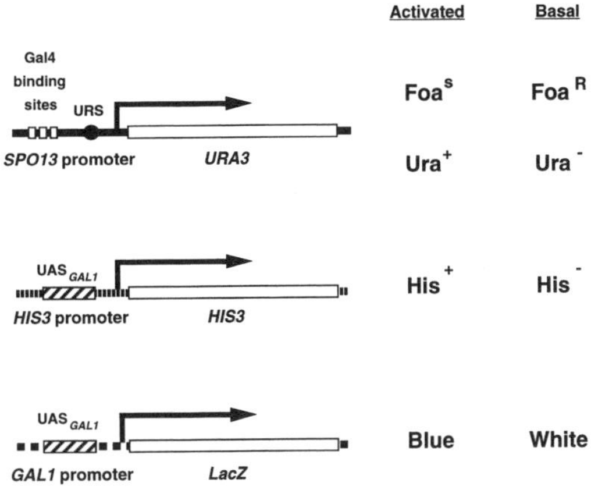

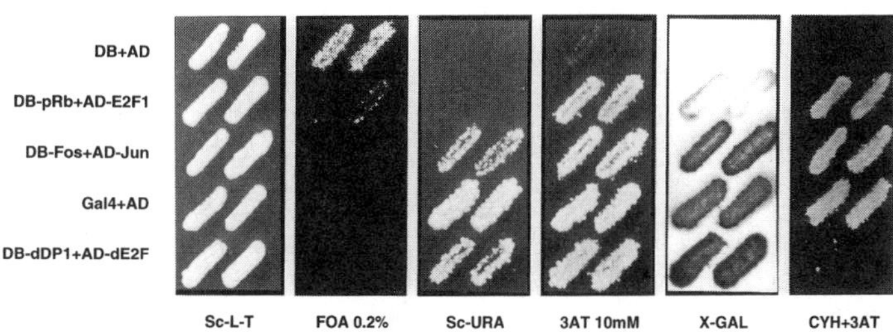

Fig. 7-4. Three reporter genes and associated phenotypes. Three GAL4-dependent reporter genes are integrated in MaV103 cells. The *SPAL10::URA3* reporter gene contains 10 GAL4 binding sites in the *SPO13* promoter fused to *URA3* and is integrated at the *URA3* locus (Vidal et al. 1996a). The *GAL1::HIS3* reporter gene contains a 125 bp $GAL1_{\mathrm{UAS}}$ in the *HIS3* promoter fused to *HIS 3* and is integrated at the *LYS2* locus (Durfee et al. 1993). The *GAL1::lacZ* reporter gene contains the *GAL1* promoter fused to *lacZ* and is integrated at an unknown locus. In the absence of Gal4p or any interacting proteins, the phenotypes are as indicated (*basal*) and shown (first row: DB+AD = control 1). In the presence of Gal4p or interacting proteins, the phenotypes are as indicated (*activated*) and shown (second row: DB-pRb+AD-E2F1 = control 2, "weak" interaction; third row: DB-Fos+AD-Jun = control 3, "strong" interaction; fourth row: Gal4 + AD = control 4, GAL4 positive control). The last row shows an example of cycloheximide/3-AT sensitivity phenotype. The DB-dDP1 fusion is expressed from a *CYH2*^{S} plasmid (control 5). Consequently cycloheximide selects for cells that have lost the corresponding plasmid and thus exhibit a 3-AT sensitive phenotype.

- deletions of the *GAL4* and *GAL80* genes encoding Gal4p and its repressor Gal80p, respectively. In the absence of Gal80p, galactose is not required for activation of Gal4p-inducible promoters.
- two recessive drug resistance mutations, *can1*^{R} and *cyh2*^{R} for plasmid shuffling (see Sikorski and Boeke 1991).
- three stably integrated single copy Gal4p-inducible reporter genes: *SPAL::URA3*, integrated at *URA3*, *GAL1::HIS3*, integrated at *LYS2*; and *GAL1::lacZ*, integrated at an unknown locus.

Strains with the above features are available for both mating types: MaV103 (*MATα leu2-3,112 trp1-901 his3Δ200 ade2-101 gal4Δ gal80Δ SPAL10::URA3 GAL1::lacZ GAL1::HIS3@LYS2 can1*^{R} *cyh2*^{R}; Vidal et al. 1996a) and MaV203 (*MATα*, same genotype). These strains are derived from a cross between two nonisogenic strains, PCY2 (Chevray and Nathans 1992) and MaV99 (Vidal et al. 1996a).

Yeast transformants expressing well-characterized protein-protein interactions have been described and are used as phenotypic controls in the experiments described below. They are the following: "control 1," pPC97 and pPC86 (Chevray and Nathans 1992) used as a negative control; "control 2," pPC97-RB and pPC86-E2F1 (Vidal et al. 1996a) which express a relatively "weak" interaction; "control 3," pPC97-Fos and pPC86-Jun (Chevray and Nathans 1992) which express a relatively "strong" interaction; "control 4," pCL1 encoding full length Gal4p (Fields and Song 1989) and pPC86 used as a positive control; "control 5," pPC97-*CYH2* -dDP and pPC86-dE2F (Du et al. 1996) which express a relatively "strong" interaction and provide a control for plasmid shuffling (Figure 7-4).

The reporter gene signal for a particular interaction should be titratable for both the negative and the positive phenotypes. For the negative phenotype, both the concentration of FOA and number of Gal4p binding sites in the *SPAL::URA3* promoter can be titrated (Vidal et al. 1996a; Figure 7-5a). Isogenic strains that contain different numbers of Gal4p binding sites (from 5 to 10) in the *SPAL::URA3* promoter have been described and are named as follows: MaV95, *SPAL5::URA3*; MaV96, *SPAL7::URA3*; MaV97, *SPAL8::URA3*; MaV99, *SPAL10::URA3*. For the positive phenotype, the concentration of 3-AT can also be titrated (Durfee et al. 1993; Figure 7-5b).

Two-Step Selections to Characterize Interactions

In order to generate models for how protein-protein interactions relate to the function of a complex, it is critical to determine how mutations affecting the interaction affect the function. Single amino acid changes are better suited for this type of analysis than truncations or deletions since they are less likely to disrupt protein function. In many cases, however, very little is known about the interacting domains, such that mutations leading to discrete amino acid substitutions that affect the interaction are relatively rare events. Consequently, a very large number of potential alleles need to be

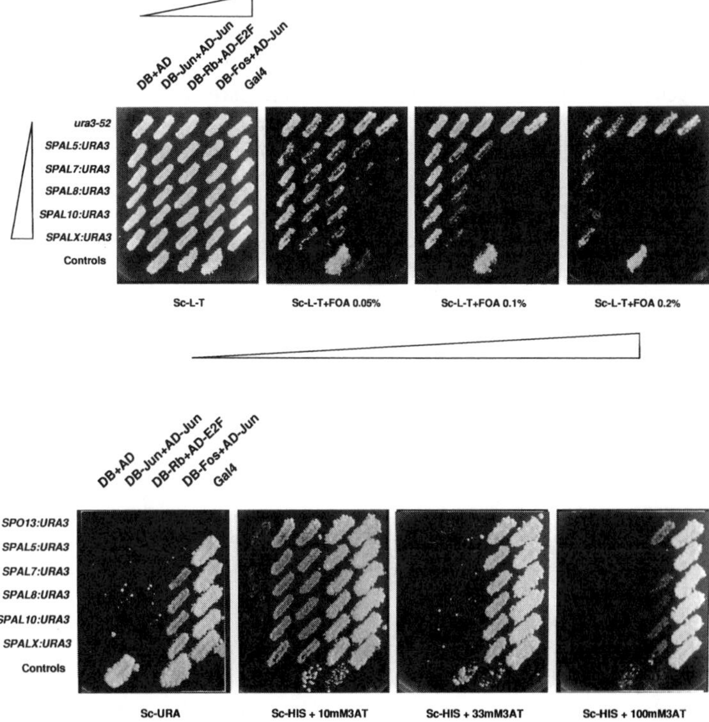

Fig. 7-5. Titration of the Foa (top) and 3AT (bottom) phenotypes. Derivatives of MaV52 (*ura3-52,* MaV52; *SPAL5::URA3,* MaV95; *SPAL7::URA3,* MaV96; *SPAL8::URA3,* MaV97; *SPAL10::URA3,* MaV99; *SPALX::URA3,* MaV94) were cotransformed with the plasmids indicated on top of the left panel. Four individual transformants were tested and one is shown here. The yeast patches were manipulated as described in the text. Figure 7-5A modified from Figure 2d in Vidal et al. (1996a), *Proc. Nat'l Acad. Sci.* 93: 10315–10320. Reprint by permission of the National Academy of Sciences.

screened in order to identify multiple informative amino acid changes. Current biochemical methods are generally time-consuming and cannot be used to identify interesting mutations in large collections of potential alleles. Hence, genetic selections are often needed.

A system is described here that combines PCR mutagenesis with yeast reverse and conventional two-hybrid systems for the genetic identification of useful mutations affecting protein-protein interactions. The system can be used to characterize in detail any interaction domain without any prior structural or functional knowledge. Conceptually, the requirements to identify functionally relevant mutations can be divided into three parts.

1. A method to produce large libraries of random mutations in the gene encoding the interacting protein of interest: The mutagenesis is accomplished by a PCR reaction under conditions that favor the misincorporation rate (mutagenic PCR). The PCR fragments are then introduced directly into yeast cells by gap repair in a linearized expression vector containing homologous sequences (Mulhard et al. 1992). These experiments are extremely efficient and many alleles can be created and introduced into yeast cells in a short time.
2. A rapid assay for protein-protein interaction capable of detecting subtle changes: A version of the two-hybrid system was designed in which the hybrids are expressed at moderate levels and the reporter gene signals are titratable.
3. Reporter genes that allow both negative and positive selection for the relevant alleles: A majority of the mutations selected in the reverse two-hybrid system consists of relatively uninformative truncations and only a minority represents more informative single amino acid changes. When combined appropriately, however, the reverse two-hybrid system and the conventional two-hybrid system provide dual selections that can be applied sequentially to identify discrete amino acid changes.

When the conditions of the mutagenic PCR reaction (see next section) are adjusted to give rise to approximately one nucleotide change per molecule, the relative ratios between the different classes of alleles are approximately as follows: (1) 99% of the molecules are wild type, mutated outside of the domain required for interaction, or mutated inside the domain but without functional consequence, (2) 0.99% of the molecules contain relatively uninformative nonsense mutations, deletions, or insertions, and (3) 0.01% of the molecules contain a missense mutation resulting in a single amino acid change that results in a more informative phenotype. Thus, as expected, the frequency of mutations that affect an interaction in the mutant library is relatively low, and the frequency of useful single amino acid changes among those is even lower. In two-step selection schemes, sequential negative and positive growth selections are used to identify interesting alleles in a large background of relatively uninformative mutations. The selection for mutations that affect an interaction is possible in the first step negative selection with *SPAL::URA3*. Subsequently, the selection for the subset of these mutations that confer interesting phenotypes is performed in a second step positive selection using *GAL1::HIS3* (Figure 7-6). Interesting mutations include: (1) amino acid changes that strongly affect the interaction in the context of a full-length protein, (2) weak mutations that do not completely abolish interaction, and (3) conditional mutations that affect the interaction only under restrictive conditions. Each class of these different mutations can be selected using different variations of the two-step selection concept. The presence of both *SPAL::URA3* and *GAL1::HIS3* reporter genes in the same yeast strains allows the testing of a large number of yeast colonies by simply replica-plating from a plate containing FOA to a plate containing 3-AT.

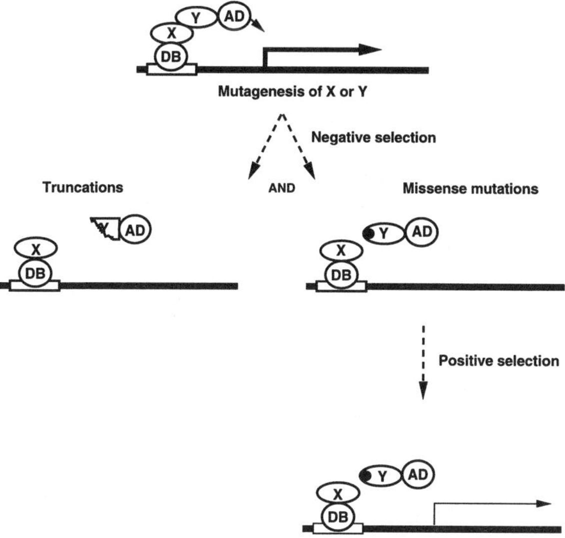

Fig. 7-6. Scheme of two-step selections. Combinations of the reverse and "forward" two-hybrid systems allows the selection of informative missense mutations from random allele libraries in two steps, a first step negative selection and a second step positive selection. Modified from Vidal et al. (1996), *Proc. Nat'l Acad. Sci.* 93:10321–10326. Reprinted by permission of the National Academy of Sciences.

Standard Protocol

The two-step selection procedure detailed below can be summarized as follows: (1) reconstitution of the interaction in the two-hybrid system, (2) PCR mutagenesis of the relevant sequences, (3) transformation of the mutagenized DNA into yeast cells, (4) sequential growth selections, (5) recovery and retesting of the mutagenized plasmids, (6) estimation of the steady-state levels of the mutant proteins by Western blot analysis, (7) testing of the binding ability of the mutant proteins in in vitro reactions, and (8) sequence analysis of the newly generated alleles.

Reconstitution of the Interaction

Recovery of interesting mutations will be optimal under selection conditions close to the growth threshold. The interaction should be titrated for both negative and positive phenotypes. For the negative phenotype, both the concentration of FOA and number of Gal4p binding sites in the *SPAL::URA3* promoter can be titrated (Vidal et al. 1996a; Figure 7-5a).

For the positive phenotype, the concentration of 3-AT can also be titrated (Durfee et al. 1993; Figure 7-5b).

The relevant plasmids should be transformed into the panel of *SPAL::URA3* strains. As controls, plasmids corresponding to controls 1 to 4 should be transformed into MaV103 (Figure 7-4). This allows comparison of the interaction of interest to signals generated by well-characterized interacting pairs. DB-X and AD-Y encoding plasmids should be transformed together as well as with plasmids encoding the activation domain (AD) and DNA-binding domain (DB), respectively, as negative controls.

The titration experiments are performed as follows:

1. Patch two colonies per transformation onto a synthetic complete plate lacking both leucine and tryptophan [Sc-L-T; drop-out media are designated by the absence (–) or presence (+) of leucine (L), tryptophan (T), histidine (H) or uracil (U)], include two patches for controls 1 to 4 (Figure 7-4), and incubate for 18 hours at 30° C.
2. Replica-plate onto Sc-L-T+U plates containing 0.05%, 0.1%, and 0.2% of FOA and onto Sc-L-T-H plates containing 10 mM, 30 mM, and 100 mM of 3-AT. Immediately replica-clean (see appendix) and incubate for ~24 hours at 30° C.
3. Replica-clean again and incubate for 2 to 3 days at 30° C.

This experiment indicates, for each interaction tested, the minimal number of Gal4p-binding sites in the *SPAL::URA3* promoter and the lowest FOA concentration that should be used, as well as the maximal concentration of 3-AT tolerated by cells expressing the interacting proteins.

PCR Mutagenesis PCR mutagenesis is advantageous over spontaneous mutations or the use of hydroxylamine since the choice of appropriate primers targets the sequence to be mutagenized (Zhou et al. 1991; Spee et al. 1993). The conditions of the PCR reaction should be titrated with different concentrations of manganese or with different ratios of nucleotides to obtain a mutation rate of about 10^{-3}. The optimal conditions will depend on the length of the fragment and the sequence to be mutagenized. The rate of misincorporation should be high enough to detect informative mutations among the yeast colonies plated; on the other hand, the rate of misincorporation should not be too high since interpretation of the results might be complicated by the presence of multiple mutations in a particular allele. In general, if single point mutations are desired, conditions giving rise to 1% Foa^R colonies are optimal.

Mutagenic PCR is performed as follows:

1. The primers are designed to generate, in the PCR product, flanking sequences that are identical to plasmid sequences such that the PCR products can be integrated in vivo by homologous recombination in yeast cells into the corresponding two-hybrid plasmids (Mulhard et al. 1992; Figure 7-7a).

   In the case where the full-length sequence introduced into pPC86 should be mutagenized, "universal primers" can be used and linearized pPC86 plasmid can be used for gap repair. The 5' primer corresponds to a sequence located in the AD coding sequence approximately 100 bp upstream of the polylinker. The sequence of the AD-specific primer should be the following: 5' CGCGTTTCCAATCACTACAGGG 3'. (In cases where DB-X is mutagenized, the sequence of the DB-specific primer should be the following: 5' GGCTTCAGTGGAGACTGATATGCCTC 3'). The 3' primer corresponds to a sequence located in the transcription termination sequence (Term) approximately 100 bp downstream of the polylinker and should be the following: 5' GGAGACTTGACCAAACCTCTGGCG 3'.

In cases where sequences corresponding to discrete domains are mutagenized, specific flanking PCR primers have to be designed appropriately. In addition, the linearized plasmid used in the transformation experiment should be a pPC86 plasmid derivative that contains the Y-encoding sequence. A unique restriction site is necessary in the mutagenized sequence for plasmid linearization (see following steps).

2. Since the PCR reaction products are transformed directly into yeast cells without prior separation, the template plasmid can generate a background of yeast transformants expressing wild type interacting protein. This background can be avoided by linearizing the pPC86-derived plasmid using a unique restriction site(s) outside of the amplified sequence (Figure 7-3).
3. PCR reactions for mutagenesis should contain the following components: 100 ng of DNA template, 1 μM "DB" or "AD" 5' primer, 1 μM "Term" 3' primer, 50 μM each dNTP, 50 mM KCl, 10 mM TrisCl pH 9.0, 0.1% Triton X-100, 1.5 mM $MgCl_2$, 1 μg/μl BSA, and 5 units of *Taq* DNA polymerase. The relative concentration of the nucleotides can be independently modified by preparing four combinations of one dNTP at 20 μM and the other three dNTPs at 100 μM.
4. Reactions are performed in 100 μl. A first set of 10 elongation cycles is performed (1' at 94°C, 1' at 45°C, 2' at 72°C), $MnCl_2$ is then added (to 100 μM, for example), and subsequently the reaction is continued for 30 elongation cycles (same conditions). Several independent PCR reactions should be performed to limit the enrichment of particular alleles arising in the early cycles.

Under the conditions of the PCR reactions described here, some amino acid substitutions cannot be recovered. Alternative methods, such as the "regional codon randomization" (Cormack and Struhl 1993) can be used. Conceptually, a library of "clustered charged-to-alanine" alleles designed to alter residues mainly on the surface of the interacting protein might also be more appropriate (Bass et al. 1991; chapter 6). In addition, the two-step selection can be used for selecting alleles conferring higher binding affinities. For this application, one could use "sexual PCR" (Stemmer 1994), as a method to generate the library of mutant alleles. In most cases, however, conventional PCR mutagenesis gives rise to sufficient numbers of interesting alleles.

Transformation The plasmid encoding the unmutagenized DB-X hybrid partner is first transformed into the optimal *SPAL::URA3* strain. The AD-Y plasmid containing the mutagenized sequence is subsequently reconsti-

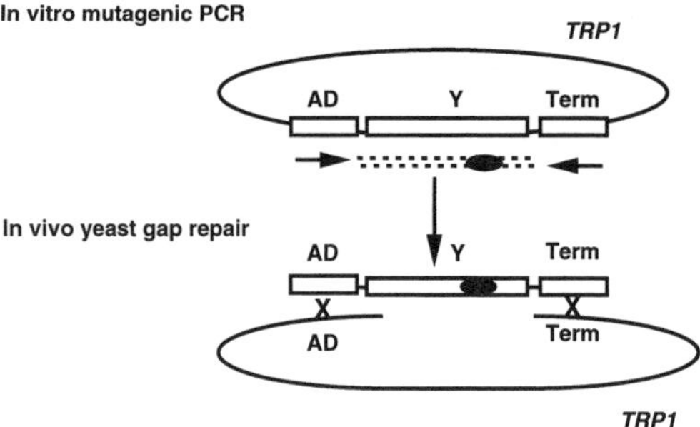

Figure 7.7 Methods for two-step selections. (A) PCR mutagenesis is performed as described in the text (AD-Y is shown here as an example). The primers are located ~100 bp upstream and downstream of the polylinker site. The DNA fragments generated in the PCR reaction are precipitated, resuspended, and directly transformed into yeast cells along with linearized pPC86 plasmid where they are expected to regenerate stable plasmids by gap repair.

tuted in the yeast cells using a gap repair transformation (Orr-Weaver et al. 1983; Mulhard et al. 1992; Figure 7-7a). The transformation/gap repair method using the PCR product from one reaction and pPC86 plasmid linearized in the polylinker (for example, *Sal*I) can be very efficient, giving rise to tens of thousands of transformants. Neither the PCR products nor the linearized plasmids are stable in yeast cells and, therefore, very low numbers of transformants should be recovered when these molecules are transformed separately. However when cotransformed into yeast cells, these molecules are expected to mediate homologous recombination or gene conversion via the two ~100 bp homologous sequences located at their extremities.

A high-efficiency protocol was established for strain MaV103 but can be extended to other *SPAL::URA3* strains (see appendix). The controls should be: (1) no DNA (negative control), (2) the PCR products and the AD-Y linearized plasmid individually (negative controls for the gap repair), and (3) pPC86 plasmid as well as the unmutagenized circular AD-Y plasmid individually (positive controls for the transformation). The pPC86- and AD-Y-containing cells are used in the growth selections described below as negative and positive controls for the two-hybrid read-outs, respectively.

Approximately 300 ng of PCR product and 150 ng of the restricted plasmid are cotransformed. The relative amounts might have to be titrated and will vary depending on the length and sequence of the PCR product. After

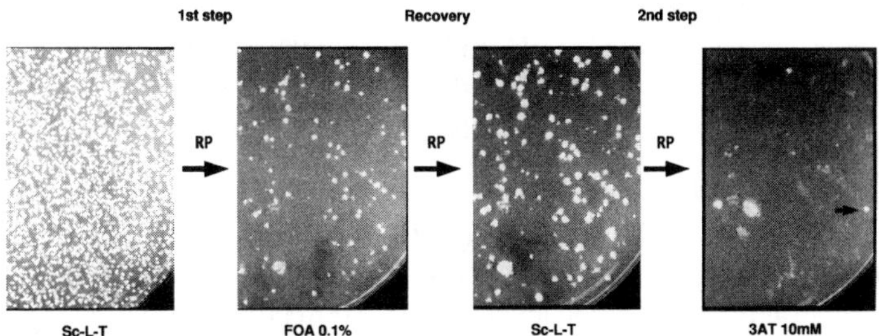

(B) In the experiment shown here, a library of PCR-generated AD-E2F1 alleles was introduced by gap-repair transformation in MaV103 cells expressing wild-type DB-DP1. Approximately 10,000 transformants were obtained on Sc-L-T plates (first panel). These plates were replica-plated (RP) onto Sc-L-T plates containing 0.1% FOA, for a first-step negative selection. After a 2-day growth, these plates were replica-plated onto Sc-L-T plates for a 24-hour recovery step and then replica-plated again onto Sc-L-T-H plates containing 10 mM 3-AT, for a second-step positive selection. For comparison, the wild-type DB-DP1/AD-E2F1 interaction provides growth on plates containing up to 100 mM3-AT (not shown). The three heavy patches at the bottom left are controls.

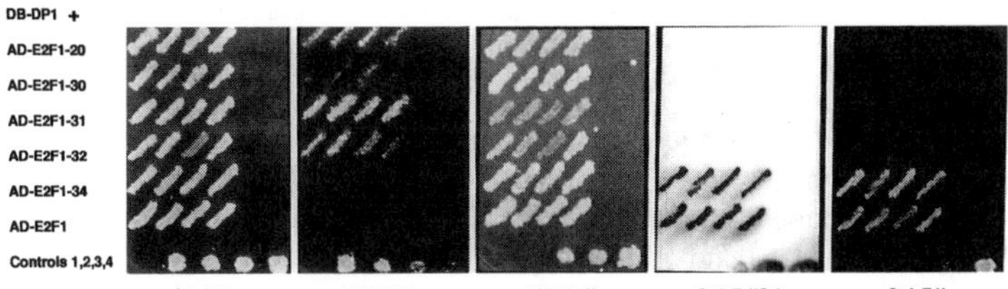

(**C**) Growth phenotypes of 5 putative E2F1 mutant alleles (the allele numbers are indicated). Colonies picked on the 3-AT plates after the second-step selection were streaked out and 4 isolated colonies were patched on a Sc-L-T plate, along with colonies containing wild-type AD-E2F1 and controls 1, 2, 3, and 4. The Sc-L-T plates were replica-plated on the indicated plates and incubated as indicated in the text. Under the conditions of this assay, the 3-AT (20 mM) phenotype is much less stringent than the Ura phenotype and therefore E2F1 alleles with weak mutations promote growth on plates containing 3-AT but not on plates lacking uracil (Sc-U). Note the range of Foa phenotypes conferred by the various alleles. In each case, these growth phenotypic differences correlated with in vitro experiments. For an unkown reason, the E2F1-34 allele did not retest in this assay and was not analyzed further.

transformation, the cells are plated onto media lacking both leucine and tryptophan to select for both the plasmid encoding the wild-type partner and the gap repair of the newly introduced plasmids containing the mutant alleles (expected numbers of transformants are presented in the following section).

Sequential Growth Selections

The length of incubation on the Sc-L-T master-plate prior to replica-plating to the selective plates is crucial, because cells approaching stationary phase exhibit different expression levels of the hybrid proteins from cells growing in exponential phase. The density of the inoculum of cells transferred to the selective plates after replica-plating is also important. As the number of cells transferred by replica-plating increases, the phenotypic differences between positive and negative controls decreases. Therefore, it is crucial to replica-plate and replica-clean (see appendix) at the suggested times.

The sequential growth selections and subsequent phenotypic analyses are performed as follows:

1. First step selection. After a 60 hour growth, replica-plate the plates containing the transformants onto Sc-L-T+U plates containing the optimal FOA concentration. Immediately replica-clean and incubate for ~24 hour at 30° C. Replica-clean again and incubate for 2 to 3 days at 30° C. Figure 7-7b shows an example of the phenotypes observed at that stage of the experiment. Also replica-plate the plates containing the controls described in the "transformation" section.
2. Recovery. The above plates are replica-plated onto Sc-L-T plates to allow recovery. This step is important because colonies containing mutations that differentially affect the interaction exhibit different levels of Foa^R phenotypes. As shown in Figure 7-7b, the reverse two-hybrid system allows the elimination of the colonies expressing wild-type interactions. Incubate for one day at 30° C.
3. Second step selection. This selection step identifies, among a large background of nonsense mutations, the alleles that affect the interaction in a selected manner. In all cases except one (Table 7-1), these alleles are selected by their ability to confer growth on plates containing 3-AT. The Foa^R colonies are replica-plated onto Sc-L-T plates containing 10 mM 3-AT, immediately replica-cleaned, incubated for ~24 hours at 30° C, and then replica-cleaned again before a 2 to 3 day incubation at 30° C. Colonies developing on these plates are indicative of the presence of an interesting allele. Figure 7-7b shows an example of the phenotypes observed at this stage in an experiment aimed at identifying weak mutations that affect the interaction only weakly. Typically, the following results can be obtained:

# of transformants	*# of FOA^R*	*# of 3AT^R*	
No DNA	0	nt	nt
PCR fragment alone	0	nt	nt
Linearized plasmid alone	100	0	
PCR fragment + linearized plasmid	10,000	500	10–20
AD-Y plasmid	10,000	2–3	0
pPC86 plasmid	10,000	10,000	0

(nt, not tested)

Table 7-1 Isolation of various types of mutations affecting protein interactions

Mutation	1st step selection	Plasmid shuffle	2nd step selection
Weak	FOA	No	Low [3-AT]
Strong	FOA	Yes	3-AT (with a reference protein)
Conditional	FOA (restrictive conditions)	No	3-AT (permissive conditions)
Altered specificity	3-AT (with non-cognate partner)	Yes	FOA (with cognate partner)

4. Colony purification. The colonies developing on Sc-L-T-H+3AT are picked and streaked out for single colonies on Sc-L-T plates.
5. Confirmation of the phenotypes. Four single colonies are patched on Sc-L-T plates for each mutant candidate and subsequently replica-plated onto media containing FOA (0.1%) and 3-AT (10 mM) and tested for β-Gal activity (see appendix). Only the colonies that retest for the three phenotypes are analyzed further (Figure 7-7c).

Recovery and Retesting of the Mutagenized Plasmids Before any further characterization of the selected alleles, the plasmids should be recovered in *Escherichia coli* and retested in fresh yeast cells (see chapters 2, 3, and 4). Candidate plasmids are reintroduced by tranformation into *GAL1::HIS3 SPAL10::URA3* DB-X containing cells. A negative control (pPC86) and a positive control wild-type AD-Y should also be transformed. Four transformants are patched onto Sc-L-T plates along with controls 1 to 4 (Figure 7-4) and subsequently replica-plated onto media containing FOA (0.1%) and 3AT (10 mM) and tested for β-Gal activity (see appendix). A semiquantitative measurement of the level to which the different alleles affect the interaction can be done by measuring β-Gal activity (see chapters 2, 3, and 4).

Biochemical Analysis of the Mutant Proteins In order to rule out the possibilities that altered phenotypes arise from altered protein stability or expression, or truncations, Western blot analysis should be performed for each potential mutant selected from two-step selection procedures.

The stability of a hybrid protein in yeast cells may be affected by complex formation with its interacting partner, in that mutants defective for interaction might consequently have lower stability in the presence of an interacting partner than their wild type counterpart. Thus, the stability of the mutant hybrid proteins and their wild type counterpart should be compared in the absence of the wild type partner.

Unambiguous results rely on good antibodies, with monoclonal antibodies best for Western blot analyses performed on yeast extracts. Cocktails of monoclonal antibodies should be used in cases where the interacting domain is unknown prior to using the two-step selection procedure, since the newly selected mutations might affect the structure of a particular epitope. Monoclonal antibodies raised against the Gal4p DB and AD domain have been shown to readily detect hybrid proteins in Western blot analyses (Printen and Sprague 1994) and should be used in cases where monoclonal antibodies raised against the protein of interest are not available.

Among the mutants obtained in the two-step selection, those genuinely affecting interaction can be identified in an independent binding assay. One approach is to compare the ability of each mutant protein to interact with the protein of interest in an in vitro binding assay. These experiments can be accomplished relatively quickly on tens of alleles without further subcloning. The sequences encoding the different alleles can be amplified in a PCR reaction using the two-hybrid plasmid recovered from the selected yeast cells as a template. The 5' primer used in the PCR reaction is designed to produce a product that contains, upstream of the Gal4p AD coding sequence, a promoter sequence recognized by the phage T7 RNA polymerase followed by a translational start codon in the context of a favorable Kozak sequence. The resulting PCR fragments can be mixed directly in an in vitro transcription/translation system to generate the corresponding proteins. The protocol of the in vitro binding reactions is outlined in the appendix.

Sequence analysis of the different alleles should indicate interacting or folding domains, and can reflect the structures or contact sites required for the interaction.

Variations of the Two-Step Selection

Different classes of mutations can be obtained using different variations of the two-step selection. In this section, the types of mutations that will be discussed include strong, weak, conditional, altered specificity, and compensatory mutations. In each case, their potential use is discussed and the variations on the growth selection schemes of the two-step selection protocol used to select them are outlined. At this time, the protocols to identify weak and strong mutations have been formally demonstrated (Vidal et al. 1996b) while the changes in these protocols to identify conditional, altered specificity, and compensatory mutations are still under development.

Strong Mutations Strong mutations can be desired for a number of reasons. First, they can yield structural information if they affect protein conformation, or if they involve residues directly involved in the interaction. Second, strong mutations can be valuable in functional studies based on overexpression of the mutant proteins in the relevant biological systems. They can be used as negative controls in dominant negative experiments or reverse genetic experiments, and they can be reintroduced into their native organism when the wild type gene product is not essential for viability.

It is possible to modify slightly the standard two-step selection procedure to select missense mutations that completely abolish a protein's ability to interact with its partner (Table 7-1). In this variation, the second step positive selection is for the alleles that maintain expression of full-length protein. The AD-Y mutagenized hybrid protein is designed to contain at its carboxy-terminus a small domain that interacts with a reference protein. After the negative selection against the interaction of interest, the plasmid encoding the DB-X wild-type partner is eliminated and, prior to the second step, a plasmid encoding the reference protein is introduced.

This system was tested with a protein that naturally contains such a domain arrangement (Vidal et al. 1996b). In the transcription factor E2F1, one domain interacts with the heterodimeric partner DP1, and C-terminal to the DP1 interaction domain, another domain of only 18 amino acids interacts with the retinoblastoma gene product (pRB; Helin et al. 1992; Helin et al. 1993). This arrangement of interacting domains allowed a double selection for missense mutations that affected DP1 binding but not pRB binding, and these mutations resulted in production of the full-length protein. Fusion of the E2F1 pRB-binding domain (RBD) with the carboxy-terminus of a protein of interest might allow similar genetic selections.

The two-step selection protocol outlined above requires four additional conditions for selection of strong mutations:

1. The hybrid protein of interest should contain a RBD at its carboxy-terminus. To generate AD-Y-RBD fusion proteins, a sequence encoding the peptide N-LDYHFGLEEGEGIRDLFD-C (Helin et al. 1992) can be cloned in the polylinker of pPC86 at the 3' end of the gene encoding the protein of interest.
2. The plasmid encoding the wild-type partner should contain a counterselectable marker, such as *CYH2*^{S}, to select for plasmid loss. Versions of the two-hybrid plasmids pPC97 and pPC86 are described above that contain the *CYH2* counterselectable marker.
3. A plasmid encoding a DB-pRB hybrid protein should be available (Vidal et al. 1996a). This plasmid should be introduced into yeast cells of the opposite mating-type, and mating procedures can be used to introduce a plasmid into many colonies in one step. The yeast strain MaV203 can be used to produce diploid cells that contain both the mutagenized plasmids and DB-pRB.
4. Variations on the growth selections of the two-step selection section are as follows:
    a. Colonies growing on recovery plates after the first step selection are replica-plated onto Sc-T plates containing cycloheximide (Sc-T+CYH). Growth on CYH medium is relatively slow and the incubation might be prolonged to 48 hours at 30° C.
    b. Approximately 5×10^9 cells of the MaV203 strain containing the DB-pRB plasmid are plated onto Sc-L plates to form a lawn of growing cells after an overnight growth. These plates are replica-plated onto YEPD plates. The CYH plates from above are then replica-plated onto these MaV203-containing YEPD plates and incubated overnight to allow mating.
    c. The colonies on YEPD plates are then replica-plated onto Sc-L-T and Sc-L-T+3AT plates and immediately replica-cleaned. The Sc-L-T plates are used as a control for the mating ability of each colony.

d. The plates are replica-cleaned again after an overnight growth at 30° C and incubated for an additional 48 hours. Growing colonies in this assay indicate that the mutant alleles express full-length protein. Depending on the concentration of FOA used in the first step selection, some alleles recovered at this stage might still represent weak mutations. In the mutagenesis of E2F1, weak alleles represented about 50% of the recovered mutations at intermediate concentrations of FOA (Vidal et al. 1996b).

e. Controls should monitor the efficiency of both plasmid loss and 3-AT selection. As a control for the efficiency of plasmid loss, plates containing lawns of MaV203 cells that contain no plasmid should be used. When plasmid loss happens with the expected efficiency, these plates should give rise to no growing diploids on Sc-L-T plates. As negative and positive controls of the 3-AT selection, a Sc-T plate should be used that contains MaV103 colonies containing pPC86 plasmid or wild-type AD-E2F1 (Vidal et al. 1996a), respectively.

It is important to note that the second step selection for strong mutations is not necessarily restricted to the pRB-binding domain. Alternative small binding domains such as the Fos/Jun leucine zipper could be used (Chevray and Nathans 1992). Methods that do not even require binding, but still assay for the intact protein, are equally possible. For example, the green fluorescent protein could be fused to the protein of interest and the second step selection would involve the search for fluorescent colonies or cells (Chalfie et al. 1994).

Weak Mutations Weak mutations that affect an interaction without completely eliminating it can be valuable for several purposes. They are useful when studying interaction domain structures because it is likely that a number of amino acids might perform redundant functions. Weak mutations can also be valuable for functional studies in reverse genetic systems in which the wild-type gene product is essential for viability or development. Since their reintroduction into their native organism containing a knock-out of the wild-type gene is expected to confer moderate phenotypes, they can be used in modifier screens aimed at identifying additional components involved in a pathway and/or a protein complex.

Weak mutations can be selected according to the standard protocol, simply by performing the second step selection on plates containing low concentrations of 3-AT (Table 7-1 and Figure 7-7).

Other mutations Conditional mutations are defined as genetic changes inducing phenotypes detectable only under certain conditions (restrictive conditions), while allowing essentially wild type phenotypes under different conditions (permissive conditions). The generation of conditional alleles is interesting for two major reasons. First, conditional phenotypes represent powerful tools to study fuction in vivo at any stage of development since temperature shift experiments can be performed. Second, modifier screens performed with these conditional organisms under intermediate conditions can be used to identify second-site suppressor mutations affecting genes acting in a pathway or in a complex.

The standard two-step selection protocol can be applied for the identification of conditional alleles by modifying the conditions of incubation in the first and second step (Table 7-1). Since the uptake rate of both FOA and 3-AT varies with the temperature of incubation, both Foa^S and His$^-$ phenotypes should be titrated carefully at the relevant temperatures. In the first step selection, yeast transformants representing a library of random alleles are incubated on FOA plates under conditions defined as restrictive. Among the defective alleles obtained after the first step selection, a few might have retained the ability to interact with the wild type partner under permissive conditions and can be selected on plates containing 3-AT.

"Altered specificity" alleles are defined here as mutant proteins that exhibit increased binding to one particular partner and decreased binding to another partner. A variation of the standard two-step selection could be used to select such altered specificity alleles (Table 7-1). In this variation, the concentration of 3-AT is titrated such that the binding of the wild type protein of interest to one partner is limiting for growth in the two-hybrid system. A library of mutant alleles of the protein of interest is first selected positively for "gain of function." After plasmid shuffling of the plasmid encoding this partner (see earlier in this section), a plasmid encoding the other partner is reintroduced by mating (see earlier in this section), and alleles with decreased binding are selected on plates containing FOA. A protocol similar to the selection of strong mutations by the two-step selection can be envisioned, with the exception that the positive selection step precedes the negative selection.

Compensatory mutations are defined here as second-site suppressor mutations located in the interacting partner of a mutant protein. A set of mutants with compensatory changes can be very useful reagents for functional studies, particularly of the relationships between partners of a complex. The loss of function alleles selected in the two-step selection procedures could be used to identify "compensatory" mutations in the interacting partner that restore the interaction. However, it is not known in advance whether a specific amino acid change in one partner of an interaction can give rise to any compensatory change in the other partner. Therefore, to obtain several pairs of compensatory changes, many mutations in each partner should be challenged simultaneously. Briefly, two independent two-step selection procedures can be performed, in two different yeast strains of opposite mating types. In one strain (e.g. MaV103), missense mutations that affect the interaction are selected in the DB-X hybrid protein and in the other strain (e.g. MaV203), missense mutations are selected in the AD-Y hybrid partner (Figure 7-6). The plasmids encoding the wild type hybrid proteins are then eliminated by plasmid shuffling in both strains (see earlier in this section). Finally, selected colonies from both strains are mated (see next section) and the resulting diploids containing potential complementary amino acid changes are selected by positive selection. Identification of such pairs of mutations might lead to a "compensatory mutation map" and could help structural understanding of protein-protein interaction domains (Gobel et al. 1994).

The Manipulation of Interactions

Small Dissociating Molecules

A strategy based on the reverse two-hybrid system might identify compounds that dissociate or prevent protein-protein interactions. It has already been shown that small stabilizing molecules can be characterized in the context of the two-hybrid system (Chiu et al. 1994; chapter 15). Analogously, small dissociating molecules might also be identified using the reverse two-hybrid system. *SPAL::URA3* cells expressing a DB-X/AD-Y interaction of interest could be plated onto FOA media and on top of these lawns of nongrowing cells, drops of solutions containing different compounds could be deposited in a matrix (Figure 7-8). Several thousand solutions could reasonably be screened on a few plates. It is known that from such drops, compounds diffuse through the surrounding medium, generating a gradient of concentration. Compounds might be identified on the basis of the fact that, inside yeast cells, they mediate dissociation. Thus, cells would form a ring of growth around the drop containing a dissociator compound. The diameter of the ring would be proportional to the concentration needed to dissociate. A potential problem inherent to this kind of assay could be the ability of yeast cells to take up the compound. Thus, yeast mutants with higher general uptake activities, such as the *erg6* mutant strains (Gaber et al. 1989; Prendergast et al. 1995) for example, might be useful.

Potential false positives are expected from such in vivo screening methods and include FOA uptake blockers, *URA3*-encoded enzyme inhibitors, Gal4p specific transcription inhibitors, Gal4p DNA-binding inhibitors, and so on. However, these nonspecific compounds could be identified in secondary screens involving unrelated proteins interacting in the context of the two-hybrid system.

The reverse two-hybrid system facilitates screening of compounds for two reasons. First, the positive phenotype conferred by dissociator molecules might allow the detection of extremely thin rings, and thus compounds that are only moderately active may be detected. Second, with negative phenotypes, threshold growth conditions can be defined by titration (see earlier in this section) to ensure identification of moderately active coumpounds. Conventional positive selectable markers cannot be modulated and, therefore, are not suitable for this kind of analysis.

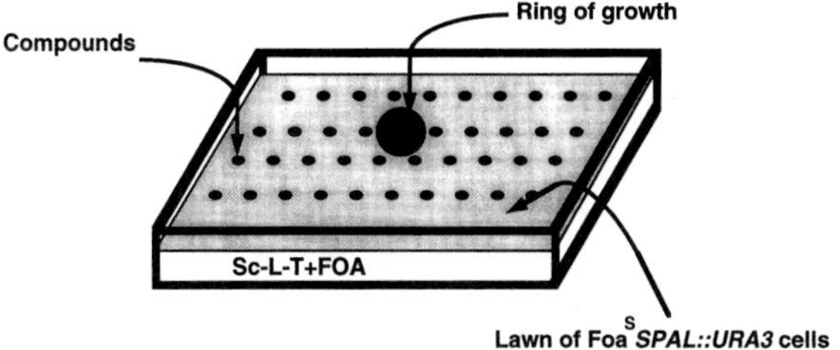

Fig. 7-8. Potential in vivo screening method for protein-protein interaction dissociator compounds. FOA sensitive *SPAL::URA3* cells expressing a DB-X/AD-Y interaction of interest are seeded on the surface of a petri plate containing Sc-L-T+FOA, resulting in a lawn of nongrowing cells. Compounds are spotted on the plate (black dots). A ring of growth of *SPAL::URA3* cells is expected around nontoxic compounds that are able to penetrate the cells and dissociate DB-X/AD-Y. The specificity controls used in such an experiment are discussed in the text.

Dissociating Peptides

Peptides are not currently useful in therapeutics due to their poor stability and the problems inherent to their delivery. However, they can be used, when their size does not exceed five or six amino acids, as lead molecules for the chemical design of small organic molecules (Huber et al. 1994). Peptides could be selected using the reverse two-hybrid system if expressed as a fusion to a carrier protein that would allows stability and structural constraint (Yang, et al. 1995; Colas, et al. 1996; chapter 16).

Dissociating peptides or proteins expressed from cDNA libraries might also be identified in the reverse two-hybrid system and used in functional studies. This concept is exemplified by the fact that the dissociation effect mediated by E1A proteins on the retinoblastoma gene product (pRB)/E2F interaction has been reconstructed in the reverse two-hybrid system (Vidal et al. 1996a).

Use of a Counterselectable Marker to Facilitate Identification of Protein-Protein Interactions

The availability of a toxic counterselectable marker extends the possibilities of the two-hybrid system in identifying protein-protein interactions in large-scale experiments. The modifications proposed here should also be helpful in experiments aimed at identifying interacting proteins of one or a few proteins of interest and, thus, several specific protocols are described.

Large-Scale Projects

The importance of protein-protein interactions suggests that a better understanding of the life of a cell might reside in global information about protein interaction networks (see chapter 12). Databases have been envisioned in which large numbers of protein-protein interactions are recorded in the form of "protein linkage maps." The elaboration of such databases relies primarily upon the construction of libraries of "pairs of interactors" from which systematic sequence analysis provides the necessary information to establish these maps (Bartel et al. 1996).

In this section, a method is described that allows the construction of libraries of "pairs of interactors," referred to here as "Bidirectional Combinatorial Libraries" (BCL). A BCL consists of yeast colonies containing two plasmids that encode proteins that interact in the context of the two-hybrid system. As described in the original two-hybrid method, a library of cDNA clones is fused to the activation domain of a transcription factor like Gal4p. In the method described here, however, a library of cDNAs clones is also fused to the DNA-binding domain. All possible pairwise combinations of DB-X and AD-Y fusion proteins in a given system are expressed in yeast cells. The few cells that express interacting proteins are selected on the basis of the expression of two-hybrid-dependent reporter genes.

Compared to conventional two-hybrid screens, the construction of BCLs is complicated by two additional difficulties. First, the existence of DB-X hybrid proteins that can activate transcription on their own, that is, without the need of an interacting protein, is a major problem. These hybrid proteins are referred to here as "self-activators." For example, approximately 1% of random *E. coli* sequences behave as transcriptional activation domains when fused to the Gal4p DB (Ma and Ptashne 1987). The availability of one (or more) toxic counterselectable markers such as *SPAL::URA3* allows the selective elimination of self-activators in each library (see end of this section). A second problem inherent to the construction of BCLs is the very low probability of finding protein-protein interactions when using large numbers of pairwise combinations. Consequently, the number of two-hybrid plasmid-containing cells that have to be generated is very high. Mating between yeast strains of opposite mating types, each containing one fusion library (Bendixen et al. 1994; Finley and Brent 1994), can be used to facilitate the generation of very large numbers of different possible pairwise combinations (Bartel et al. 1996).

Practically, BCLs are generated using yeast cells of opposite mating type containing different two-hybrid-dependent reporter genes including a toxic counterselectable marker such as *SPAL::URA3*. The DB-X and AD-Y cDNA fusion libraries are transformed independently, one library in cells of one mating type, and the other in cells of the opposite mating type. The self-activators present in the DB-X and AD-Y libraries are eliminated by transferring the transformants to FOA-containing media. The "cleared"

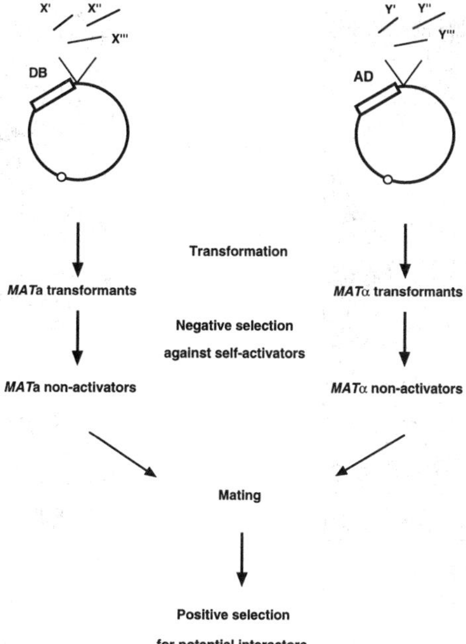

Fig. 7-9. Construction of bidirectional combinatorial libraries. The two-hybrid system can be applied to large-scale protein-protein interaction identification projects. Two libraries are generated, one is fused to a DNA-binding domain (DB) and the other to an activation domain (AD), and introduced into yeast cells of opposite mating-type. Each library is cleared from self-activator clones that activate reporter genes on their own, by using the negative selection possible with a counterselectable marker. The two libraries are then combined by mating and potential interactors are identified by conventional positive selection.

DB-X and AD-Y libraries are then challenged by mating and selecting diploids that contain plasmids encoding interacting DB- and AD-hybrid proteins (Figure 7-9).

More than 30 two-hybrid screens using a variety of unrelated baits and cDNA libraries have been performed with the modified version presented here. (Figure 7-10) Among these different screens, the technical feasibility of each step proposed to generate BCLs has been assessed. In all cases, the number of false positives was extremely low and could be identified conve-

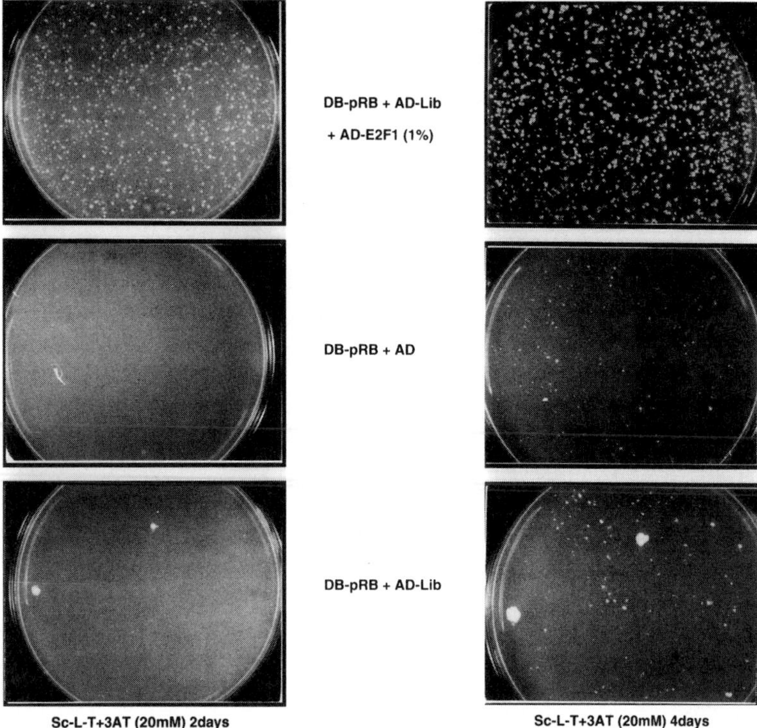

Fig. 7-10. Two-hybrid 3-AT positive selection. In the experiment shown here, a DB-pRB fusion was used to screen a mammalian AD-cDNA-fusion library. After transformation of the indicated plasmids, approximately 70,000 transformants developed on Sc-L-T plates. These plates were transferred to 3-AT plates as indicated in the text. The left and right panels show the phenotypes observed after incubation at 30°C for 2 and 4 days, respectively. In the two top panels (positive control), a plasmid expressing AD-E2F1, a known DB-pRB-interacting protein, was diluted to ~1/100th in the AD-cDNA library prior to transformation. In the two middle panels (negative control), the empty plasmid pPC86 was transformed. The two colonies visible in the two lower panels after a two day growth express putative DB-pRB-interacting proteins.

niently by phenotypic analysis (Sardet et al. 1995; Du et al. 1996; and unpublished observations).

The "Swapped" Two-Hybrid System

Some proteins exhibit very high levels of transcriptional activation when fused to DB and this activity cannot be counteracted by addition of 3-AT (see earlier in this section). Consequently, these proteins have been refractory to two-hybrid cloning with conventional systems. However, several aspects of the construction of BCLs can be applied to this situation.

It is possible to "swap" the two-hybrid system and express the bait as an AD-Y fusion protein and the cDNA library clones as DB-X fusions (Du et al. 1996). As described previously, it is expected that such libraries will contain approximately 1% of self-activating DB-X clones. Two genetic strategies described below can be applied to solve that problem. At this time, the protocols of the "shuffling" strategy have been formally demonstrated (Du et al. 1996) while those of the "clearing and mating" procedure are still under development.

Two-hybrid Screen and Plasmid Shuffling The first strategy relies upon plasmid shuffling (see previous section). Since plasmid shuffling methods can be applied on very large numbers of colonies, colonies containing genuine interacting pairs can be identified from numerous self-activator colonies. The AD-Y bait hybrid is expressed from a *CYH2*-containing plasmid and, hence, dependence of the two-hybrid phenotypes on the expression of the bait can be tested conveniently.

The "swapped" two-hybrid system can be performed using the conventional methods described above with the modifications in the growth selections as follows. After a 60 hour growth, the plates containing the transformants are replica-plated onto Sc-L-T-H plates containing the optimal 3-AT concentration, immediately replica-cleaned and incubated for 48 hours at 30° C. Several thousands of growing colonies are expected. They are then replica-plated and immediately replica-cleaned onto two Sc-L-H+3-AT plates, one containing 1 μg/ml cycloheximide and the other lacking cycloheximide. These plates are incubated for 2 to 3 days at 30° C. The colonies not growing on the plate containing cycloheximide while growing in the absence of cycloheximide are indicative of a plasmid expressing a genuine interacting protein (Du et al. 1996).

Two-Hybrid Screens by Clearing and Mating Another version of the "swapped" two-hybrid system can also be used to identify interacting proteins of baits containing strong activation domains. Here, the DB-X cDNA library is first cleared of its self-activators and subsequently challenged with the bait in the context of a mating selection. (Figure 7-11) In a clearing experiment, the growth selections will eliminate self-activator clones from a DB-X library by

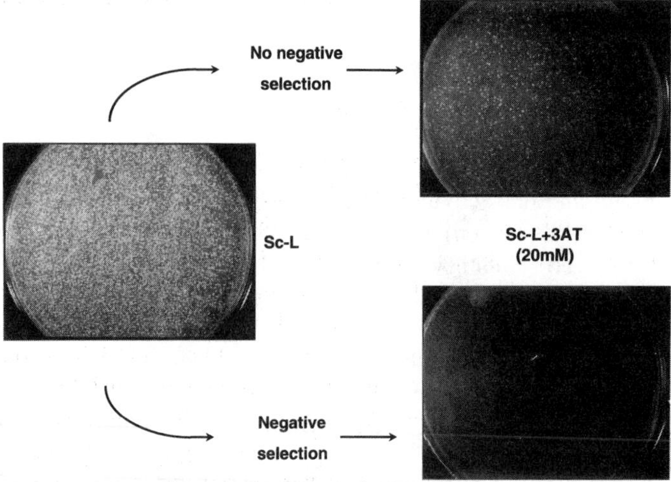

Fig. 7-11. To demonstrate the feasibility of clearing DB-X libraries from their self-activating clones, a well-characterized self-activator encoding plasmid was diluted to 1/100th in a pPC97 plasmid solution. That solution was subsequently transformed into MaV103 cells and plated on a Sc-L plate (left panel). This master-plate was replica-plated to both a Sc-L plate and a Sc-L plate containing 0.2% FOA (not shown on the figure). On the Sc-L plate, there is "no negative selection" and the self-activator-expressing colonies are expected to be viable. However, when using Sc-L+FOA plates, the self-activator expressing colonies are challenged with "negative selection" and are expected to be eliminated. This is verified on Sc-L-H plates containing 20 mM 3-AT. The 3-AT resistant colonies indicate that most self-activator-expressing colonies remain viable when no negative selection is applied (top panel). However, during the negative selection "passages" these clones have been eliminated and no colony shows any 3-AT resistant phenotype (negative selection, bottom panel).

transferring *SPAL::URA3* transformants to plates containing FOA. An advantage of this procedure in the context of the "swapped" two-hybrid is that the clearing experiment needs to be performed only once for a given library.

Conclusions

Genetic selections are required for experiments in which very large numbers of reactions are monitored in order to identify particular molecules of interest. For example, macromolecular binding reactions can be assessed in genetic selections schemes. In vitro genetic selections can be applied to identify RNA molecules that bind to a molecule of interest (Szostack

1992). In vivo genetic selections based on phage presentation techniques can be applied to identify peptides that bind to a molecule of interest (Barbas 1993). Finally, the yeast two-hybrid system and its variations allow genetic selections to be applied to identify protein-protein interactions and other macromolecular binding reactions. In all cases, the selections can be extremely powerful at discovering elements that *favor* binding (positive selection).

However, the characterization of binding partners often requires the identification of elements that *prevent* binding and here again, genetic selections are needed (negative selection). Such a negative selection can be performed in the context of the "reverse" two-hybrid system described in this chapter. The reverse two-hybrid system can be used to characterize protein-protein, DNA-protein, and potentially RNA-protein interactions. Two-step selection strategies were presented that allow the identification of discrete amino acid changes that affect protein-protein interactions by combining positive and negative selections in the same yeast strains. Mutant alleles generated in two-step selection procedures might be helpful to characterize the structure/function of protein complexes. It is also likely that the reverse two-hybrid system will be helpful in identifying molecules that dissociate or prevent interactions.

NOTE An alternative version of the reverse two-hybrid system described here (Shih et al. 1996 Proc Natl. Acad. Sci. USA 93: 13896-13901) was published while this book was in press.

ACKNOWLEDGMENTS Most of the concepts described in this chapter were imagined and developed in the laboratory of Ed Harlow. I am extremely grateful to him for his constant financial and scientific support as well as his encouragement. His "touch" tremendously improved this manuscript.

Many thanks to Jef D. Boeke and the members of his laboratory for their help at an early stage of this work. I also greatly appreciate the contributions of P. Chevray, S. Elledge, R. Esposito, A. Fattaey, S. Fields, J. LaBaer, D. Nathans, R. Strich, C. Sardet, and R. Weinberg for providing yeast strains, plasmids and/or libraries; J. Benson, N. Dyson, W. Du, A. Fattaey, S. Fields, L. Hartwell, H.-L. Ji, B. Kempkes, J. LaBaer, F. McCormick, and C. Sardet for stimulating discussions; E. Chen, P. Braun, and D. Lindstrom for their help; and M. Brasch, L. Cironi, C. Englert, B. Kennedy, B. Kempkes, L. Ko, and B. Schulman for reading the manuscript. Special thanks to J. Vidal and N. Piton. I am the recipient of an American Cancer Society (Mass. Div.) Senior Fellowship.

References

Anderson, S. F., Steber, C. M., Esposito, R. E., and Coleman, J. E. (1995). UME6, a negative regulator of meiosis in *Saccharomyces cerevisiae*, contains a C-terminal Zn2Cys6 binuclear cluster that binds the URS1 DNA sequence in a zinc-dependent manner. Protein Sci. 4:1832–1843.

Barbas, C. F. (1993). Recent advances in phage display. Curr. Opin. Biotech. 4:526–530.

Bartel, P. L., Roecklein, J. A., SenGupta, D., and Fields, S. (1996). A protein linkage map of *Escherichia coli* bacteriophage T7. Nature Genet. 12:72–77.

Bass, S. H., Mulkerrin, M. G., and Wells, J. A. (1991). A systematic mutational analysis of hormone-binding determinants in the human growth hormone receptor. Proc. Natl. Acad. Sci. USA 88:4498–4502.

Bendixen, C., Gangloff, S., and Rothstein, R. (1994). A yeast mating-selection scheme for detection of protein-protein interactions. Nucl. Acids Res. 22:1178–1179.

Boeke, J. D., LaCroute, F., and Fink, G. R. (1984). A positive selection for mutants lacking orotidine-5'-phosphate decarboxylase activity in yeast: 5-fluoro-orotic acid resistance. Mol. Gen. Genet. 197:345–346.

Bowdish, K. S., and Mitchell, A. P. (1993). Bipartite structure of an early meiotic upstream activation sequence from *Saccharomyces cerevisiae*. Mol. Cell. Biol. 13:2172–2181.

Buckingham, L. E., Wang, H.-T., Elder, R. T., McCarroll, R. M., Slater, M. R., and Esposito, R. E. (1990). Nucleotide sequence and promoter analysis of *SPO13*, a meiosis-specific gene of *Saccharomyces cerevisiae*. Proc. Natl. Acad. Sci. USA 87:9406–9410.

Chalfie, M., Tu, Y., Euskirchen, G., Ward, W. W., and Prasher, D. C. (1994). Green Fluorescent Protein as a marker for gene expression. Science 263:802–805.

Chevray, P., and Nathans, D. (1992). Protein interaction cloning in yeast: Identification of mammalian proteins that react with the leucine zipper of Jun. Proc. Natl. Acad. Sci. USA 89:5789–5793.

Chien, C.-T., Bartel, P. L., Sternglanz, R., and Fields, S. (1991). The two-hybrid system: a method to identify and clone genes for proteins that interact with a protein of interest. Proc. Natl. Acad. Sci. USA 88:9578–9582.

Chiu, M. I., Katz, H., and Berlin, V. (1994). RAPT1, a mammalian homolog of the yeast Tor, interacts with the FKBP12/rapamycin complex. Proc. Natl. Acad. Sci. USA 91:12574–12578.

Choi, K. Y., Satterberg, B., Lyons, D. M., and Elion, E. A. (1994). Ste5 tethers multiple protein kinases in the MAP kinase cascade required for mating in *S. cerevisiae*. Cell 78:499–512.

Colas, P., Cohen, B., Jessen, T., Grishina, I., McCoy, J., and Brent, R. (1996). Genetic selection of peptide aptamers that recognize and inhibit cyclin-dependent kinase 2. Nature 380:548–550.

Cormack, B. P., and Struhl, K. (1993). Regional codon randomization: defining a TATA-binding protein surface required for RNA polymerase III transcription. Science 262:244–248.

Dalton, S., and Treisman, R. (1992). Characterization of SAP-1, a protein recruited by Serum Response Factor to the c-*fos* Serum Response Element. Cell 68:597–612.

Du, W., Vidal, M., Xie, J.-E., and Dyson, N. (1996). A homologue of the retinoblastoma family of proteins regulates E2F activity in *Drosophila*. Genes Dev. 10:1206–1218.

Durfee, T., Becherer, K., Chen, P.-L., Yeh, S.-H., Yang, Y., Kilburn, A. E., Lee, W.-H., and Elledge, S. J. (1993). The retinoblastoma protein associates with the protein phosphatase type 1 catalytic subunit. Genes Dev. 7:555–569.

Fields, S., and Song, O.-K. (1989). A novel genetic system to detect protein-protein interactions. Nature 340:245–246.

Finley, R. L., and Brent, R. (1994). Interaction mating reveals binary and ternary connections between *Drosophila* cell cycle regulators. Proc. Natl. Acad. Sci. USA 91:12980–12984.

Gaber, R. F., Copple, D. M., Kennedy, B., Vidal, M., and Bard, M. (1989). The yeast gene *ERG6* is required for normal membrane function but is not essential for biosynthesis of the cell-cycle-sparking sterol. Mol. Cell. Biol. 9:3447–3456.

Giniger, E., Varnum, S. M., and Ptashne, M. (1985). Specific DNA binding of GAL4, a positive regulatory protein of interest. Cell 40:767–774.

Gobel, U., Sander, C., Schneider, R., and Valencia, A. (1994). Correlated mutations and residue contacts in proteins. Proteins 18:309–317.

Gyuris, J., Golemis, E., Chertkov, H., and Brent, R. (1993). Cdi1, a human G1 and S phase protein phosphatase that associates with Cdk2. Cell 75:791–803.

Helin, K., Lees, J. A., Vidal, M., Dyson, N., Harlow, E., and Fattaey, A. (1992). A cDNA encoding a pRB-binding protein with properties of the transcription factor E2F. Cell 70:337–350.

Helin, K., Wu, C., Fattaey, A., Lees, J., Dynlacht, B., Ngwu, C., and Harlow, E. (1993). Heterodimerization of the transcription factors E2F-1 and DP-1 leads to cooperative transactivation. Genes Dev. 7:1850–1861.

Huber, H. E., Koblan, K. S., and Heimbrook, D. C. (1994). Protein-protein interactions as therapeutic targets for cancer. Current Medicinal Chemistry 1:13–34.

Irniger, S., Piatti, S., Michaelis, C., and Nasmyth, K. (1995). Genes involved in sister chromatid separation are needed for B-type cyclin proteolysis in budding yeast. Cell 81:269–277.

Ito, H., Fukada, Y., Murata, K., and Kimura, A. (1983). Transformation of intact yeast cells treated with alkali cations. J. Bacteriol. 153:163–168.

Katcoff, D. J., Yona, E., Hershkovits, G., Friedman, H., Cohen, Y., and Dgany, O. (1993). SIN1 interacts with a protein that binds the URS1 region of the yeast *HO* gene. Nucl. Acids Res. 21:5101–5109.

King, R. W., Peters, J.-M., Tugendreich, S., Rolfe, M., Hieter, P., and Kirschner, M. W. (1995). A 20S complex containing CDC27 and CDC16 catalyzes the mitosis-specific conjugation of ubiquitin to cyclin B. Cell 81:279–288.

Kishore, G. M., and Shah, D. M. (1988). Amino acid biosynthesis inhibitors as herbicides. Ann. Rev. Biochem. 57:627–663.

Koleske, A. J., and Young, R. A. (1994). An RNA polymerase II holoenzyme responsive to activators. Nature 368:466–469.

Le Douarin, B., Pierrat, B., vom Baur, E., Chambon, P., and Losson, R. (1995). A new version of the two-hybrid assay for detection of protein-protein interactions. Nucl. Acids Res. 23:876–878.

Lopes, J. M., Schulze, K. L., Yates, J. W., Hirsch, J. P., and Henry, S. A. (1993). The *INO1* promoter of *Saccharomyces cerevisiae* includes an upstream repressor sequence (URS1) common to a diverse set of yeast genes. J. Bacteriol. 175:4235–4238.

Luche, R. M., Smart, W. C., and Cooper, T. G. (1992). Purification of the heterodimeric protein binding to the URS1 transcriptional repression site in *Saccharomyces cerevisiae*. Proc. Natl. Acad. Sci. USA 89:7412–7416.

Luche, R. M., Smart, W. C., Marion, T., Tillman, M., Sumrada, T. A., and Cooper, T. G. (1993). *Saccharomyces cerevisiae* BUF protein binds to sequences participating in DNA replication in addition to those mediating transcriptional repression (URS1) and activation. Mol. Cell. Biol. 13:5749–5761.

Luche, R. M., Sumrada, R., and Cooper, T. G. (1990). A *cis*-acting element present in multiple genes serves as a repressor binding site for the yeast *CAR1* gene. Mol. Cell. Biol. 10:3884–3895.

Ma, J., and Ptashne, M. (1987). A new class of yeast transcriptional activators. Cell 51:113–119.

Mulhard, D., Hunter, R., and Parker, R. (1992). A rapid method for localized mutagenesis of yeast genes. Yeast 8:79–82.

Munder, T., and Fürst, P. (1992). The *Saccharomyces cerevisiae CDC25* gene product binds specifically to catalytically inactive Ras proteins in vivo. Mol. Cell. Biol. 12:2091–2099.

Nakanishi, T., Shimoaraiso, M., Kubo, T., and Natori, S. (1995). Structure-function relationship of yeast S-II in terms of stimulation of RNA polymerase II, arrest relief, and suppression of 6-Azauracil sensitivity. J. Biol. Chem. 270:8991–8995.

Orr-Weaver, T. L., Szostack, J. W., and Rothstein, R. J. (1983). Genetic applications of yeast transformation with linear and gapped plasmids. Methods Enzymol. 101:228–245.

Park, H.-O., and Craig, E. A. (1991). Transcriptional regulation of a yeast *HSP70* gene by heat shock factor and an Upstream Repression Site-binding factor. Genes Dev. 5:1299–1308.

Prendergast, J. A., Singer, R. A., Rowley, N., Johnston, G. C., Danos, M., Kennedy, B., and Gaber, R. (1995). Mutations sensitizing yeast cells to the start inhibitor nalidixic acid. Yeast 11:357–547.

Printen, J. A., and Sprague, G. F. (1994). Protein-protein interactions in the yeast pheromone response pathway: Ste5p interacts with all members of the MAP kinase cascade. Genetics 138:609–619.

Sardet, C., Vidal, M., Cobrinik, D., Geng, Y., Onufryk, C., Chen, A., and Weinberg, R. A. (1995). E2F-4 and E2F-5, two novel members of the E2F family, are expressed in the early phases of the cell cycle. Proc. Natl. Acad. Sci. USA 92:2403–2407.

Sikorski, R. S., and Boeke, J. D. (1991). *In vitro* mutagenesis and plasmid shuffling: from cloned gene to mutant yeast. Meth. Enzymol. 194:302–318.

Spee, J. H., de Vos, W. M., and Kuipers, O. P. (1993). Efficient random mutagenesis method with adjustable mutation frequency by use of PCR and dITP. Nucl. Acids Res. 21:777–778.

Stemmer, W. P. (1994). Rapid evolution of a protein in vitro by DNA shuffling. Nature 370:389–391.

Strich, R., Surosky, R. T., Steber, C., Dubois, E., Messenguy, F., and Esposito, R. E. (1994). *UME6* is a key regulator of nitrogen repression and meiotic development. Genes Dev. 8:796–810.

Szostack, J. W. (1992). *In vitro* genetics. Trends Biochem. Sci. 17:89–93.

Tugendreich, S., Tomkiel, J., Earnshaw, W., and Hieter, P. (1995). CDC27Hs colocalizes with CDC16Hs to the centrosome and mitotic spindle and is essential for the metaphase to anaphase transition. Cell 81:261–268.

Vidal, M., Brachmann, R., Fattaey, A., Harlow, E., and Boeke, J. D. (1996a). Genetic selection for the dissocation of protein-protein and DNA-protein interactions. Proc. Natl. Acad. Sci. USA 93:10315–10320.

Vidal, M., Braun, P., Chen, E., Boeke, J. D., and Harlow, E. (1996b). Genetic characterization of a mammalian protein-protein interaction using a reverse two-hybrid system. Proc. Natl. Acad. Sci. USA 93:10321–10326.

Vidal, M., Buckley, A. M., Yohn, C., Hoepner, D. J., and Gaber, R. F. (1995). Identification of essential nucleotides in an upstream repressing sequence of *Saccharomyces cerevisiae* by selection for increased expression of *TRK2*. Proc. Natl. Acad. Sci. USA 92:2370–2374.

Vojtek, A. B., Hollenberg, S., and Cooper, J. (1993). Mammalian Ras interacts directly with the serine/threonine kinase Raf. Cell 74:205–214.

Wang, H.-T., Frackman, S., Kowalisyn, J., Esposito, R. E., and Elder, R. (1987). Developmental regulation of *SPO13*, a gene required for separation of homologous chromosomes at meiosis I. Mol. Cell. Biol. 7:1425–1435.

Yang, M., Wu, Z., and Fields, S. (1995). Protein-peptide interactions analyzed with the yeast two-hybrid system. Nucl. Acids Res. 23:1152–1156.

Young, K. H., and Ozenberger, B. A. (1995). Investigation of a ligand binding to members of the cytokine receptor family within a microbial system. Ann. N. Y. Acad. Sci. 766:279–286.

Zhou, Y., Zhang, X., and Ebright, R. H. (1991). Random mutagenesis of gene-sized DNA molecules by use of a PCR with TAQ DNA polymerase. Nucl. Acids Res. 19:6052.

Appendix

Although most of the methods described below have already been published, adapted versions are described with modifications for the particular strains, plasmids, and strategies presented here.

Preparation of Sc, Sc+3AT, and Sc+FOA Media

Prepare a powder of amino acids and purine by mixing equal weight (for example 10 g for each compound) of the following Sigma products (the catalog number is indicated in the parentheses): adenine sulfate (A3159), alanine (A7627), arginine (A3784), aspartic acid (A4409), asparagine (A8381), cysteine (C1276), glutamic acid (G5889), glutamine (G3126), glycine (G7126), isoleucine (I2752), lysine (L5626), methionine (M9625), phenylalanine (P5030), proline (P0380), serine (S4500), threonine (T8625), tyrosine (T1020), and valine (V0500).

Approximately twenty 15 cm Petri dishes can be poured with 2 L of agar-containing media. The liquid medium and the agar are autoclaved in two separate 2-liter flasks. In the first 2-liter flask, add 3 g of Yeast Nitrogen Base without amino acids and ammonium sulfate (Difco # 03335-15-9), 10 g of ammonium sulfate (ICN 80229), and 2 g of the amino acid powder mix. Add a clean stirring bar, resuspend in 1 L of distilled water and adjust the pH to 5.9 with NaOH. The stirring bar should be kept in the flask to stir the media after autoclaving. In the second 2-liter flask, add 40 g of agar (Difco # 03335-15-9) in 900 mls of H_2O. The agar will be solubilized during autoclaving.

After autoclaving, pour the content of the agar-containing flask into the media-containing flask. Incubate in a 50°C water bath for about 1 hour. Add 100 ml of 40% glucose and, depending on the the different auxotrophies needed to be tested with the drop-out medium, 16 ml of 20 mM uracil (Sigma # U-0750), 100 mM histidine-HCl (Sigma # H-9386), 100 mM leucine (Sigma # L-1512), 40 mM Tryptophan (Sigma # T-0271). Add 5-fluoroorotic acid (FOA) or 3-aminotriazole (3-AT) as powders. FOA can be purchased for a reasonable price from Diagnostic Chemicals Limited (catalog # 1555). 3-AT can be purchased from Sigma (catalog # A8056). Stir at room temperature (a few minutes for 3-AT and ~ 30' for FOA). Do not adjust the pH at this stage and pour approximately 100 mls of media in 15 cm plates.

Yeast Transformations

During the transformation experiment (Ito *et al*, 1983), yeast cells should always be maintained at room temperature (RT). The transformation efficiency should always be controlled using a well-characterized plasmid as a positive control and no DNA as a negative control. The high-efficiency protocol is as follows:

1. Patch cells expressing the wild-type hybrid DB-X on half of a 10 cm Sc-L plate and incubate overnight at 30°C.
2. Resuspend these cells in 500 ml liquid YEPD medium in a one liter flask such that the optical density (OD) is ~ 0.1.
3. Incubate with agitation at 30°C for ~4 hours until OD is 0.3.
4. Spin in a tabletop centrifuge in two conical Corning 250 ml tubes at 1800 rpm at 20°C for 5 min.
5. Pour off the supernatants and resuspend each pellet in 100 ml sterile H_20 stored at RT.
6. Spin again at 1800 rpm at 20°C for 5 min.
7. Pour off the supernatants and resuspend in 50 ml *fresh* TE/LiAc (a solution made of 10 mM Tris.HCl pH 7.5, 1 mM EDTA, 0.1 M LiAc).
8. Spin again at 1800 rpm at 20°C for 5 min.
9. Pour off the supernatants, resuspend each pellet in 1.5 ml of fresh TE/LiAc and pool both suspensions.
10. In 50 μl of cells, add 5 μl freshly boiled carrier (10 μg/μl), 300 ml PEG (40% PEG4000, 10 mM Tris.HCl pH 7.5, 1 mM EDTA, 0.1 M LiAc) and the relevant DNAs (0.5 μg for cDNA libraries in cloning experiments and approximately 200 ng of PCR product and 200 ng of restricted plasmid in two-step selection experiments).
11. Incubate at 30°C for 30 min.
12. Incubate at 42°C for 15 min.

13. Spin in microfuge for 20 sec. Discard the supernatant by careful aspiration. Spin again briefly and resuspend in 500 μl H_20.
14. After transformation, the cells are plated with 3 mm sterile glass beads (Fisher Scientific # 11-312A) onto media lacking both leucine and tryptophan to select for transformants.

The low-efficiency protocol is as follows (for six transformations):

1. Grow yeast cells on a YEPD plate for 1 to 4 days and, using a sterile toothpick, resuspend the amount of cells corresponding to a match head in 300 μl TE/LiAc in a sterile Eppendorf tube.
3. Spin, resuspend in 300 μl TE/LiAc.
4. In 50 μl of cells, add 1μg plasmid, 5 μl freshly boiled carrier (10 μg/ul) and 300 μl PEG.
5. Incubate at 30°C for 30 min.
6. Incubate at 42°C for 15 min.
7. Spin in microfuge for 20 sec. Discard the supernatant by careful aspiration. Spin again briefly and resuspend in 200 μl H_2O and plate on a 10 cm plate.

Replica-Plating

The velvets used should contain 100% cotton velveteen without rayon and should be cut in 220 × 220 mm pieces. The velvets can be recycled by washing and autoclaving such that a stock of about 100 velvets will allow convenient turnovers. After replica-plating, autoclave the velvets in a plastic bucket with a 25-minute cycle without drying. This step eliminates remaining viable yeast cells containing recombinant DNA prior to the washing in a laundry machine. Wash under a "color" cycle; it is very important not to add soap. Dry under a "color" cycle. Pack the velvets flat in aluminum foil, approximately 30/pack, and autoclave with a 45-minute autoclaving cycle and 99-minute drying cycle.

The 15 cm replica-plating blocks can be purchased from Replica-Tech, Inc.

There are two extremely important parameters to keep in mind while replica-plating: (1) the number of cells inoculated onto the selective plates after replica-plating. As the number of cells transferred by replica-plating increases, the phenotypic differences between positive and negative controls decreases. Therefore, it is crucial to transfer a minimal number of cells to the selective plates, and (2) the time of incubation of the master plate. This is probably due to the fact that yeast cells approaching stationary phase exhibit different expression levels of the hybrid proteins from cells growing in exponential phase.

Replica-plating should be performed the following way:

1. After a 60 h or 18 h incubation depending on the experiments (see chapter 7 text for details), transfer the colonies or patches from the master plate to a sterile velvet and subsequently to the selective plate(s) of interest (replica-plating).

2. The inoculum on these selective plates is then "diluted" by transferring to a new sterile velvet (replica-cleaning).

3. In some cases, the plates should be "replica-cleaned" again after a first 24 h incubation.

4. In order to control the efficiency of these different steps, controls 1 to 4 should be patched on every single master plate, and their phenotypes verified (Figure 7-5).

X-Gal Overlay Assay

1. Replica-plate the patches directly onto a 137 mm "MagnaGraph" Nylon membrane (MSI # NJ4HY13750) placed on a YEPD plate and incubate for 18 hours at 30°C. Include a patch for control 1 to 5 on each master-plate.
2. For one membrane, prepare a solution with 10 mg X-Gal (Fisher Biotech BP1615-1) dissolved in 100 μl DMF N,N-dimethyl formamide (Sigma D-42554), 60 μl of β-mercaptoethanol and 10 ml of Z buffer [6.1g $Na_2HPO_4 \cdot 7H_2O$ (or 8.52 g anhydrous), 5.5g $NaH_2PO_4 \cdot H_2O$ (or 4.8 g anhydrous), 0.75g KCL, 0.246 g $MgSO_4 \cdot 7H_2O$ (or 0.12 g anhydrous), dissolved in 1 l and adjusted to pH 7.0].
3. Add the above mix onto a stack of two round 125 mm Whatman 541 filter papers (Whatman # 1541 125) placed on the lid of a petri plate.
4. Dip the nitrocellulose filters in liquid nitrogen for approximately 20 sec. and lay onto the Whatman filters.
5. Cover and incubate for 24 hours, at 37°C.

No conclusion can be drawn from this experiment unless the controls show the expected phenotypes (Figure 7-5).

Western Blotting

1. Patch the yeast cells on 1/6 of a 15 cm Sc-L-T plate and incubate overnight at 30°C.
2. Resuspend in 100 mls in Sc-L-T liquid medium for an additional 12 hours at 30°C.
3. Harvest, wash, and resuspend the cells in 500 μl "breaking" buffer (100 mM Tris pH 8.0, 20% glycerol, 1 mM EDTA, 0.1% Triton X-100, 5 mM $MgCl_2$, 10 mM β-mercaptoethanol, 1 mM phenylmethylsulfonylfluoride (PMSF), 1 μg/ml leupeptin, 1 μg/ml pepstatin). The cell density should be approximately $2-3 \times 10^9$ cells/ml.
4. Add approximately 400 μl of glass beads (Sigma # G-8772).
5. Vortex at full speed for 10 min at 4°C.
6. Pellet the debris by centrifugation at 12,000 RPM for 15 min at 4°C.
7. Estimate the protein concentrations in the supernatants by using the Bradford assay.
8. Subject 50-100 mg of proteins to SDS-Polyacrylamide Gel Electrophoresis (SDS-PAGE) and subsequent Western-blot analysis using the appropriate antibodies.

In vitro Binding Reactions

The protocol of the in vitro binding reactions can be divided in three sections: (1) the PCR amplification of the DB-X or AD-Y fusion sequences, (2) the in vitro translation of the PCR product, and (3) the binding reaction itself.

1. The PCR amplification is performed using the following reagents and conditions:

   a. 200 ng of the pPC97 or pPC86 derivative mutant plasmid as the DNA template.

   b. a 5' primer (1 μM) with a "hybrid" sequence which consists of 25 nucleotides corresponding to the T7 promoter (underlined below) followed by 20 nucleotides that anneal with the AD or DB sequence (italics below) right at the junction with the insert in the polylinker. A T7-DB primer used in these experiments should be: 5'CCAAGCTTCTAATACGACTCACTATAGGGAAGATG*GGTCAAAGACAGTTGACTGTATCG*3'. A T7-AD primer used in these experiments should be: 5'CCAAGCTTCTAATACGACTCACTATAGGGAAGATG*AACCCAAAAAAAGAGGGTGGGTCG*3'.

   c. 3' primer (1 μM) corresponding to a sequence located in the transcriptional terminating (TERM) sequence located approximately 100bp downstream of the both the pPC97 and pPC86 derived plasmids: 5'GGAGACTTGACCAAACCTCTGGCG3'.

   d. A cocktail containing 50 μM of each dNTP, 50 mM KCl, 10 mM Tris pH 9.0, 0.1% Triton X-100, 1.5 mM $MgCl_2$, 1 μg bovine serum albumin (BSA), and 5 units of *Taq* DNA polymerase in 100 μl.

   e. Runs of 30 cycles of 1 min at 94°C, 1 min at 50°C, and 3 min at 72°C.

   The yield and the length of the PCR products should be estimated by electrophoresis on an agarose gel. The PCR products should be precipitated and resuspended in 30ul.
2. The in vitro transcription and translation (IVT) of the PCR products can be performed using the TNT T7-coupled reticulocyte lysate system from Promega (Catalog number L4950) using the manufacturer's protocol.
3. The binding reactions are performed using conventional methods.

8

Dissecting the Biological Relevance of Multiple Interactions

Michael A. White

The advent of powerful biochemical and genetic techniques that allow the screening of libraries to identify proteins that interact with a protein of interest has resulted in an explosion of information concerning the protein-protein interactions that mediate biological processes. However, a major problem confronting any biologist studying these interactions is to determine their physiological relevance. Any screen in which proteins are identified solely by the criterion of physical association with another protein is by nature susceptible to artifacts resulting from nonphysiologically relevant association. This problem is compounded when multiple partners are identified for a particular protein. Careful analysis must be undertaken to determine which interactions occur in vivo, and, of these, which interactions are required or responsible for mediating a particular biological process. Defining the biological relevance of identified protein-protein interactions, rather than their identification, has become the rate-limiting step in deciphering molecular mechanisms.

The complications just described have been encountered in studies of the cellular function of *Ras* protooncogenes. Ras-mediated signaling events have been implicated in the control of a variety of cellular processes, including gene expression, cell cycle progression, and cytoskeletal rearrangements (reviewed in Egan et al. 1993; Moodie et al. 1994). Mutationally activated Ras is associated with numerous human tumors (Bos 1989), and can itself induce growth and morphological transformation of many cell lines (Barbacid 1987). Thus, interest has centered on the determination of Ras function and the identification of the molecules that mediate its effects (Ras effectors). A growing roster of molecules has been identified that physically interact with Ras, both in the two-hybrid system and in vitro, in a manner consistent with their classification as potential Ras effectors (reviewed in Feig et al. 1994). However, the biological role of the majority of these proteins is unclear. In addition, with few exceptions, it is not known which if any Ras-induced phenotypes these potential effectors mediate in cells.

A means of dissecting the involvement of Ras-binding proteins in Ras function, which should be generally applicable to any protein with multiple partners, is the generation of Ras mutants that are defective in different interactions. Examination of the phenotypes induced upon expression of these mutants in vivo should lead to information regarding the function of specific interactions. Furthermore, a particular partner can be designated as required for mediating Ras function if a complementary mutation can be generated in the partner that restores interaction with a Ras mutant and rescues Ras function when coexpressed.

In the past, the identification both of mutants that affect protein binding and of complementary mutants in the partner that restore binding has not been easy. Often, extensive analysis of the domains and residues involved in a particular protein-protein interaction is required before mutants can be produced that appropriately affect binding. Such was the case in the generation of the specific complementary mutants of Ras and Cdc25, a Ras-activating protein (Park et al. 1994), and in the identification of mutants demonstrating that the oncogenic activity of the c-Myc protein requires dimerization with the Max protein (Amati et al. 1993). Other approaches involve genetic screens for allele-specific extragenic suppressors, which require beginning with an appropriate mutant in a genetically tractable organism. The advent of the two-hybrid system, combined with yeast genetics, allows the direct selection from pools of randomly generated mutants of those mutations that affect specific protein interactions. Because large numbers of mutants can be quickly and efficiently screened, the prior identification of domains of interaction and of the residues that contribute binding specificity is not necessary. I describe here the procedures developed to use the two-hybrid system to isolate mutations in Ras that separate its ability to interact with different downstream effectors, and to select for complementary mutants in the effectors that restore interaction with binding-defective Ras mutants (White et al. 1995). These methods are generally applicable to any proteins whose interaction can be detected with the two-hybrid system.

Mutant Library Construction

To identify Ras mutants defective in protein interactions, we first produced a pool of randomly generated mutants expressed as fusions to the Gal4p DNA-binding domain (GBD). *Ha-ras* was randomly mutagenized along its entire length by PCR amplification with the error-prone *Taq* DNA polymerase as described by Zhou et al. (1991). Essentially standard reaction conditions were used with a very dilute template concentration (2 femtomoles in 200 μl) and 30 cycles of amplification. Given a 600 bp template, and a misincorporation frequency of approximately 10^{-5} errors per nucleotide synthesized, this protocol results in a single base substitution in approximately 35% of the product. Higher or lower mutation frequencies can be obtained by altering the amount of starting template and the number of cycles of amplification (Zhou et al. 1991). Mutation frequencies that are too high may be disadvantageous due to the production of large numbers of mutants with multiple substitutions. The use of standard PCR conditions produces both transition and transversion substitutions. In addition, Zhou et al. (1991) reported finding no evidence for mutational hotspots, although the presence of such hotspots may be template specific. The PCR-generated pool of randomly mutated *Ras* genes was ligated into the two-hybrid expression vector pHP5 (White et al. 1995) such that Ras mutants would be expressed as GBD-fusions. Mutant libraries containing more than 1.5×10^4 clones were quickly and easily produced by this method.

Isolation of Ras Mutants that Discriminate between Partners

We first generated Ras mutants that discriminate between the bona fide Ras effectors Raf1 and byr2. Considerable genetic and biochemical evidence indicates that both of these proteins interact directly with Ras, and that Raf1 functions immediately downstream in mammalian cells (reviewed in Moodie et al. 1994) and byr2 in *Schizosaccharomyces pombe* (Wang et al. 1991). The association of Ras with Raf1 or byr2 can be detected in the two-hybrid system (Van Aelst et al. 1993; Vojtek et al. 1993). Although byr2 is a yeast protein and has no recognizable homology to Raf1 (a mammalian Ras effector), human Ras can productively interact with byr2 in vivo (Nadin-Davis et al. 1986). This conservation of function suggests that human Ras interacts with another effector molecule in mammalian cells that has similar binding determinants on Ras as byr2.

The mutant Ras library was separately screened with a Raf1-Gal4p activation domain fusion (GAD) and a byr2-GAD fusion, in the reporter strain YPB2, to identify Ras mutants that no longer interact with these targets. One hundred fifty mutants were isolated from each screen. The Ras mutants defective in Raf1 interaction were in turn screened for those that retained binding to byr2. Similarly, the Ras mutants defective in byr2 interaction were screened for those that retained interaction with Raf1. Positives from the second round of screening that fulfilled these criteria were isolated and sequenced. In this way, we found that a substitution of glycine for glutamic acid at position 37 (RasG37) disrupted interaction with Raf1 but not byr2, and a substitution of serine for threonine at position 35 (RasS35) interfered with binding to byr2 but not Raf1 (White et al., 1995).

Expression of activated Ras proteins containing these mutations (RasV12G37 and RasV12S35) revealed that both are defective in the induction of cellular transformation and exhibited different patterns of activation of downstream components that were consistent with the observed two-hybrid interactions (White et al. 1995). The mutants had complementary activity, however, as demonstrated by a synergistic induction of cellular transformation upon coexpression, suggesting that each mutant was defective in different interactions required for Ras to transform cells. The introduction of the activating mutation (V12) did not affect the profile of two-hybrid interactions.

RasV12G37 and RasV12S35 discriminate not only between Raf1 and byr2, but also among other Ras-binding proteins that have been identified (Table 8-1). These putative partners fall into two classes: those that bind RasV12G37 and those that bind RasV12S35. This classification may be due to similarities in the binding determinants on Ras of the proteins in each group. It is possible, however, to isolate Ras mutants with different profiles of interactions. RasV12C40 was isolated in a two-hybrid screen for Ras mutants that do not interact with Raf1 but do interact with AF6, a Ras-binding protein involved in myeloid leukemia (Prasad et al. 1993; Van Aelst et al. 1994; Kuriyama et al. 1995). RasV12C40 interacts exclusively with AF6 among all Ras partners tested in the two-hybrid system. This mutant retains the ability to induce membrane ruffling in REF52 cells (Joneson et al. 1996), a Ras-induced response lacking in RasV12G37 and RasV12S35. Because RasV12C40 does not bind or activate the Ras effector Raf1 in vivo (Joneson et al. 1996; Koshravi-Far and Der, personal communication), it has been instrumental in uncovering the involvement of a novel effector molecule responsible for mediating Ras-induced membrane ruffling. This effector may be an as yet unidentified protein that interacts with RasV12C40 but not RasV12S35 or RasV12G37. Alternatively, AF6 may mediate Ras-induced membrane ruffling. Although RasV12G37 does not induce this response but does interact with AF6, this interaction may be enzymatically unproductive. The two-hybrid system cannot discriminate between productive and nonproductive protein-protein interactions.

Isolation of Complementary Mutations

RasV12G37 is most likely defective in the induction of cellular transformation due to its defective interaction with Raf1, an important component

Table 8-1. Two-hybrid interactions between Ras mutants and Ras partners.

	Raf1	byr2	Ral GDS	Rin1	RIP36	AF6
rasV12	+	+	+	+	+	+
rasV12,G37	–	+	+	+	+	+
rasV12,S35	+	–	–	–	–	–
rasV12,C40	–	–	–	–	–	+
rasA35	–	–	–	–	–	–
references	(19,21)	(19)	(9,11,17)	(8)	(21)	(12,20)

in Ras-induced transformation. To determine if this was the case, we used the two-hybrid system to generate mutations in Raf1 that suppress the Raf1-binding defect of RasV12G37. The Ras-binding domain in Raf1 had been localized to the amino-terminal half of the protein (Vojtek et al. 1993). A different protein, MEK (mitogen activated/extracellular signal-regulated kinase kinase), interacts with the C-terminal half of Raf1 (Van Aelst et al. 1993). A library of mutations in the N-terminal half of Raf1 (amino acids 1-281) was produced by PCR by the same protocol used to generate the mutant Ras library (Zhou et al. 1991). This library was fused to the wild-type half of Raf1 in the vector pGADGH to produce a library of full-length Raf1 mutants expressed as fusions to the Gal4p activation domain.

The mutant Raf1 library was screened using RasG37 for proteins that had acquired the ability to interact with the Ras mutant. Positives were in turn tested for interaction with MEK to ensure that full-length Raf1 protein was expressed. This secondary screen was important since some Raf1 mutants, isolated as binding to RasG37, were truncations that destroy the in vivo activity of the Raf1 enzyme. Three different Raf1 mutants were isolated that could interact both with RasG37 and MEK. All three resulted from single amino acid substitutions in the same region of the Raf1 protein: a change from arginine to glycine at position 256, or from serine to either leucine or proline at position 257. These mutations did not affect the interaction of Raf1 with wild-type Ras. A surprising result was that none of these mutations mapped to the previously defined minimal Ras-interaction domain (Vojtek et al. 1993), suggesting that regions of proteins outside of domains minimally required for interaction can be important in the context of the interaction of full-length proteins.

Coexpression of the Raf1 mutants with activated Ras carrying the E37G mutation suppressed the transformation defect of the Ras mutant, demonstrating that the in vivo defects of RasV12G37 were at least in part due to its lack of interaction with Raf1.

Variant-Specific Complementary Mutants

It is important to note that the complementary Raf1 mutants we isolated to restore binding to RasG37 cannot be used to provide information on the actual residues mediating Ras-Raf1 contact. These mutants, which bind wild-type Ras, may function by altering the conformation of the full-length Raf1 protein such that RasG37 can bind without contributing binding energy to the interaction. The production of truly variant-specific complementary mutants, Raf1 proteins that interact only with specific mutant Ras proteins and vice versa, should be more useful in identifying the amino acid residues directly involved in Ras-Raf1 binding. Theoretically, the isolation of such mutants in Ras and Raf1, or any two interacting proteins, could be accomplished by repetition of the processes of mutant library production and two-hybrid screening as outlined in Figure 8-1.

Mutagenesis Revisited

Our use of PCR-based mutagenesis to produce mutant libraries was selected due to its speed, simplicity, and efficiency at producing a large pool of templates with single base changes. The protocol had been developed to avoid selection for transitions versus transversions, and results in a wide variety of alterations at most positions (Zhou et al. 1991). There are limitations to this method, however, that may make it unsuitable for some applications. For example, because the majority of templates contain less than one alteration, there are limitations to the amino acid substitutions that can be made at any specific position. It would be difficult, for instance, to change a Phe to a Pro, as this would require altering two nucleotides in the same codon. Strategies employing oligo-directed mutagenesis, while more tedious, can be more successful at generating all possible amino acid changes at every position of interest. Stone et al. (1991) employed a cassette mutagenesis strategy in which oligonucleotides containing various alterations were used to replace wild-type sequence in the Ras effector domain. This type of procedure is most efficient when saturation mutagenesis of a small region of a protein is desired.

Reporter Strains Revisited

The generation of new two-hybrid reporter strains will simplify the search for mutations in proteins that separate interactions with multiple partners, as well as the identification of mutations that are allele-specific. It is possible to construct a yeast strain with two different reporter genes controlled by two different promoter elements. For example, an auxotrophic yeast strain could be generated with the *URA3* gene under the control of a promoter containing the Gal4p binding sites, and the *HIS3* gene under the control of a promoter containing the LexA binding sites. This strain would

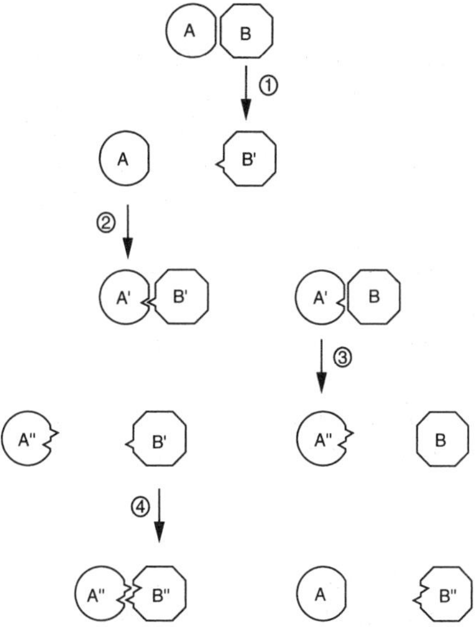

Fig. 8-1. Production of variant-specific complementary mutants in interacting proteins. The process begins with a pair of proteins A and B whose physical interaction can be detected using the yeast two-hybrid system. Step 1: A mutant B (B') is identified that disrupts interaction with A. This mutant may be isolated by screening a library expressing randomly mutated B, or by site-directed mutagenesis if information about the domain of interaction is available. Preferably, the selected B mutant will retain a positive interaction with a third protein if available. This additional criterion will increase the probability that the alteration is not too severe (for example, a truncation or frameshift) to be complemented by a mutation in A. Step 2: A library of randomly mutated A is screened for mutants (A') that can interact with B'. These complementary mutants are then tested for interaction with the wild-type B. Step 3: If A' interacts with B, then a second mutation in A' (resulting in A") is identified that disrupts this interaction. This is accomplished by constructing a mutant library using A' as the template, and screening for proteins that do not interact with B'. Again, it is important to select an A" that does not severely disrupt the structure of the protein. This selection is most easily accomplished if other proteins are available that interact with A but have different binding determinants (for example, they bind to a different domain of A) than B, which can be tested for interaction with A". A" is then tested for the ability to interact with B'. Step 4: If the second mutation in A" disrupts interaction with B', then a compensatory mutation in B' is selected to restore interaction with A". Again, this is accomplished by constructing a mutant library using B' as the template, and selecting a B" that interacts with A". A B" is selected that does not interact with A, culminating in the isolation of the variant-specific complementary mutants A" and B".

allow mutant protein libraries to be tested against two partners simultaneously. The advantage of using *URA3* as a reporter is that it allows both positive and negative selection (see chapter 7) through the use of defined media, either lacking uracil or containing 5-fluoroorotic acid, which is toxic to yeast expressing the *URA3* gene product (Boeke et al. 1984). For example, a mutant library of protein X could be expressed as Gal4p activation domain fusions, and positive selection could be applied to identify mutations that fail to interact with protein Y fused to the Gal4p DNA-binding domain but that retain interaction with protein Z fused to the LexA DNA-binding domain. A similar strategy could be used to select for complementary mutations in a partner that will specifically restore interaction with a mutant protein without being able to interact with the wild-type form.

References

Amati, B., Brooks, M. W., Levy, N., Littlewood, T. D., Evan, G. I., and Land, H. (1993). Oncogenic activity of the c-Myc protein requires dimerization with Max. Cell 72: 233–245.

Barbacid, M. (1987). *ras* genes. Annu. Rev. Biochem. 56: 779–827.

Boeke, J. D., Lacroute, F., and Fink, G. R. (1984). A positive selection for mutants lacking orotidine-5'-phosphate decarboxylase activity in yeast. Mol. Gen. Genet. 197: 345–346.

Bos, J. L. (1989). Ras oncogenes in human cancer: a review. Cancer Res. 49: 4682–4689.

Diaz-Meco, M., Lozano, J., Municio, M. M., Berra, E., Frutos, S., Sanz, L., and Moscat, J. (1994). Evidence for the in vitro and in vivo interaction of Ras with Protein Kinase C ζ. J. Bio. Chem. 50: 31706–31710.

Egan, S. E., and Weinberg, R. A. (1993). The pathway to signal achievement. Nature 365: 781–783.

Feig, L. and Schaffhausen, B. (1994). Signal Transduction. The hunt for Ras targets. Nature 370: 508–509.

Han, L., and Colicelli, J. (1995). A human protein selected for interference with Ras function interacts directly with Ras and competes with Raf1. Mol. Cell. Biol. 15: 1318–1323.

Hofer, F., Fields, S., Schneider, C., and Martin, G. S. (1994). Activated Ras interacts with the Ral guanine nucleotide dissociation stimulator. Proc. Natl. Acad. Sci. USA 91: 11089–11093.

Joneson, T., White, M. A., Wigler, M., and Ber-Sagi, D. (1996). Stimulation of membrane ruffling and MAP kinase activation by distinct effectors of Ras. Science 271: 810–812.

Koide, H., Satoh, T., Nakafuku, M., and Kaziro, Y. (1993). GTP-dependent association of raf-1 with Ha-Ras: Identification of Raf as a target downstream of Ras in mammalian cells. Proc. Natl. Acad. Sci. USA 90: 8683–8686.

Kuriyama, M., Harada, N., Kuroda, S., Yamamoto, T., Nakafuku, M., Iwamatsu, A., Yamamoto, D., Prasad, R., Croce, C., Canaani, E., and Kaibuchi, K. (1995). Identification of AF-6 and canoe as putative targets for Ras. J. Biol. Chem. 271: 607–610.

Moodie, S. A., and Wolfman, A. (1994). The 3Rs of life: Ras, Raf, and growth regulation. Trends Genet. 10: 14–18.

Nadin-Davis, S. A., Nasim, A., and Beach, D. (1986). Involvement of *ras* in sexual differentiation but not in growth control in fission yeast. EMBO J. *5*: 2963–2971.

Park, W., Mosteller, R. D., and Broek, D. (1994). Amino acid residues in the CDC25 guanine nucleotide exchange factor critical for interaction with Ras. Mol. Cell. Biol. 14: 8117–8122.

Prasad, R., Gu, Y., Alder, H., Nakamura, T., Canaani, O., Saito, H., Hubner, K., Gale, R. P., Nowell, K., Kuriyama, K., Miyazaki, Y., Croce, C. M., and Canaani, E. (1993). Cloning of the *ALL-1* fusion partner, the *AF-6* gene, involved in acute myeloid leukemias with the t(6;11) chromosome translocation. Cancer Res. 53: 5624–5628.

Spaargaren, M., and Bischoff, J. R. (1991). Identification of the guanine nucleotide dissociation stimulator for Ral as a putative effector molecule of R-ras, H-ras, K-ras, and Rap. Proc. Natl. Acad. Sci. USA 91: 12609–12613.

Stone, J. C., and Blanchard, R. A. (1991). Genetic definition of ras effector elements. Mol. Cell. Biol. 11: 6158–6165.

Van Aelst, L., Barr, M., Marcus, S., Polverino, A., and Wigler, M. (1993). Complex formation between RAS and RAF and other protein kinases. Proc. Natl. Acad. Sci. USA 90: 6213–6217.

Van Aelst, L., White, M., and Wigler, M. (1994). Ras Partners. Cold Spring Harbor Symp. Quant. Biol. 59: 181–186.

Vojtek, A. B., Hollenberg, S. M., and Cooper, J. A. (1993). Mammalian Ras interacts directly with the serine/threonine kinase Raf. Cell 74: 205–214.

Wang, Y., Xu, H.-P., Riggs, M., Rodgers, L., and Wigler, M. (1991). *byr2*, a Schizosaccharomyces pombe gene encoding a protein kinase capable of partial suppression of the *ras1* mutant phenotype. Mol. Cell. Biol. 11: 3554–3563.

White, M. A., Nicolette, C., Minden, A., Polverino, A., van Aelst, L., Karin, M., and Wigler, M. H. (1995). Multiple Ras functions can contribute to mammalian cell transformation. Cell 80: 533–541.

Zhou, Y., Zhang, X., and Ebright, R. H. (1991). Random mutagenesis of gene-sized DNA molecules by use of PCR with TAQ DNA polymerase. Nucl. Acids Res. 19: 6052.

9

Investigation of Ligand/Receptor Interactions and the Formation of Tertiary Complexes

Bradley A. Ozenberger
Kathleen H. Young

Protein-protein interactions occurring at the mammalian cell surface, such as the binding of circulating peptide ligands to transmembrane receptors, are of paramount importance to cellular functions and biological responses. The two-hybrid system has been used extensively in studies of intracellular protein interactions, such as those occurring among signal pathway components and transcription complex proteins. However, protein-protein interactions that normally occur at the mammalian cell surface were generally considered precluded from two-hybrid analysis since the specific interaction of the fusion proteins must occur in the nucleus of a yeast cell. Extracellular proteins generally have complex tertiary structures stabilized by disulfide bonds which would not be predicted to form readily in the reductive environment of the cell interior. However, we have found that two-hybrid methods are applicable to many diverse extracellular protein-protein associations. To illustrate this application, we describe the interaction of specific peptide ligands with their cognate receptor protein(s).

The mammalian peptides growth hormone (GH), prolactin (PRL), and vascular endothelial growth factor (VEGF) and their cognate extracellular ligand-binding proteins were expressed in a two-hybrid system. Specific interactions between GH and the GH receptor (GHR), and PRL and the PRL receptor (PRLR), were coupled to an easily scorable change in yeast cell phenotype (growth on selective medium). Permutations devised to express a third protein enabled examination of reversible and specific ligand/receptor interactions, thereby establishing pharmacology in yeast. The expression of a third heterologous protein facilitated the investigation of ligand-dependent receptor dimerization for GH and VEGF, demonstrating the utility of these methods for characterizing higher order protein complexes. This easily manipulable system creates a unique biological forum for the rapid investigation of peptide hormone/receptor interactions.

Reagents

Gal4p activation domain fusions were assembled in the vector pACT2, and Gal4p DNA-binding domain fusions were assembled in the vector pAS2. These two-hybrid expression plasmids are described in chapter 2. The plasmid used to express nonfusion proteins is denoted pCUP (Ozenberger and Young 1995). Briefly, this vector was constructed by inserting the *CUP1* promoter region (Butt et al. 1984) plus the 3' end of the yeast *PGK1* gene (Kang et al. 1990) into the cloning region of the centromeric vector pRS316 (Hill et al. 1986) to provide transcriptional initiation and termination signals, respectively. In strains described in this report, constitutive expression from the *CUP1* promoter in the absence of inducer (Cu^{2+}) was sufficient to attain desired levels of protein. Specific DNA sequences encoding mammalian peptide ligands and receptors are described elsewhere (Ozenberger and Young 1995). Receptor fusion proteins were engineered to truncate the protein just prior to the transmembrane region, leaving only the amino-terminal extracellular segment. Ligands expressed as nonfusion proteins were not modified.

All yeast strains were derived from strain Y190 (Harper et al. 1993). Yeast were grown in YEPD medium or, if maintaining plasmids, in synthetic drop-out media as described in Rose et al. (1990). To incorporate a third plasmid, a uracil auxotroph of strain Y190 was identified by passage on 5-fluoroorotic acid (Rose et al. 1990). This Ura- derivative was designated CY770. This modification likely resulted in the deletion of a UAS_{GAL}-*lacZ* reporter gene. Strains were generated by transforming expression plasmids into either Y190 or CY770 by the lithium acetate method (Rose et al. 1990). To counteract background expression of the UAS_{GAL}-*HIS3* reporter gene, the imidazoleglycerol phosphate dehydratase inhibitor 3-amino-1,2,4-triazole was added to 60 mM in bioassay medium.

Applications

Ligand/Receptor Interactions in Yeast

Complementary DNA sequences encoding porcine GH or the extracellular region of the rat GHR were cloned into Gal4p-fusion protein expression plasmids. The yeast strain Y190 (Harper et al. 1993), containing a UAS_{GAL}-*HIS3* reporter gene, was transformed with both fusion plasmids or

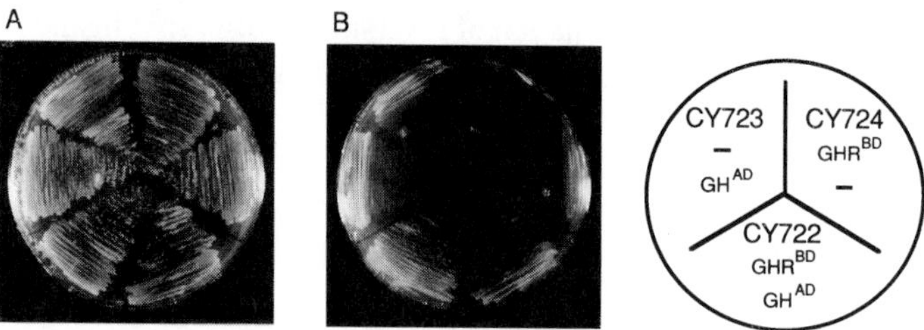

Fig. 9-1. Growth of strains expressing GHR and GH fusion proteins. Strains expressing the indicated proteins are described in the text. The position and the expressed heterologous proteins of each strain are indicated in the diagram. Superscripts indicate that a protein was expressed as a fusion to the transcriptional activation domain (AD) or DNA-binding domain (BD) of Gal4p. A dash indicates that the strain contained an unmodified vector (pAS2 or pACT2). Two independent isolates of each strain were streaked on synthetic medium deficient in leucine, tryptophan and uracil (nonselective; Plate A) or on the same medium lacking histidine (selective; Plate B). Both plates contain 60 mM 3-amino-1,2,4-triazole. Plates were incubated for 4 days at 30°C.

with a single fusion construct plus the opposing vector containing no heterologous DNA. All strains grew vigorously on medium that did not require two-hybrid interaction (nonselective medium) (Figure 9-1A). These strains were then tested for histidine prototrophy (selective medium). Only the strain (CY722) containing both ligand and receptor fusion proteins grew, while strains containing either the ligand or receptor fusion alone were not able to grow (Figure 9-1B). These results suggest that GH and GHR can mediate the Gal4p-dependent activation of the reporter gene in an interaction suggestive of ligand-receptor binding.

To further investigate these findings, we developed a similar system using the peptide hormone PRL and its receptor. PRL is structurally related to GH. The PRLR, like the GHR, is a member of the cytokine receptor family. Porcine PRL and the extracellular domain of the porcine PRLR were expressed as Gal4p-fusion proteins. As in the GH/GHR experiment, the strain expressing both the PRL- and PRLR-fusion proteins was able to grow on selective medium while strains containing either the ligand or receptor fusion alone could not (Table 9-1). These results mirror those observed in the GH/GHR expression strains and suggest that this methodology has general utility for examination of ligand binding to members of this receptor superfamily.

Competitive Interactions Determine Specificity

To substantiate the apparent binding of GH to its receptor in the foreign environment of a yeast nucleus, we modified the system to express "free"

Table 9-1 Summary of growth and competition assays.

Strain	pACT2	pAS2	pCUP	Growth
Y190	—	—	—	0
CY722	GH	GHR	–	+++
CY723	vector	GHR	—	0
CY724	GH	vector	–	0
CY726	PRL	PRLR	–	++
CY770	—	—	–	0
CY781	GH	GHR	GH	+
CY784	GH	GHR	vector	+++
CY785	GH	GHR	PRL	+++
CY786	PRL	PRLR	PRL	+
CY787	PRL	PRLR	vector	++
CY788	PRL	PRLR	GH	++

All yeast strains were derived from strain Y190 (Harper et al. 1993). Strains with number designations equal to or greater than 770 do not contain the *URA3::GAL-lacZ* gene. pACT2 carries the *LEU2* selectable marker and expresses the Gal4p transcriptional activation domain as a protein fusion with the indicated peptide. pAS2 carries the *TRP1* selectable marker and expresses the Gal4p DNA-binding domain as a protein fusion with the indicated receptor. pCUP carries the *URA3* selectable marker and is used to express nonfusion peptide. A dash indicates that a strain does not contain the denoted plasmid. Strains were incubated on selective medium for 4 days at 30° and scored for growth. A zero indicates that no growth was observed. The most robust growth was observed with the GH/GHR-fusion protein interaction (strains CY722 and CY784), indicated by three pluses. Growth of strains (CY726 and CY787) expressing PRL/PRLR fusion proteins was less robust as indicated by two pluses. Specific competition of GH/GHR or PRL/PRLR interactions with nonfusion protein (CY781 or CY786, respectively) significantly reduced, but did not eliminate, the growth response as indicated by one plus.

nonfusion ligand from a third plasmid (pCUP) carrying a *URA3* selectable marker. The GH peptide should compete with the GH-fusion protein, reversing the interaction of the fusion proteins (Figure 9-2A). Strain CY770, a Ura$^-$ derivative of Y190, was transformed with the GH- and GHR-fusion expression plasmids plus a plasmid expressing nonfusion GH, to produce strain CY781. Concurrent expression of GH with the GH- and GHR-fusion proteins in CY781 substantially attenuated cell proliferation on selective medium in comparison to strain CY722 (Figure 9-2B). This experiment typifies an in vivo competition assay and illustrates the reversibility of the ligand/receptor interaction. We refer to two-hybrid systems employing three heterologous proteins as PRIMES (Peptide/Receptor Interactions in a Microbial Expression System).

Additional strains were developed to assess ligand/receptor specificity. Strains expressing GH- and GHR-fusion proteins or PRL- and PRLR-fusion proteins were transformed with pCUP, pCUP-GH, or pCUP-PRL. As demonstrated previously, a strain expressing the GH- and GHR-fusions in the absence of competitor grew on selective medium, and this growth was abrogated with coexpression of free GH (Figure 9-2B, Table 9-1). The strain containing the GH- and GHR-fusions and also expressing PRL exhibited a strong growth phenotype (CY785; Table 9-1). Strain CY787, con-

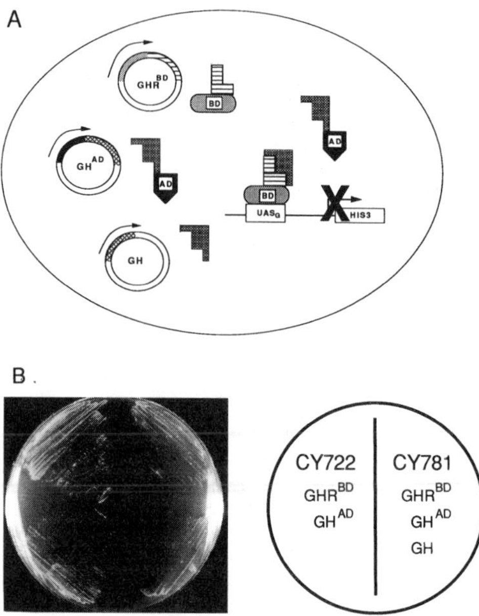

Fig. 9-2. Growth inhibition by coexpression of nonfusion GH. (A) Monomer model—competitive ligand/receptor interaction. The extracellular domain of the rat GHR is expressed as a fusion to the DNA-binding domain (BD) of the Gal4p transcriptional activator protein; mature porcine GH is expressed as a fusion to the Gal4p activation domain (AD). Additionally, GH is expressed from pCUP as a nonfusion protein. In this strain, GH is free to associate with the GHR-fusion protein thereby blocking the GH-fusion from interacting with the GHR-fusion. This competitive interaction abrogates the functional reconstitution of DNA binding and activation domains of Gal4p, limiting induction of the UAS_{GAL}-*HIS3* reporter gene. (B) Strains expressing the indicated proteins are described in the text. The position and expressed heterologous proteins of each strain are indicated in the diagram. Ligand without designating superscript denotes expression as a nonfusion protein from pCUP. Two independent isolates of each strain were assayed on selective medium as described in Fig. 9-1.

taining PRL and PRLR fusion proteins, grew on selective medium, and this growth was abrogated by expression of free PRL but not GH (CY786 and CY788; Table 9-1). These data suggest that ligand binding to its cognate receptor in this system retains the pharmacological specificity of the native interaction.

Ligand-Induced Receptor Dimerization

The demonstration of a specific ligand/receptor interaction in yeast by utilizing three expression plasmids presented an opportunity to examine more complex associations of proteins. Differential receptor protein interactions dependent on ligand binding could be observed by concomitantly expressing a receptor extracellular domain with both the Gal4p DNA-binding and activation domain as fusion proteins, and the ligand as a nonfusion protein (Figure 9-3A). Experiments were conducted using GH/GHR to investigate ligand-dependent receptor dimer formation. These required the construction of an additional plasmid to express the extracellular domain of GHR as a fusion with the Gal4p activation domain. Yeast strains (CY899, CY901) that expressed any single receptor fusion plasmid with GH failed to grow on selective medium (Figure 9-3B). The strain that expressed both receptor fusions in the absence of GH (CY887) grew poorly on selective medium as did a strain that expressed both GHR fusions and PRL (data not shown). Only the strain (CY886) that expressed both receptor fusions plus GH exhibited substantial growth (Figure 9-3B).

The growth of strains expressing both GHR fusion proteins was clearly stimulated by coexpression of GH but there was some growth of the strain (CY887) that did not express ligand (Figure 9-3B). These effects were confirmed and quantitated by determining the maximal growth rates of strains in selective liquid medium (minimal synthetic medium supplemented with 60 mM 3-amino-1,2,4-triazole) (see Ozenberger and Young 1995). To summarize these data, expression of GHR-fusion proteins resulted in a growth rate appreciably greater than controls, even in the absence of GH. Coexpression of nonfusion GH further enhanced cell proliferation. These results corroborate the observations of GH-dependent GHR dimerization as shown in Figure 9-3b. The apparent avidity between GHR extracellular domains in the absence of ligand was not anticipated. However, the crystal structure of the ligand/receptor complex (DeVos et al. 1992) does reveal contacts between membrane-proximal sequences that would be present in the Gal4p fusion proteins. In sharp contrast to the results observed in the GHR dimerization assays, PRL-mediated receptor dimerization could not be observed. Expression of PRLR fusion proteins produced no induction of growth, and concomitant expression of PRL with the receptor fusion proteins produced no enhancement of growth rate (see Ozenberger and Young 1995). These data suggest that PRL, although structurally related to GH, may interact

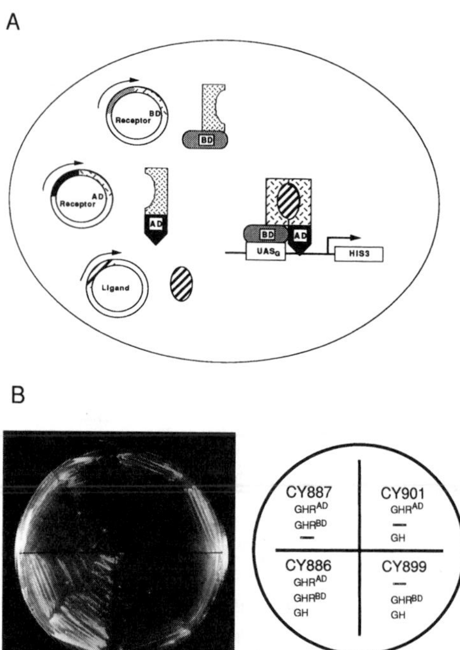

Fig. 9-3. Ligand-dependent receptor dimerization. (A) Dimer model. The extracellular domain of a receptor is fused to both the DNA-binding (BD) and activation (AD) domains of Gal4p. The mature protein for the cognate ligand is expressed from pCUP as a nonfusion protein. Maximal induction of reporter gene transcription is dependent on ligand-promoted dimerization of receptor fusion proteins. (B) Strains expressing the indicated GHR-fusion proteins and GH are described in the text. The position and the expressed heterologous proteins of each strain are shown in the diagram. A dash indicates that the strain contained an unmodifed vector (pAS2, pACT2 or pCUP). Two independent isolates of each strain were assayed as described in Fig. 9-1.

with its cognate receptor in a manner not conducive to receptor homodimer formation.

An alternative example of ligand-mediated receptor dimerization in this system was provided by the mitogenic peptide VEGF (Gospodarowicz et al. 1989) and its receptor, flt1/KDR (Shibuya et al. 1990; Matthews et al. 1991; Terman et al. 1992). VEGF-induced KDR dimer formation is important for receptor activation (Matthews et al. 1991; DeVries et al. 1992) and occurs with a different (2:2) stoichiometry (Pötgens et al. 1994) than that for GH/GHR (Cunningham et al. 1991). KDR fusions were generated by cloning sequences encoding the extracellular region of KDR into both fusion protein expression vectors. VEGF was expressed as a nonfusion protein from the pCUP expression vector. Yeast strains (CY867, CY869) that expressed either single KDR fusion protein in the presence of VEGF failed to grow, and the strain (CY847) that expressed both KDR-fusion proteins in the absence of VEGF grew poorly on selective medium (Figure 9-4). Only the strain (CY846) that expressed both fusion proteins plus VEGF exhibited substantial growth on selective medium (Figure 9-4).

Reporter gene activation, observed as growth on selective medium, occurred in strain CY847 expressing KDR fusion proteins in the absence of VEGF (Figure 9-4), a finding similar to that observed in GHR dimerization assays (Figure 9-3B). Measurement of maximal growth rates in selective liquid medium produced data consistent with the results shown in Figure 9-4. KDR fusion proteins induced a specific two-hybrid response in the absence of ligand but coexpression of VEGF further stimulated cell proliferation (Ozenberger and Young 1995). These data suggest that interaction with ligand enhances the formation and/or stabilization of KDR dimers.

Essential Disulfide Bonds Do Not Obviate Expression in Yeast

The peptide ligands and receptors used in these studies have structural motifs stabilized by essential disulfide bonds. The redox potential within most cells is not conducive to disulfide bond formation. However, the observations of various ligand/receptor interactions in a two-hybrid system suggest that the interior of yeast may be more suitable for the formation of disulfide bonds than predicted. The acidic environment preferred by yeast may be an important distinction in comparison to other eucaryotes. The growth media used in our research had a pH of 4.3, standard for optimal yeast growth. Regardless of environmental pH, yeast cytosol is maintained at neutral pH via the action of a membrane proton-ATPase (Vallejo and Serrano 1989). GH/GHR and VEGF/KDR/KDR interactions (as shown in Figures 9-1 and 9-4) were examined on media with elevated pH. Two-hybrid responses in both ligand/receptor assays were attenuated as pH approached neutral (Figure 9-5). There was no inhibition of cell growth on nonselective (+ histidine) medium adjusted to a pH of 6.9 (Figure 9-5). The interaction of two

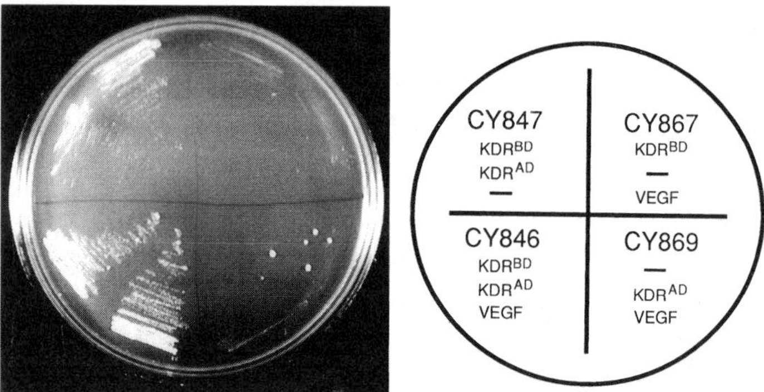

Figure 9-4. VEGF interaction with KDR-fusion proteins. Strains expressing KDR fusion proteins and VEGF are described in the text. The position and expressed heterologous proteins of each strain are shown in the diagram. A dash indicates that the strain contained an unmodified vector (pAS2, pACT2, or pCUP). Two independent isolates of each strain were assayed as described in Fig. 9-1.

yeast intracellular proteins, Snf1p and Snf4p (Fields and Song 1989), which are unlikely to require disulfide bonds for structural integrity, appeared unaffected by changes in the extracellular pH in these experiments. These findings were further validated by cysteine substitution analysis of VEGF that demonstrated that cysteines essential for in vivo activity in a mammalian expression system (Pötgens et al. 1994) are also essential for activity in a two-hybrid system (data not shown; Ozenberger et al., 1995).

Discussion

The two-hybrid system has been successfully employed to study numerous intracellular protein associations. We demonstrate that this method, which depends on gene activation within the yeast nucleus, also can be used to examine protein associations that normally occur at the mammalian cell surface. When expressed in yeast, the peptide hormones GH and PRL were found to interact with their respective receptors. Importantly, ligand/receptor binding was reversible and specific as demonstrated by coexpressing a third protein as a competitive inhibitor. Use of a third heterologous component also enabled the examination of ligand-induced receptor dimerization by expressing two receptor fusion proteins plus free peptide ligand. Additionally, the diversity of these proteins demonstrates that complex structural motifs such as four helix bundles (Abdel-Meguid et al. 1987), cysteine knots (McDonald and Hendrickson 1993), and immunoglobulin-like do-

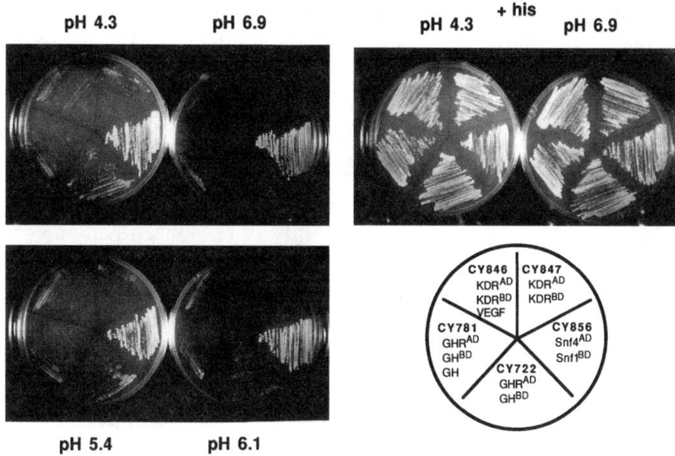

Fig. 9-5. Attenuation of ligand/receptor interactions at elevated pH. Strains expressing the indicated proteins are described in the text. The position of each strain is shown in the diagram. The GH/GHR strains are the same strains shown in Fig. 9-1; VEGF/KDR strains are also shown in Fig. 9-4. Each strain was streaked on synthetic complete medium deficient in leucine, tryptophan, uracil and histidine with the addition of 3-aminotriazole to 60 mM. The initial pH of this medium was 4.3. pH was adjusted to the indicated value by adding potassium hydroxide prior to autoclaving. Plates shown on the right were supplemented with histidine.

mains (Williams and Barclay 1988) are not restrictive for protein-protein interactions in this system.

Biochemical and pharmacological measurements of ligand/receptor interactions have often been elucidated through production of recombinantly expressed ligand, followed by evaluation in receptor binding assays under various experimental conditions. Reducing conditions, which attenuate disulfide bond formation, have a detrimental effect on binding for most ligand/receptor pairs. This observation suggests that structural motifs imparted or stabilized by disulfide bonds are important for the integrity of these ligand and receptor proteins. Each of the ligand and receptor proteins described in this chapter is structurally complex and contains essential disulfide bonds in its native form (Rozakis-Adcock and Kelly 1992; Vaisman et al. 1990; Pötgens et al. 1994; Bignon et al. 1994; Fuh et al. 1990). The proper generation of these structural elements was an important consideration in the described experiments because the redox potential within cells is generally not conducive to disulfide bond formation. The observations of functional interactions between various ligands and receptors in yeast suggest that the interior of this organism may, in fact, be suitable for the formation of tertiary protein structures requiring disulfide bonds. The acidic environment (laboratory growth medium pH ≈ 4.3) preferred by yeast is one factor associated with this phenomenon. The large outward movement of protons across the yeast cell membrane, necessary to maintain a neutral cytosolic environment, might result in an enhanced capacity for disulfide bond formation through localized variations in redox potential. More likely, levels of glutathione, the major thiol-disulfide redox buffer in the cytosol, can vary significantly in response to changes in the physiological state of cells (Gilbert 1990). Yeast grown under standard laboratory conditions may have lower levels of reduced glutathione, resulting in a more oxidative cell interior. When the pH of assay medium was elevated, interactions were attenuated between proteins, such as GH/GHR and VEGF/KDR, that require disulfide bonds for activity. In contrast, there was no apparent effect on the interaction between two yeast intracellular proteins that are not expected to contain structural disulfide bonds. These findings were further validated by cysteine substitution analysis of VEGF, which demonstrated that cysteines essential for in vivo activity in a mammalian expression system (Pötgens et al. 1994) are also essential for activity in a two-hybrid system (Ozenberger et al. 1995). These data suggest that the characteristic physiology of *Saccharomyces cerevisiae* may enable the examination of peptide ligand/receptor interactions inside yeast cells.

Another post-translational modification, glycosylation, may warrant consideration prior to analysis of ligand/receptor interactions in yeast. Proteins modified by N-linked glycosylation are unlikely to be found in the yeast nucleus. The peptide hormones and receptors investigated in this report contain N-linked glycosylation sites and have associated carbohydrate moieties in their native forms (Vaisman et al. 1990; Kelly et al. 1993). However, glycosylation was less of a concern in these studies because the nonglycosylated receptors bind ligand with little or no change in affinity (Fuh et al. 1990; Cunningham et al. 1991; Bignon et al. 1994). Therefore, protein glycosylation was not a factor in the assays described in this report.

The observation that ligand-dependent receptor dimerization can be examined in yeast should catalyze further investigation of higher order protein interactions using two-hybrid methods. The peptide hormone GH was shown to functionally interact with its receptor in both monomer and dimer models. These findings agree with stoichiometric and kinetic studies of GH association with its receptor in vitro or in mammalian cell systems (DeVos et al. 1992; Fuh et al. 1992). PRL displays a high degree of similarity to GH at both the ligand and receptor level, so a 1:2 ligand/receptor stoichiometry was predicted for PRL. However, PRL-induced receptor dimerization was not observed although a specific interaction between ligand and receptor fusion proteins was observed in a monomer model. Perhaps PRL does not promote receptor-receptor homodimer formation; rather, it may stimulate heterodimerization with an alternative receptor, as is prevalent among members of the cytokine receptor family (Young, 1992). GH clearly stimulated GHR dimerization in a two-hybrid system. Interestingly, there was a significant stimulation of growth in a strain expressing only GHR fusion proteins (CY887), suggesting that there is some avidity between these receptors in the absence of ligand. Interaction between GHRs may occur via the receptor stem region (DeVos et al. 1992) or via inter-receptor disulfide bond formation at the unpaired membrane-proximal cysteine. The frequency of ligand-independent GHR dimerization in natural systems is unknown. The VEGF/KDR assays present a similar paradigm. Strain CY847, expressing only the two KDR fusion proteins, exhibited substantial growth above controls. The affinity between KDRs in the absence of VEGF may be indicative of receptor-receptor association via an Ig-loop, as has been observed for the related platelet-derived growth factor receptor (Herren et al. 1993) and c-kit (Lev et al. 1993). VEGF expression significantly enhanced KDR dimer formation or stabilization. In contrast to the predicted sequential binding of GH to one receptor protein followed by interaction with a second receptor (Wells et al. 1993), VEGF functions as a homodimeric ligand and may simultaneously bind paired receptor proteins, stabilizing dimer formation and consequent receptor activation.

Future Considerations

We have described novel methods to examine reversible and specific binding of peptide ligands and receptors and the formation of specific multiprotein complexes. Importantly, the data suggest that two-hybrid methodologies can be generally applied to the multitude of proteins that normally interact at cell surfaces to control cellular activities. Also, as described elsewhere in this book, powerful genetic tools available in the two-hybrid system can be applied to identify new ligands and receptors via cDNA library screening, to perform rapid structure/function analyses, and to characterize receptor protein partners. These applications should contribute to the elucidation of the complex molecular associations involved in ligand/receptor multimerization and potentially assist in the identification of novel therapeutics acting on this important family of molecules.

ACKNOWLEDGMENTS The authors thank S. Hellings for important contributions to the research. Portions of the work are excerpted from "B.A. Ozenberger and K.H. Young, Functional Interaction of Ligands and Receptors of the Hematopoietic Superfamily in Yeast, Molecular Endocrinology 9: 1321-1329, 1995"; copyright The Endocrine Society.

References

Abdel-Meguid, S., Shieh, H., Smith, W., Dayringer, H., Violand, B., and Bentle, L. (1987). Three-dimensional structure of a genetically engineered variant of porcine growth hormone. Proc. Natl. Acad. Sci. USA 84:6434–6437.

Bignon, C., Sakal, E., Belair, L., Chapnik-Cohen, N., Djiane, J., and Gertler, A. (1994). Preparation of the extracellular domain of the rabbit prolactin receptor expressed in Escherichia coli and its interaction with lactogenic hormones. J. Biol. Chem. 269:3318–3324.

Butt, T. R., Sternberg, E. J., Gorman, J. A., Clark, P., Hamer, D., Rosenberg, M., and Crooke, S. T. (1984). Copper metallothionein of yeast, structure of the gene and regulation of expression. Proc. Natl. Acad. Sci. *USA* 81:3332–3336.

Chein C.-T., Bartel, P. L., Sternglanz, R., and Fields, S. (1991). A method to identify and clone genes for proteins that interact with a protein of interest. Proc. Natl. Acad. Sci. USA 88:9578–9582.

Cunningham, B. C., Ultsch, M., DeVos, A. M., Mulkerrin, M. G., Clauser, K. R., and Wells, J. A. (1991). Dimerization of the extracellular domain of the human growth hormone receptor by a single hormone molecule. Science 254:821–825.

Cunningham, B., and Wells, J. (1993). Comparison of a structural and a functional epitope. J. Mol. Biol. 234:554–563.

DeVos, A. M., Ultsch, M., and Kossiakoff, A. A. (1992). Human growth hormone and extracellular domain of its receptor: crystal structure of the complex. Science 255:306–312.

DeVries, C., Escobedo, J. A., Ueno, H., Houck, K., Ferrara, N., and Williams, L. T. (1992). The fms-like tyrosine kinase, a receptor for vascular endothelial growth factor. Science 255:989–991.

Fields, S., and Song, O. (1989). A novel genetic system to detect protein-protein interactions. Nature 340:245–246.

Fuh, G., Cunningham, B. C., Fukunaga, R., Nagata, S., Goeddel, D. V., and Wells, J. A. (1992). Rational design of potent antagonists to the human growth hormone receptor. Science 256:1677–1680.

Fuh, G., Mulkerrin, M. G., Bass, S., McFarland, N., Brochier, M., Bourell, J. H., Light, D. R., and Wells, J. A. (1990). The human growth hormone receptor. Secretion from *Escherichia coli* and disulfide bonding pattern of the extracellular binding domain. J. Biol. Chem. 265:3111–3115.

Gilbert, H. (1990). Molecular and cellular aspects of thiol-disulfide exchange. Adv. Enzymol. 63:69–172.

Gospodarowicz, D., Abraham, J. A., and Schilling, J. (1989). Isolation and characterization of a vascular endothelial cell mitogen produced by pituitary-derived folliculo stellate cells. Proc. Natl. Acad. Sci. USA 86:7311–7315.

Harper, W. J., Adami, G. R., Wei, N., Keyomarsi, K., and Elledge, S. J. (1993). The p21 Cdk-interacting protein Cip1 is a potent inhibitor or G1 cyclin-dependent kinases. Cell 75:805–816.

Herren, B., Rooney, B., Weyer, K. A., Iberg, N., Schmid, G., and Pech, M. (1993). Dimerization of extracellular domains of platelet-derived growth factor receptors. J. Biol. Chem. 268:15088–15095.

Hill, J. E., Myers, A. M., Koerner, T. J., and Tzagoloff, A. (1986). Yeast/E. coli shuttle vectors with multiple unique restriction sites. Yeast 2:163–167.

Kang, Y.-S., Kane, J., Kurjan, J., Stadel, J. M., and Tipper, D. J. (1990). Effects of expression of mammalian Gα and hybrid mammalian-yeast Gα proteins on the yeast pheromone response signal transduction pathway. Mol. Cell. Biol. 10:2582–2590.

Kelly P. A., Ali, S., Rozakis, M., Goujon, M., Nagano, M., Pellegrini, I., Gould, D.,

Djiane, J., Edery, M., Finidori, J., and Postel-Vinay, M.C. (1993). The growth hormone/prolactin receptor family. Rec. Prog. Horm. Res. 48:123–164.

Leung, D. W., Spencer, S. A., Cachianes, G., Hammonds, R. G., Collins, C., Henzel, W. J., Barnard, R. Waters, M. J., and Wood, W. I. (1987). Growth hormone receptor and serum binding protein: purification, cloning and expression. Nature 330:537–543.

Lev, S., Blechman, J., Nishikawa, S.-I. , Givol, D., and Yarden, Y. (1993). Interspecies molecular chimeras of Kit help define the binding site of the stem cell factor. Mol. Cell. Biol. 13:2224–2234.

Matthews, W., Jordan, C. T., Gavin, M., Jenkins, N. A., Copeland, N. G., and Lemischka, I. R. (1991). A receptor tyrosine kinase cDNA isolated from a population of enriched primitive hematopoietic cells and exhibiting close genetic linkage to *c-kit.* Proc. Natl. Acad. Sci. USA 88:9026–9030.

McDonald, N. Q., and Hendrickson, W. A. (1993). A structural superfamily of growth factors containing a cystine knot motif. Cell 73:421–424.

Ozenberger, B. A., and Young, K. H. 1995. Functional interaction of ligands and receptors of the hematopoietic superfamily in yeast. Mol. Endocrinol. 9:1321–1329.

Ozenberger, B. A., Hellings, S., Kajkowski, E. M., and Young, K. H. (1995). Interactions between mammalian ligands and cell surface receptors expressed in a two-hybrid system. *Program of the 77th Annual Meeting of the Endocrine Society.* Washington, D.C. Abstract, p. 533.

Pötgens, A. J. G., Lubsen, N. H., van Altena, M. C., Vermeulen, R., Bakker, A., Schoenmakers, J. G. G., Ruiter, D. J., and de Waal, R. M. W. (1994). Covalent dimerization of vascular permeability factor/vascular endothelial growth factor is essential for its biological activity. J. Biol. Chem. 269:32879–32885.

Rose, M. D., Winston, F., and Hieter, P. (1990). In *Methods in Yeast Genetics.* Cold Spring Harbor Laboratory Press. New York, Cold Spring Harbor.

Rozakis-Adcock, M., and Kelly, P. A. (1992). Identification of ligand binding determinants of the prolactin receptor. J. Biol. Chem. 267:7428–7433.

Shibuya, M., Yamaguchi, S., Yamane, A., Ikeda, T., Tojo, A., Matsushime, H., and Sato, M. (1990). Nucleotide sequence and expression of a novel human receptor-type tyrosine kinase gene (flt) closely related to the fms family. Oncogene 5:519–525.

Staten, N. R., Byatt, J. C., and Krivi, G. G. (1993). Ligand-specific dimerization of the extracellular domain of the bovine growth hormone receptor. J. Biol. Chem. 268:18467–18473.

Terman, B. I., Dougher-Vermanzen, M., Carrion, M. E., Dimitrov, D., Armellina, D. C., Gospodarowicz, D., and Bohlen, P. (1992). Identification of the KDR tyrosine kinase as a receptor for vascular endothelial cell growth factor. Biochem. Biophys. Res. Comm. 187:1579–1586.

Tischer, E., Mitchell, R., Hartman, T., Silva, T., Gospodarowicz, D., Fiddes, J. C., and Abraham, J. A. (1991). The human gene for vascular endothelial growth factor. J. Biol. Chem. 266:11947–11954.

Vaisman, N., Gospodarowicz, D., and Neufeld, G. (1990). Characterization of the receptors for vascular endothelial growth factor. J. Biol. Chem. 265:19461–19466.

Vallejo, C. G., and Serrano, R. (1989). Physiology of mutants with reduced expression of plasma membrane H+-ATPase. Yeast 5:307–319.

Wells, J. A., Cunningham, B. C., Fuh, G., Lowman, H. B., Bass, S. H., Mulkerrin, M. G., Ultsch, M., and DeVos, A. M. (1993). The molecular basis for growth hormone-receptor interactions. Rec. Prog. Horm. Res. 48:253–275.

Williams, A., and Barclay, A. (1988). The immunoglobulin superfamily—Domains for cell surface recognition. Ann. Rev. Immunol. 6:381–405.

Young, P. R. (1992). Protein hormones and their receptors. Curr. Opin. Biotech. 3:408–421.

10

Two-Hybrid Analysis of Protein-Protein Interactions in the Yeast Pheromone Response Pathway

George F. Sprague, Jr.
John A. Printen

Overview of the Pheromone Response Pathway

The pheromone response pathway is a multistep pathway that enables haploid yeast cells (*Saccharomyces cerevisiae*) to communicate and prepare for mating. As a result of activation of the pathway by binding of pheromone to receptor, several physiological changes occur: transcription of genes whose products catalyze mating is elevated, cell cycle progression halts in the G1 phase, and cell polarity is reoriented toward the perceived mating partner. Hence, a vegetative cell differentiates into a mating competent cell.

The components of the pathway have been identified by the isolation of mutants with altered pathway activity, including both mutants that cannot respond to pheromone and mutants that exhibit a constitutively-activated pathway. Double mutant studies utilizing nonresponsive and constitutive mutants have determined the relative order of action of the gene products. Moreover, the cloning and sequencing of the genes has suggested specific biochemical functions, which in many cases have been verified. Together, these studies have given a sophisticated picture of the pathway, but substantial questions remain. Do sequential proteins in the deduced linear pathway indeed interact in vivo, or are there as yet unidentified components? What is the biochemical nature of signal propagation at each step? Once the pathway is activated, what mechanisms serve to attenuate the sig-

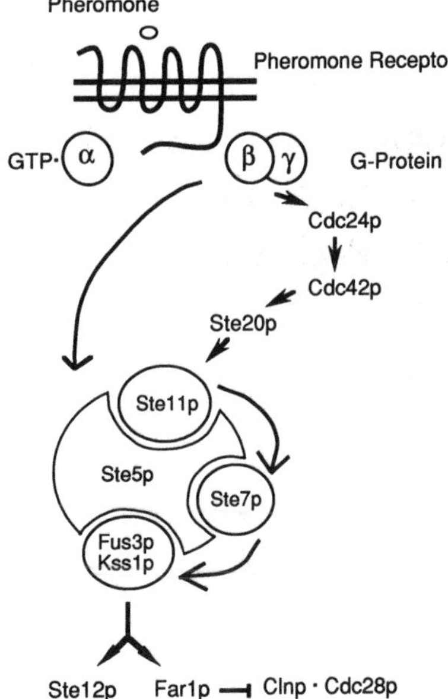

Fig. 10-1. A schematic of the pheromone response pathway. A discussion of the known or presumed biochemical function of each pathway component is given in the text. Ste5p is represented as a scaffold facilitating formation of a complex involving each tier of the MAP kinase cascade. Each arrow represents an interaction that is known or presumed to be important for signal propagation. Interactions that may lead to signal attenuation are not depicted.

nal so that the diploid zygote can initiate vegetative growth? In at least one case, the gene sequence did not suggest a biochemical function for its encoded product. What is the role of that gene product? Two-hybrid analysis has shed light on each of these questions and, thereby, considerably enhanced the understanding of the pheromone response pathway. Below, we first summarize the current picture of the pheromone response pathway (Figure 10-1; for review, see Sprague and Thorner 1992; Kurjan 1994; Elion 1995; Herskowitz 1995; Kron and Gow 1995), incorporating insights gained from the two-hybrid analysis, and then discuss some specific contributions from two-hybrid analysis. In this latter discussion, we will point out new reagents we have developed that may be of general utility to those carrying out two-hybrid analysis; we will summarize our strategy and findings; and we will highlight new biological perspectives that have come from the analysis.

The pheromone receptors are members of the 7-transmembrane receptor family and are coupled to a heterotrimeric G protein. Binding of pheromone to receptor presumably causes a conformational change that is propagated across the plasma membrane to the G protein. GTP is exchanged for GDP on the Gα subunit, leading to release of the G$\beta\gamma$ subunit, which propagates the signal. However, the mechanism by which G$\beta\gamma$ propagates the signal is not known and may be complex. This possibility follows from the observation, made initially by two-hybrid analysis, that the Gβ subunit (Ste4p) interacts with at least two proteins required for pheromone response: Ste5p, a scaffolding protein (Whiteway *et al.* 1995; see later in this paragraph), and Cdc24p (Zhao *et al.* 1995). This latter interaction forms part of an argument, emerging from a series of recent findings, that a member of the Ras superfamily, Cdc42p, is a component of the pheromone response pathway and serves as an activator of Ste20p, a serine/threonine protein kinase (Manser *et al.* 1994; Simon *et al.* 1995; Stevenson *et al.* 1995; Zhao *et al.* 1995). Cdc24p is the guanine nucleotide exchange factor for Cdc42p. Each of these proteins is required for activation of a three-tiered MAP kinase cascade. Moreover, the findings that Gβ (Ste4p) interacts with both Ste5p and Cdc24p, and that Ste20p interacts with both Ste5p (revealed by immunoprecipitation studies; Leeuw *et al.* 1995) and Cdc42p, imply that activation of the MAP kinase pathway involves several events. The MAP kinase cascade is composed of Ste11p, an MEKK homolog, Ste7p, an MEK homolog, and Fus3p and Kss1p, two related and partially redundant MAP kinase homologs. The pathway bifurcates after the MAP kinases, one branch leading to transcription induction and the other branch leading to cell cycle arrest. Both Fus3p and Kss1p are thought to phosphorylate Ste12p, a transcription factor for pheromone-responsive genes and, thereby, promote transcription induction. Fus3p, but not Kss1p, can also phosphorylate Far1p. Phosphorylated Far1p binds to and inhibits the activity of the Clnp•Cdc28p complex, providing one means by which pheromone treatment leads to arrest of the cell division cycle. Thus, it is possible to trace a signal from the receptor through a number of components to target proteins that govern two of the physiological responses to pheromone. In addition, the beginnings of a molecular link between pheromone response and the reorientation of cell polarity is provided by the identification of Cdc42p as a component of the response pathway. Cdc42p is required for polarity establishment in vegetative cells; perhaps its interaction with Cdc24p activates it at the site of pheromone-bound receptors, leading to the establishment of a new polarity axis.

Protein-Protein Interactions Revealed by the Two-Hybrid System

As stated earlier, two-hybrid analysis has had a profound impact on the current understanding of the pheromone response pathway. The first, and perhaps most surprising, finding from this analysis is that Ste5p interacts with protein kinases that operate at each step of the MAP kinase cascade. This finding suggests that one role of Ste5p is to serve as a scaffold to facilitate interactions among members of the kinase cascade. In this role as a facilitator, Ste5p may serve at least three functions: (1) it may make both signal propagation and signal attenuation more efficient; (2) it may help minimize crosstalk with other MAP kinase cascades and, thus, ensure the integrity of the pheromone response pathway; and (3) it may serve to localize the kinase cascade to a particular subcellular destination. Second, two-hybrid analysis has revealed that activation of the MAP kinase cascade by the G protein may be complex, requiring several independent biochemical events. Third, two-hybrid analysis has revealed interactions between proteins that are not thought to act sequentially in the pathway. The potential implications of these interactions can be pursued by further genetic and biochemical analyses. Fourth, once an interaction between two proteins is detected, it has been possible in some cases to identify a protein domain responsible for this interaction. More refined hypotheses as to how signaling is propagated or attenuated can then be developed. Thus, two-hybrid analysis has greatly enriched our perspective on and understanding of the pheromone response pathway. In its absence, the relationships among pathway components would be based largely on formal genetic analysis, and there would be little evidence indicating whether the relationships are direct or indirect.

Beginning from the formal genetic framework, we and others used the two-hybrid system to investigate protein-protein interactions among essentially all components of the pheromone response pathway (Choi *et al.* 1994; Marcus *et al.* 1994; Printen and Sprague, 1994; Simon *et al.* 1995; Whiteway *et al.* 1995; Zhao *et al.* 1995). The overall strategy and the salient findings are summarized below.

Materials

Initially, we made hybrid genes that fused essentially the entire open reading frame of Ste4p, Ste20p, Ste5p, Ste11p, Ste7p, Ste12p, Fus3p, and Kss1p to the DNA-binding domain (BD), the transcription activation domain (AD), or both of Gal4p (Printen and Sprague, 1994). The original system of Fields and Song (1989) created hybrid proteins with the Gal4p AD at the N-terminus. To increase the flexibility of this system, we created vectors that enable the Gal4p AD to be encoded at the C-terminus of the hybrid protein. Indeed, in some cases we were able to detect an interaction when the transcription AD was located at the C-terminus of the hybrid protein, but not at the N-terminus of the equivalent protein. We verified that all hybrid genes indeed directed the synthesis of the appropriate fusion protein, using monoclonal antibodies developed against Gal4p BD or Gal4p AD. In addition, we verified that each of the fusion proteins was functional by examining the ability of the hybrid genes to complement a deletion of the corresponding pathway gene. The hybrids were then expressed in all pair-wise combinations in an appropriate yeast strain containing a *GAL1-lacZ* reporter to detect potential interactions between the hybrid proteins.

Matrix

The results of these pair-wise tests can be compiled into a matrix (Table 10-1) and organized by a series of summary statements. First, several interactions that were expected based on biochemical or genetic studies were indeed detected. For example, Ste7p interacted with both Fus3p and Kss1p, and Kss1p interacted with Ste12p. Second, Ste11p interacted with both MAP kinases, Fus3p and Kss1p. These interactions were unexpected because Ste11p is at the first tier in the kinase cascade, whereas Fus3p and Kss1p are at the third tier. Third, we observed interaction of Ste5p with Ste11p, Ste7p, and Fus3p. This finding is particularly striking because although genetic epistasis experiments had suggested that Ste5p and Ste11p might act sequentially in the pathway, there was no evidence that Ste5p might interact with the more distal members of the MAP kinase cascade. Choi *et al.* (1994) and Marcus *et al.* (1994) have also detected interaction between Ste5p and each member of the MAP kinase cascade (including Kss1p, as well) by two-hybrid analysis. More important, Choi *et al.* (1994) have obtained independent biochemical support for these interactions. Using both immunoprecipitation and glycerol gradient methodologies, they detected a complex that contains Ste5p, Ste11p, Ste7p, and Fus3p. In principle, the interactions detected by two-hybrid analysis need not reflect direct interaction between the two proteins being tested—a bridging protein could mediate the interaction. This possibility seems particularly germane given the evidence for a complex involving a number of the pathway components. We therefore reexamined the ability of pairs of hybrid proteins to interact in strains carrying deletions at *STE4, STE5, STE7, STE11,* or *FUS3*. These deletions had little or no effect on interaction between the hybrids, and in no case was the interaction abolished. Thus, although we cannot exclude the possibility that unknown proteins mediate the observed interactions, this experiment supports the idea that the tested gene products interact directly.

Fourth, we did not detect interaction between a number of pairs of proteins. In some cases, there is no reason to suppose that the tested pair in fact interacts, but in other cases, subsequent studies provided evidence for interaction not detected in these initial experiments. A discussion of two examples is worthwhile because it may be instructive about the ways in which

Table 10-1 Summary of two-hybrid interactions involving components of the pheromone response pathway. This matrix includes results from Choi et al. (1994), Marcus et al. (1994), Printen and Sprague (1994), and Whiteway et al. (1995). A plus (+) indicates that the protein pair interacted in the two-hybrid system; a minus (–) indicates that no interaction was detected.

DNA-Binding Domain Hybrid	Transcription Activation Domain Hybrid							
	Ste4p	Ste20p	Ste5p	Ste11p	Ste7p	Fus3p	Kss1p	Ste12p
Ste20p	–	–	–	–	–	–	–	–
Ste5p	+	–	–	+	+	+	–	–
Ste11p	–	–	+	+	–	–	+	–
Ste7p	–	–	+	–	–	+	+	–
Fus3p	–	–	+	+	+	–	–	–
Kss1p	–	–	+	+	+	–	–	+

the two-hybrid system can be manipulated to examine more thoroughly the possibility that two particular proteins interact. One example is provided by Ste11p and Ste7p. We failed to detect an interaction between these two proteins. However, other investigators were able to detect an interaction if Ste5p was overexpressed in the same cells (Choi *et al.* 1994; Marcus *et al.* 1994). Hence, in this case, a bridging protein (Ste5p) may greatly enhance or be required for the interaction between these two protein kinases. Ste4p (Gβ) and Ste5p provide a second example. In this case, interaction was detected when the Ste5p hybrid protein contained residues 1–214, but not when it contained residues 25–917 (Whiteway *et al.* 1995). Perhaps residues 1–24 of Ste5p are critical for interaction with Ste4p, or perhaps a domain of Ste5p gives a stronger signal in the two-hybrid system than does complete Ste5p, as has been observed for other interactions (Vojtek *et al.* 1993). These examples illustrate that the failure to detect an interaction cannot be taken as evidence that it does not occur. Conversely, however, it is important to remember that success at detecting an interaction cannot establish its physiological role or even its biological relevance. For example, the interaction could reflect an event in signal propagation or an event in signal attenuation. Subsequent genetic and biochemical experiments are needed to support or refute the possible relevance.

Domains of Interaction and Analysis of Interactions Using Mutant Hybrid Proteins

The experiments summarized above identify a number of protein-protein interactions that are likely to be important for the functioning of the pheromone response pathway. In some cases, it may be possible to gain further insight into the physiological role of the interaction by determining if the interaction can be ascribed to a particular segment (domain?) of each of the proteins. Ideally, such domain analysis will be complemented by other biochemical or genetic experiments that examine the function of the protein from a different perspective. The interaction of Ste5p with Ste7p and Ste11p and the interaction of Ste11p with Fus3p provide examples.

The DNA sequences of *STE7* and *STE11* predict that each encodes a protein composed of two domains: a carboxyl-terminal catalytic domain with protein kinase activity, and an amino-terminal, presumably regulatory, domain. A regulatory role has been documented most clearly for the Ste11p amino-terminal domain. Truncations and amino acid substitutions in this region of Ste11p cause constitutive activation of the pheromone response pathway, arguing that this domain functions as a negative regulator of the kinase activity (Cairns *et al.* 1992; Stevenson *et al.* 1992). The Ste7p regulatory domain may have a positive role.

We examined the interaction of hybrids containing each of these domains with the set of proteins that showed interaction with full-length Ste11p and Ste7p. We also examined the effect on protein-protein interactions caused by a Ste11p amino acid substitution that leads to the constitutive activation of pheromone response. This latter set of experiments was done only in the context of full-length Ste11p. The amino-terminal regulatory domain of Ste11p interacted with itself and with Ste5p. An amino acid substitution, P279S, in the amino-terminal segment, had dramatic effects on interaction with other pathway components. Specifically, hybrids containing this substitution showed 15-fold greater activation of the reporter when coexpressed with the Ste5p hybrid than did wild-type Ste11p hybrids coexpressed with the Ste5p hybrid. Conversely, interactions with Fus3p and Kss1p were decreased to 1% and 16%, respectively, of the reporter activation seen with wild-type Ste11p. The P279S substitution does not affect the steady-state expression of the fusion protein. Thus, the changes in expression of the reporter appear to reflect a change in the affinity of the hybrid bearing the substitution for Ste5p, Fus3p, and Kss1p.

Together, the experiments involving the Ste11p amino-terminal hybrids and the Ste11p amino-acid substitution hybrids suggest some interesting possibilities for regulation of signal transmission. First, the interaction of the Ste11p amino-terminal domain with Ste5p suggests that Ste5p may have a role in pheromone response, in addition to its role as a scaffolding protein. For example, perhaps an upstream component modifies or alters the conformation of Ste5p such that Ste5p becomes competent to relieve the inhibitory effect of the Ste11p amino-terminal domain. The analysis of dominant constitutive alleles of *STE5*, such as the one reported by Hasson *et al.* (1993), may help to address this possibility. Second, the interaction of Ste11p with the MAP kinases Fus3p and Kss1p, which are not adjacent to Ste11p in the pathway, may reflect part of the mechanism whereby the pathway signal is attenuated. Support for this possibility comes from the properties of the partially constitutive Ste11p P279S mutant. The inability of this mutant to interact with Kss1p or Fus3p could, in part, account for the partial constitutivity of this version of Ste11p, if Fus3p and Kss1p negatively influence Ste11p activity. Indeed, it is conceivable that the absence of this negative influence on the mutant Ste11p could account for its greater ability to associate with Ste5p. Even in a cell that has not been stimulated with pheromone, there is some active Kss1p and Fus3p. Perhaps phosphorylation of wild-type Ste11p by the MAP kinases affects its ability to interact with Ste5p.

Experiments to investigate the interaction properties of the regulatory and catalytic domains of Ste7p also suggest possible regulatory events that can be pursued by other experimental approaches. The amino-terminal regulatory domain of Ste7p interacts with both Fus3p and Kss1p. As argued earlier for Ste11p, perhaps this interaction reflects attenuation of the pathway. Indeed, preliminary evidence indicates that the amino-terminal Ste7p domain is phosphorylated by Fus3p and mutation of the sites of phosphorylation in otherwise wild-type Ste7p lessens the ability of cell to attenuate the signal and recover from pheromone treatment (B. Errede, personal communication). The carboxyl-terminal catalytic domain of Ste7p interacts with Ste5p. This interaction may simply indicate that the binding site on Ste7p for the scaffold protein is within the catalytic domain. More interestingly, it might indicate that Ste5p is a substrate for Ste7p. Because Ste7p functions after Ste5p in the response pathway, the putative phosphorylation of Ste5p by Ste7p may be another part of the attenuation mechanism.

The domain structure of Ste11p and Ste7p is suggested by the sequence of the proteins, and that information directed the interaction experiments summarized above. The sequence of Ste5p, in contrast, predicts no particular domain structure. To gain insight into how Ste5p interacts with each of the protein kinases, Choi *et al.* (1994) tested the ability of segments of Ste5p to interact. A rather short segment of Ste5p, from residues 240–336, was found to be necessary and sufficient for interaction with the two MAP kinases. On the other hand, Ste11p and Ste7p appear to interact with different regions in the carboxyl-terminal two-thirds of Ste5p, although segments sufficient for interaction were not identified. Thus, this information implies that Ste5p can interact simultaneously with kinases at each tier of the MAP kinase cascade, a possibility substantiated by subsequent biochemical experiments. In addition, this still crude definition of interaction domains will certainly guide future thinking about regulation of signal transmission, for example, as the sites of phosphorylation on Ste5p by the MAP kinases (Kranz *et al.* 1994) or by Ste7p (as suggested by two-hybrid interactions) are determined.

Summary

We have described analysis of the yeast pheromone response pathway using the two-hybrid system, offering this analysis as a case study of how a thorough matrix approach examining all possible pair-wise interactions can yield important new insights even in a well-studied pathway. Indeed, the payoff has been substantial and our discussion emphasized results that provided new perspective. The hypothesis that Ste5p, a protein whose sequence provides no clues as to its biochemical function, serves as a scaffold for the MAP kinase cascade, is one such new perspective. It will be interesting to learn whether scaffolds are a common feature of other MAP kinase cascades, both in yeast and in other organisms. The pheromone pathway is well studied by standard genetic and biochemical approaches, and those studies provided the framework for a second new perspective that emerged from two-hybrid analysis. In several cases, interactions were detected between proteins not thought to control adjacent steps in the pathway. By mapping the domains responsible for these interactions and by examining the effects of mutations of known physiological consequence on the interactions, we were able to hypothesize that some of the interactions between nonadjacent components are part of the mechanism that serves to attenuate signal transmission. As this case study illustrates, two-hybrid analysis is a powerful approach that can provide new insights and hypotheses that, otherwise, might be slow to emerge by more traditional approaches.

References

Cairns, B.R., Ramer, S.W., and Kornberg, R.D. (1992). Order of action of components in the yeast pheromone response pathway revealed with a dominant allele of the STE11 kinase and multiple phosphorylation of the STE7 kinase. Genes Dev. 6:1305–1318.

Choi, K.-Y., Satterberg, B., Lyons, D.M., and Elion, E.A. (1994). Ste5 tethers multiple protein kinases in the MAP kinase cascade required for mating in S. cerevisiae. Cell 78:499–512.

Elion, E.A. (1995). Ste5: A meeting place for MAP kinases and their associates. Trends Cell Biol. 5:322–327.

Fields, S., and Song, O. (1989). A novel genetic system to detect protein-protein interactions. Nature 340:245–246.

Hasson, M.S., Blinder, D., Thorner, J., and Jenness, D.D. (1993). Mutational activation of the *STE5* gene product bypasses the requirement for G protein β and γ subunits in the yeast pheromone response pathway. Mol. Cell. Biol. 14:1054–1065.

Herskowitz, I. (1995). MAP kinase pathways in yeast: For mating and more. Cell 80:187–97.

Kranz, J.E., Satterberg, B., and Elion, E.A. (1994). The MAP kinase Fus3 associates with and phosphorylates the upstream signaling component Ste5. Genes Dev. 8:313–327.

Kron, S.J., and Gow, N.A.R. (1995). Budding yeast morphogenesis: Signalling, cytoskeleton and cell cycle. Curr. Opin. Cell Biol. 7:845–855.

Kurjan, J. (1993). The pheromone response pathway in *Saccharomyces cerevisiae*. In: *Annual Review of Genetics*, Campbell, A., ed. Palo Alto, California, Annual Reviews, Inc. pp. 147–179.

Leeuw, T., Fourest-Lieuvin, A., Wu, C., Chenevert, J., Clark, K., Whiteway, M., Thomas, D.Y., and Leberer, E. (1995). Pheromone response in yeast: Association of Bem1p with proteins of the MAP kinase cascade and actin. Science 270:1210–1213.

Manser, E., Leung, T., Salihuddin, H., Zhao, Z.S., and Lim, L. (1994). A brain serine/threonine protein kinase activated by Cdc42 and Rac1. Nature 367:40–46.

Marcus, S., Polverino, A., Barr, M., and Wigler, M. (1994). Complexes between STE5 and components of the pheromone-responsive mitogen-activated protein kinase module. Proc. Natl. Acad. Sci. USA 91:7762–7766.

Printen, J.A., and Sprague, Jr., G.F. (1994). Protein-protein interactions in the yeast pheromone response pathway: Ste5p interacts with all members of the MAP kinase cascade. Genetics 138:609–619.

Simon, M.-N., De Virgilio, C., Souza, B., Pringle, J.R., Abo, A., and Reed, S.I. (1995). Role for the Rho-family GTPase Cdc42 in yeast mating pheromone signal pathway. Nature 376:702–705.

Sprague, G.F., Jr., and Thorner, J. (1992). Pheromone response and signal transduction during the mating process of *Saccharomyces cerevisiae*. In: *The Molecular and Cellular Biology of the Yeast Saccharomyces: Gene Expression*, Jones, E.W., Pringle, J.R., and Broach, J.R., eds. New York: Cold Spring Harbor Press. pp. 657–744.

Stevenson, B.J., Rhodes, N., Errede, B., and Sprague, Jr., G.F. (1992). Constitutive mutants of the protein kinase STE11 activate the yeast pheromone response pathway in the absence of the G protein. Genes Dev. 6:1293–1304.

Stevenson, B.J, Ferguson, B., De Virgilio, C., Bi, E., Pringle, J.R., Ammerer, G., and Sprague, Jr., G.F. (1995). Mutation of *RGA1*, which encodes a putative GTPase-activating protein for the polarity-establishment protein Cdc42p, activates the pheromone-response pathway in the yeast *Saccharomyces cerevisiae*. Genes Dev. 9:2949–2963.

Vojtek, A.B., Hollenburg, S.M., and Cooper, J.A. (1993). Mammalian Ras interacts directly with the serine/threonine kinase Raf. Cell 74:205–214.

Whiteway, M.S., Wu, C., Leeuw, T., Clark, K., Fourest-Lieuvin, A., Thomas, D.Y., and Leberer, E. (1995). Association of the yeast pheromone response G protein βγ subunits with the MAP kinase scaffold Ste5. Science 269:1572–1575.

Zhao, Z-S., Leung, T., Manser, E., and Lim, L. (1995). Pheromone signaling in *Saccharomyces cerevisiae* requires the small GTP-binding protein Cdc42p and its activator *CDC24*. Mol. Cell. Biol. 15:5246–5257.

11

Two-hybrid Screening and the Cell Cycle

Gregory J. Hannon

Cell Cycle Control in Mammalian Cells

Unlike many other reviews in this volume, this chapter does not present technical advice on the execution of the two-hybrid screen or its variants. Instead, it focuses on the impact that the two-hybrid approach has had on a single area of research. The mechanics of cell cycle control have been intensively studied, both because of the ubiquitous importance of this process in biology and because of the probable connection to human disease. This was one of the first fields of inquiry in which the two-hybrid screen found widespread use and had broad impact.

Progression through the cell cycle is controlled by a family of evolutionarily conserved enzymes known as the cyclin-dependent kinases or CDKs (see Sherr 1994; Hunter and Pines 1994). These proteins are not active on their own but instead require association with a positive regulatory subunit called a cyclin (see Draetta 1990). Cyclin-dependent kinases were first discovered by genetic methods in the fission yeast, *Schizosaccharomyces pombe*, and in the budding yeast, *Saccharomyces cerevisiae* (see Draetta 1990). In these organisms, a single cyclin-dependent kinase, cdc2 in *S. pombe* or Cdc28p in *S. cerevisiae*, controls progress through the entire division cycle. In mammalian cells, the situation is more complex. The CDK family in human cells consists of at least seven members: CDC2

(CDK1) and CDK2-CDK7 (see Sherr 1993, 1994). Each of these enzymes is thought to execute a specialized function, and it is their ordered activation that promotes the proper sequence of cell cycle events (see Sherr 1993).

The cyclin family contains multiple members in both lower and higher eukaryotes (see Lew and Reed 1993), and the specific configuration of cyclin and CDK subunits is proposed to determine the substrate specificity of cyclin/CDK complexes. In mammals, the combination of seven different CDK subunits with at least as many cyclins could conceivably generate a vast array of cell cycle regulatory kinases. However, each CDK prefers a limited subset of cyclin partners. For example, CDC2 complexes mainly with cyclins A and B to control events during the G2 and M phases (see Sherr 1993). CDK2 associates with cyclins A and E, and these kinases regulate late G1 and S phases. CDK4 and CDK6 bind exclusively to D-type cyclins, and these complexes control commitment to the division cycle during early and mid G1.

Since inappropriate execution of cell cycle events would be disastrous for either the individual cell or for the organism as a whole, the activation of cyclin/CDK complexes is regulated at multiple levels. Activity requires association with the cyclin subunit and is controlled, to some extent, by the availability of this positive regulator. Abundance of some cyclins (for example, A, B, E) fluctuates in a cell-cycle-dependent manner (see Sherr 1993). In these cases, the presence of the cyclin generally correlates with the cell cycle phase in which that protein functions. Changes in the abundance of cyclin subunits is controlled both by regulation of their synthesis and by regulation of their destruction (see King et al. 1994). For example, cyclin B synthesis is increased during G2 and M phases; however, at the end of M-phase, cyclin B becomes unstable and is degraded. This process leaves CDC2 in a monomeric form which is reused during the next division cycle. In the case of cyclin D1, protein levels are relatively constant throughout the cell cycle (see Sherr 1994). The abundance of this protein is instead regulated by extracellular signals. Under conditions that inhibit cell growth (for example, low serum), cyclin D levels fall. However, when cells are shifted to a growth-permissive environment (for example, serum rich), cyclin D synthesis increases (see Sherr 1993, 1994).

Even after they form, cyclin/CDK complexes are still inert. Activity requires an additional modification that is accomplished by CDK-activating kinase (CAK). This enzyme phosphorylates a threonine residue that lies near position 160 in all known CDK family members (see Solomon 1993). This phosphorylation is necessary to promote a conformational change which results in the relocation of an inhibitory domain (the T-loop) such that it no longer blocks substrate entry into the active site (Debondt et al. 1993; Jeffrey et al. 1995). Although activation by CAK is essential, it has not yet been demonstrated to be a cell cycle regulatory mechanism. Instead, the CAK enzyme appears to be constitutively active under all conditions that have been examined to date.

Potentially active complexes can be further regulated by inhibitory phosphorylation of tyrosine and/or threonine residues near the ATP binding site of the CDK subunit (see Draetta 1990). This mode of regulation has been best demonstrated for the cdc2/cyclin B complex in fission yeast. Fully formed, CAK-phosphorylated cyclin B/cdc2 enzyme is held in an inactive state throughout G2 until the onset of mitosis. At this point, there is an abrupt activation of the kinase that correlates with dephosphorylation of the Thr14 and Tyr15 residues. Control over these inhibitory phosphorylations is exerted by the balance between the activity of inhibitory kinases, wee1 and mik1, and an activating phosphatase, cdc25 (see King et al. 1994). Similar regulatory strategies probably also apply to the diverse cyclin/CDK enzymes in mammalian cells as human homologs of the fission yeast cdc25 and wee1 enzymes have been identified (Galaktionov et al. 1991; Sadhu et al. 1990; Parker et al. 1995).

The final layer of regulation over cyclin/CDK activity (at least for the moment) has emerged with the discovery of CDK-inhibitory proteins. In general, these proteins control cell cycle progression in response to inputs that are extrinsic to the basic cell cycle machine. The two-hybrid approach has played a particularly large role in the identification of this class of regulators, which will be discussed in detail later in the chapter.

The two-hybrid screen has been most often successful in the identification of stable, protein-protein interactions. Perhaps because such interactions are prevalent among components of cell cycle control, cell cycle regulatory proteins have proven amenable to the two-hybrid approach. In the following sections, I will discuss three aspects of cell cycle control in which the two-hybrid technique has been of particular importance. These are the regulation of the G1/S transition by phosphorylation of pRb, global control of cell cycle progression by the p21/p27 family of CDK inhibitors and the role of CAK and KAP in the metabolism of threonine ~160 phosphorylation.

The Rb Family and the Control of G1 Progression.

It has long been clear that the critical decisions concerning continued proliferation, growth arrest, and differentiation are made during the G1 phase of the cell cycle. Since these controls are invariably disrupted, at some level, in tumor cells, regulation of G1 progression has been the subject of intense scrutiny.

At the nexus of the circuitry controlling G1 progression is the Rb protein, the product of the retinoblastoma susceptibility gene (for a detailed review, see Weinberg 1995). In its hypophosphorylated form, as it exists in early G1, Rb acts as a growth inhibitor, and blocks the G1/S phase transition. As cells progress through G1, Rb becomes increasingly phosphorylated and the G1/S blockade is lifted. Rb is then maintained in a phosphorylated form throughout the remainder of the cell cycle until the following

G1 phase, when it is again present in its hypophosphorylated state. Thus, understanding the control of Rb phosphorylation seemed one of the keys to unraveling proliferation control in mammalian cells.

A major step forward came from the realization that phosphorylation of Rb during the G1 phase is accomplished by cyclin-dependent kinases, particularly the cyclin D-associated enzymes, CDK4 and CDK6. Initial studies revealed that a fragment of the Rb protein is an in vitro substrate for these complexes (Kato et al. 1993), and subsequent in vivo experiments have demonstrated that Rb phosphorylation is the only critical role for cyclin D-associated enzymes. Microinjection of cyclin D antibodies could block the progress of normal cells through the division cycle, but only if they were injected prior to the point at which Rb becomes fully phosphorylated. Furthermore, Rb-deficient tumor cells are immune to the effects of the cyclin D antibody (see Weinberg 1995).

The control of Rb phosphorylation by the cyclin D-associated enzymes was intellectually satisfying since it had been previously shown that the abundance of cyclin D is not linked to progress through the cell cycle, as is the case with many of the other cyclins. Instead, the abundance of cyclin D1 responds to the presence of growth stimulatory factors in the extracellular environment that affect the decision to enter the division cycle.

It had been supposed that the loss of proliferation control that accompanies the transformation of a normal cell into a tumor cell must somehow translate into changes in the control of the cell division cycle. The involvement of the cyclin D/CDK4(6)-Rb pathway in tumorigenesis has lent credence to this hypothesis. Rb is a tumor suppressor protein whose inactivation plays a role in a wide variety of tumors (see Weinberg 1995). Also, cyclin D1 is the product of an oncogene; in fact, cyclin D1 was isolated (in one case) as the product of the PRAD1 oncogene (see Sherr 1993, 1994; Hunter and Pines 1994). An additional link between the cyclin D-Rb pathway and tumor suppression stemmed initially from a comparison of cyclin D/CDK4 complexes present in normal and tumor cells. In normal fibroblasts, CDK4 is associated with cyclin D and two additional proteins, PCNA and p21. However, in virally transformed fibroblasts, the exclusive partner of CDK4 is a small protein of ~16 kDa, p16 (Xiong, Zhang, and Beach 1993a). A cDNA clone corresponding to p16 was isolated from a HeLa cell cDNA library using a two-hybrid screen in which CDK4 was the interaction target (Serrano et al. 1993). The testing of the interaction of p16 with a number of DNA-binding domain fusions demonstrated that p16 binds specifically to CDK4 (and CDK6) (Serrano et al. 1993).

In vitro reconstitution experiments revealed that p16 is an inhibitor of CDK4 and CDK6 kinases (Serrano et al. 1993). This finding was unexpected since the p16/CDK4 complex is prominent in rapidly growing, transformed cells (Xiong, Zhang, and Beach 1993a). The resolution to this problem came with the understanding that Rb is the only critical substrate for CDK4(CDK6) kinases. In the cells in which p16 is abundant, Rb is invariably inactivated (see Hirama and Koeffler 1995).

The gene encoding p16 maps to 9p21, a site of frequent chromosomal rearrangement in human tumors and the position of a gene that had been implicated in familial predisposition to melanoma (Kamb et al. 1994; Nobori et al. 1994). Subsequent work has demonstrated that p16 function is absent in a high percentage of tumors derived from a wide variety of cell types (see Hirama and Koeffler 1995). An inactive p16 allele also tracks with early-onset melanoma in susceptible families (Ranade et al. 1995). p16 function may be lost in a number of ways. In some tumors, p16 is homozygously deleted. In others, expression is prevented by methylation of the p16 promoter region. The p16 gene may also contain inactivating point mutations or small deletions or insertions (see Hirama and Koeffler 1995).

The link between loss of p16 function and neoplastic transformation has been cemented by an analysis of the mutant p16 proteins present in tumor cells. Here again, the two-hybrid approach has been exploited. Inhibition of CDK4 and CDK6 by p16 requires the ability of p16 to bind to these enzymes (see, for example, Wick et al. 1995). A number of tumor-specific p16 mutants are impaired for CDK interaction as measured by the two-hybrid assay (Reymond and Brent 1995; Yang et al. 1995). In vitro reconstitution studies with some of these same mutants have confirmed a defect in their ability to inhibit kinase activity (Koh et al. 1995), thus validating the more rapid genetic approach.

p16 is a member of a family of CDK4 and CDK6 inhibitors known as the INK4 family. Presently, this family contains three additional members, p15INK4B, p18INK4C and p19INK4D. Each of these family members was also isolated by the use of the two-hybrid screen. p15 and p18 were found in a screen for proteins that interact with CDK6 (Guan et al. 1994) while p18 and p19 were found in a separate study as CDK4 interactors (Hirai et al. 1995). Finally, p19 was found in a two-hybrid screen in which nur77, an orphan steroid receptor that is required for activation-induced apoptosis in T-cells, was the target (Chan et al. 1995). Of the entire INK4 family, only p15 was also isolated independent of the two-hybrid approach (Hannon and Beach 1994).

Although all members of the INK4 family are biochemically indistinguishable in their ability to inhibit CDK4 and CDK6 kinases, the inhibitors are likely to play different roles in vivo. p15 is clearly one downstream effector of the growth inhibitory cytokine, TGF-β (Hannon and Beach 1994). The abundance of p16 is not affected by TGF-β but is instead greatly increased in Rb-negative tumor cells (see Sherr 1994; Hirama and Koeffler 1995). This result might suggest that p16 acts in a checkpoint mechanism that responds to some aspect of neoplastic transformation; however, the specific signals that cause p16 induction are unknown. The biological roles of p18 and p19 are even less clear at present.

Although the cyclin D-associated kinases are clearly the key mediators of Rb phosphorylation during G1, Rb is also a substrate for other cyclin dependent kinases (see Weinberg 1995). These are viewed as the probable candidates for the kinases that maintain the Rb phosphorylation state throughout the S, G2, and M phases. The fact that Rb must be recycled at the end of each division cycle into an unphosphorylated, growth-inhibitory state predicts the existence of a phosphatase that opposes the Rb kinases. A two-hybrid screen in which Rb was the target has produced the only good candidate for this activity (Durfee et al. 1993). This type 1 protein phosphatase (PP1-α2) binds preferentially to the under-phosphorylated form of Rb in vitro. In vivo, interaction of the phosphatase with Rb occurs during the period extending from mitosis to early G1. This is precisely the time in which Rb dephosphorylation must be accomplished so that the Rb pathway can be reset for the next cell cycle.

Like the INK4 proteins, Rb is only one member of a multigene family. The Rb-related proteins p107 and p130 were first identified by their interaction with the adenoviral oncoprotein, E1A (see Weinberg 1995). p107 was subsequently cloned by a reverse-genetics approach (Ewen et al. 1991) and p130 was isolated both by a similar strategy and by a two-hybrid approach (Hannon et al. 1993; Li et al. 1993). p107 and p130 share many features with Rb but are, on the whole, most similar to each other. Unlike Rb, p107 and p130 have not yet been shown to function as tumor suppressors, although in vitro, all three family members can act as growth suppressors.

Undoubtedly, we have not yet discovered all of the mechanisms through which Rb family members exert control over cell proliferation. However, one mode involves the regulation of genes that are essential for progress through the division cycle (see Weinberg 1995). Rb and its relatives bind to a family of transcription factors known collectively as E2Fs. This binding prevents E2F family members from activating gene expression. Phosphorylation of Rb disrupts Rb/E2F complexes, thus freeing the active transcription factor. The E2F family currently consists of five members: E2F1–E2F5. E2F1, E2F2, and E2F3 bind preferentially to Rb (see Weinberg 1995) while E2F4 and E2F5 prefer p107 and p130. In fact, E2F4 and E2F5 were cloned in two-hybrid screens for proteins that interact with p130 (Hijmans et al. 1995; Vairo et al. 1995; Sardet et al. 1995; Ginsberg et al. 1994; Beijersbergen et al. 1994).

Control of the G1/S transition by the Rb pathway is clearly one of the most critical aspects of proliferation control in normal cells. This importance is attested to by the fact that the pathway is disrupted at some point (for example, Rb, p16, cyclin D1, CDK4) in virtually all human tumors. It should be clear from the preceding section that the two-hybrid screen played a large role in identifying the constituents of this regulatory pathway and in providing clues to their biochemical functions.

Checkpoint Control and the General CDK Inhibitors.

Inhibitors of cyclin dependent kinases can be divided into two classes (see Sherr 1994). INK4 family members specifically inhibit the CDK4 and CDK6 kinases, while the p21 family of CDK inhibitors displays much broader specificity.

The discovery of the general CDK inhibitors represented a remarkable convergence of a number of fields. Arguably, the first notice of p21 was taken during a comparison of CDK-associated proteins in normal and tumor cells (Xiong, Zhang, and Beach 1993a). p21 was seen in association with multiple CDKs in normal fibroblasts but was absent from CDK complexes in virally transformed cells. This observation led to the cloning of p21 by a reverse genetics approach (Xiong et al. 1993). During the same period, a number of other groups worked toward the isolation of the same protein from different perspectives. One group cloned p21 (sdi1) in a search for growth inhibitors that are abundant in senescent cells (Noda et al. 1994). Another isolated p21 (waf1) as a transcriptional target of the p53 tumor suppressor (el-Deiry et al. 1993). Finally, p21 was isolated in a two-hybrid search for proteins that interact with CDK2 (Harper et al. 1993).

In vitro reconstitution experiments quickly revealed that p21, like p16, is a CDK inhibitor, although p21 shows a broader specificity both in vitro and in vivo (Xiong et al. 1993b, Harper et al. 1993). Addition of p21 to CDK2, CDK4, or CDC2 kinases abolishes their activity. Unlike with p16, this inhibition is accomplished without destabilizing the interaction between cyclin and CDK subunits. The precise mechanism of CDK inhibition by p21 is unknown; however, it is clear that p21 binding can prevent the essential activating CDK phosphorylation by CAK (Aprilekova et al. 1995; Harper et al. 1995). p21 can also inhibit CAK-activated complexes, suggesting that the p21 family of inhibitors can block CDK function through multiple mechanisms (Zhang et al. 1994; Harper et al. 1995).

In vivo, p21-bound cyclin/CDK complexes also contain an additional subunit, PCNA (Xiong et al. 1993a). In vitro, PCNA can bind to cyclin/-CDK complexes only in the presence of p21, suggesting a direct p21-PCNA interaction. This interaction was confirmed both by two-hybrid studies in yeast and by biochemical approaches (Waga et al. 1994; G. H., unpublished). PCNA is an essential accessory factor for DNA polymerase and plays two roles in DNA synthesis. First, PCNA recognizes the primer/template junction and facilitates binding of DNA polymerase δ. Second, PCNA enhances the processivity of the polymerase. Direct interaction of p21 with PCNA in vitro inhibits DNA replication by interfering with the role of PCNA as a processivity factor (Waga et al. 1994; Flores-Rozas et al. 1994; Li et al. 1994). In vivo, this might allow p21 to specifically inhibit replicative DNA synthesis without interfering with DNA repair.

Just as the cloning of p21 through its interaction with CDK2 gave clues to the biochemical function of p21, the isolation of the p21 gene as a growth inhibitor from senescent cells and as a p53 target provided insight into the biological role of p21. p53 is a central mediator of the cellular response to DNA damage (see Lee and Bernstein 1995). Damage-induced increases in the abundance of p53 can lead either to apoptosis or to growth arrest, depending upon the context. p53 has also been implicated as one component of the irreversible cell cycle arrest that defines senescence. The fact that a CDK inhibitor is a transcriptional target of p53 provided a satisfactory biochemical explanation for the growth inhibitory aspects of p53 function. Definitive evidence for the involvement of p21 in p53-induced growth arrest came from studies of mouse and human cells with targeted disruptions of the p21 gene (Brugarolas et al. 1995; Waldman et al. 1995; Deng et al. 1995).

The second member of the p21 family of CDK inhibitors was also isolated by both biochemical and two-hybrid approaches. p27kip1 was first sought as a CDK2 inhibitor that is abundant in quiescent cells or in cells that had been arrested following treatment with the growth-inhibitory cytokine, TGF-β (Slingerland et al. 1994; Koff et al. 1993). Subsequent purification and cloning led to the conclusion that p27 is a broad-specificity CDK inhibitor (Polyak et al. 1994). p27 was concurrently isolated in a two-hybrid screen that was designed to identify proteins that interact with cyclin D1 (Toyoshima and Hunter 1994). A third member of the p21 family, p57Kip2, has subsequently been found through a low-stringency hybridization approach (Matsuoka et al. 1995).

As with the INK4 family, members of the p21 family show similar biochemical properties but have distinct biological functions. p27 levels are unaffected following induction of p53; however, interaction of p27 with CDK2 is prominent following treatment of cells with a number of extracellular growth inhibitors (see Sherr 1994). It is as yet unclear precisely how each of these extracellular factors affect cell cycle progression through their effect on p27.

Progress through the division cycle is monitored by oversight mechanisms that respond to cellular damage, extracellular signals, and to intracellular signals which ensure that each cell cycle phase has been successfully completed before the next phase is executed. These controls are broadly categorized as "checkpoints." p21 is clearly one component of the checkpoint mechanism that responds to DNA damage. p27 may function as one part of a checkpoint mechanism that allows entry into the cell cycle only when extracellular conditions are appropriate. With the p21 family, both conventional reverse genetics and two-hybrid approaches have played complementary roles in the discovery of this group of cell cycle regulators.

CAK, KAP, and the Balance of Threonine ~160 Phosphorylation

Structural studies indicated that monomeric cyclin-dependent kinases exist in a conformation that is incompatible with activity (DeBondt et al. 1993). Assumption of an active conformation requires major structural shifts which occur upon cyclin binding and upon phosphorylation of a threonine residue near amino acid 160 in all known CDK family members (DeBondt et al. 1993; Jeffrey et al. 1995). The conserved threonine lies within an inhibitory domain called the T loop which, in the inactive, unphosphorylated state, blocks entry of the substrate into the active site. Phosphorylation of threonine ~160 is thought to cause a conformational change that displaces the inhibitory domain.

The essential CDK phosphorylation is accomplished by an enzyme known as CDK-activating kinase or CAK (see Solomon 1993). This activity was first detected in *Xenopus* oocytes, and a subsequent reverse-genetics approach yielded a kinase subunit that had been previously dubbed MO15. The MO15 kinase subunit is not capable of phosphorylating CDKs on its own, but this finding was not surprising since MO15 is itself a member of the CDK family. This result predicted the existence of a CAK-cyclin subunit; however, this hypothetical protein eluded pursuers for a number of years. The complete human CAK complex finally yielded simultaneously to reverse genetics and two-hybrid approaches (Fisher and Morgan 1994; Makela et al. 1994). Cyclin H was concurrently isolated by two groups who found it to be a member of a growing family of mammalian "C-type" cyclins.

A single CAK (CDK7)-cyclin H complex is capable of phosphorylating all human cyclin/CDK enzymes in vitro, suggesting that a single CAK might activate the entire spectrum of mammalian cell cycle regulatory kinases in vivo (see Sherr 1994). Consistent with this notion, neither cyclin H abundance nor apparent levels of CAK activity vary during the cell cycle.

Individual cyclin/CDK enzymes complete their jobs during a single cell cycle. Following execution of these functions, the enzymes are inactivated, in part, by the destabilization of their cyclin subunits. Although most cyclins are synthesized afresh with each cycle, CDK subunits are recycled in a monomeric, unphosphorylated state. This observation predicted the existence of a phosphatase that would remove the activating phosphate from threonine ~160. A candidate for this enzyme was found as a dual-specificity phosphatase that was initially isolated in two-hybrid screens based upon its ability to interact with human CDC2 and CDK2 (Hannon et al. 1994; Gyuris et al. 1993; Poon and Hunter 1995). This KAP/CDI1 phosphatase appears to serve mainly a recycling function rather than as a foil to CAK, since the phosphatase is only active against monomeric CDK subunits and not against cyclin/CDK complexes (Poon and Hunter 1995).

Table 11-1 Cell cycle regulators cloned by two-hybrid screening.

Function	Gene	Target	Reference
CDK4/CDK6 inhibitors			
	p16	CDK4	Serrano et al. 1993
	p15	CDK6	Guan et al. 1994
	p18	CDK4	Hirai et al. 1995
		CDK6	Guan et al. 1994
	p19	CDK4	Hirai et al. 1995
		Nur77	Chen et al. 1995
Rb family member			
	p130	CDK2	Hannon et al. 1993
Rb phosphatase			
	PP1-α2	Rb	Durfee et al. 1993
Rb-binding transcription factors			
	E2F-4	p130	Sardet et al. 1995
			Beijersbergen et al. 1995
	E2F-5	p130	Hijmans et al. 1995
			Sardet et al. 1995
General CDK inhibitors			
	p21	CDK2	Harper et al. 1993
	p27	cyclin D1	Toyoshima and Hunter, 1994
CAK cyclin			
	cyclin H	CAK	Makela et al. 1994
CDK Thr 161 phosphatase			
	KAP	CDK2	
		CDC2	Hannon et al. 1994
	CDI1	CDK2	Gyuris et al. 1993

Conclusions

In this review, I have detailed three pathways in which the two-hybrid approach has contributed to our understanding of the mechanisms that control the cell division cycle (see Table 11-1). In some cases, the relevant proteins were isolated almost simultaneously by multiple approaches (for example, p21, p130). In others, the two-hybrid screen provided our only access to key regulators (for example, p16, KAP/CDI1). In the past, the identification of new cell cycle regulators required the forethought, patience, and precision of a fly-fisherman. With the development of the two-hybrid system, it sometimes seems that new components of the cell cycle control machinery are being pulled from the depths in bulk. With its relative simplicity, its minimal requirement of time and effort, and its ability to lead in surprising directions, the two-hybrid approach promises to be a useful tool in our ongoing quest for new pieces of the cell cycle puzzle.

References

Aprelikova, O., Xiong, Y., and Liu, E. T. (1995). Both p16 and p21 families of cyclin-dependent kinase (CDK) inhibitors block the phosphorylation of cyclin-dependent kinases by the CDK-activating kinase. J. Biol. Chem. 270:18195–18197.

Beijersbergen, R. L., Kerkhoven, R. M., Zhu, L., Carlee, L., Voorhoeve, P. M., and Bernards, R. (1994). E2F-4, a new member of the E2F gene family, has oncogenic activity and associates with p107 in vivo. Genes Dev. 8:2680–2690.

Brugarolas, J., Chandrasekaran, C., Gordon, J. I., Beach, D., Jacks, T., and Hannon, G. J. (1995). Radiation-induced cell cycle arrest compromised by p21 deficiency. Nature 377:552–557.

Chan, F. K., Zhang, J., Cheng, L., Shapiro, D. N., and Winoto, A. (1995). Identification of human and mouse p19, a novel CDK4 and CDK6 inhibitor with homology to p16INK4. Mol. Cell. Biol. 15:2682–2688.

DeBondt, H. L., Rosenblatt, J., Jancarik, J., Jones, H. D., Morgan, D. O., and Kim, S. H. (1993). Crystal struture of cyclin-dependent kinase 2. Nature 363:595–602.

Deng, C., Zhang, P., Harper, J. W., Elledge, S. J., and Leder, P. (1995). Mice lacking p21CIP1/WAF1 undergo normal development, but are defective in G1 checkpoint control. Cell 82:675–684.

Draetta, G. (1990). Cell cycle control in eukaryotes: molecular mechanisms of cdc2 activation. Trends Biochem. Sci. 15:378–383.

Durfee, T., Becherer, K., Chen, P. L., Yeh, S. H., Yang, Y. Kilburn, A. E., Lee, W. H., and Elledge, S. J. (1993). The retinoblastoma protein associates with the protein phosphatase type 1 catalytic subunit. Genes Dev. 7:555–569.

el-Deiry, W. S., Tokino, T., Velculescu, V.E., Levy, D. B., Parsons, R., Trent, J. M., Lin, D., Mercer, W. E., Kinzler, K. W., and Vogelstein, B. (1993). WAF1, a potential mediator of p53 tumor suppression. Cell 75:817–825.

Ewen, M. E., Xing, Y. G., Lawrence, J. B., and Livingston, D. M. (1991). Molecular cloning, chromosomal mapping, and expression of the cDNA for p107, a retinoblastoma gene product-related protein. Cell 66:1155–1164.

Fisher, R. P., and Morgan, D. O. (1994). A novel cyclin associates with MO15/CDK7 to form the CDK-activating kinase. Cell 78:713–724.

Flores-Rozas, H., Kelman, Z., Dean, F. B., Pan, Z. Q., Harper, J. W., Elledge, S. J., O'Donnell, M., and Hurwitz, J. (1994). Cdk-interacting protein 1 directly binds with proliferating cell nuclear antigen and inhibits DNA replication catalyzed by the DNA polymerase delta holoenzyme. Proc. Natl. Acad. Sci. USA 91:8655–8659.

Galactionov, K., and Beach D. (1991). Specifc activation of cdc25 tyrosine phosphatases by B-type cyclins: evidence for multiple roles of mitotic cyclins. Cell 67:1181–1194.

Ginsberg, D., Vairo, G., Chittenden, T., Xiao, Z. X., Xu, G., Wydner, K. L., DeCaprio, J. A., Lawrence, J. B., and Livingston, D. M. (1994). E2F-4, a new member of the E2F transcription factor family, interacts with p107. Genes Dev. 8:2665–2679.

Guan, K. L., Jenkins, C. W., Li, Y., Nichols, M. A., Wu, X., O'Keefe, C. L., Matera, A. G., and Xiong, Y. (1994). Growth suppression by p18, a p16INK4/MTS1- and p14INK4B/MTS2-related CDK6 inhibitor, correlates with wild-type pRb function. Genes Dev. 8:2939–2952.

Gyuris, J., Golemis, E., Chertkov, H., and Brent, R. (1993). Cdi1, a human G1 and S phase protein phosphatase that associates with Cdk2. Cell 75:791–803.

Hannon, G. J., and Beach, D. (1994). p15INK4B is a potential effector of TGF-beta-induced cell cycle arrest. Nature 371:257–261.

Hannon, G. J., Casso, D., and Beach, D. (1994). KAP: a dual specificity phosphatase that interacts with cyclin-dependent kinases. Proc. Natl. Acad. Sci. USA 91:1731–1735.

Hannon, G. J., Demetrick, D., and Beach, D. (1993). Isolation of the Rb-related p130 through its interaction with CDK2 and cyclins. Genes Dev. 7:2378–2391.

Harper, J.W., Adami, G. R., Wei, N., Keyomarsi, K., and Elledge, S. J. (1993). The p21 Cdk-interacting protein Cip1 is a potent inhibitor of G1 cyclin-dependent kinases. Cell 75:805–816.

Harper, J. W., Elledge, S. J., Keyomarsi, K., Dynlacht, B., Tsai, L. H., Zhang, P., Dobrowolski, S., Bai, C., Connell-Crowley, L., Swindell, E., Fox, M. P., and Wei, N. (1995). Inhibition of cyclin-dependent kinases by p21. Mol. Biol. Cell 6:387–400.

Hijmans, E. M., Voorhoeve, P. M., Beijersbergen, R. L., vant Veer, L.J., and Bernards, R. (1995). E2F-5, a new E2F family member that interacts with p130 in vivo. Mol. Cell. Biol. 15:3082–3089.

Hirai, H., Roussel, M. F., Kato, J. Y., Ashmun, R. A., and Sherr, C. J. (1995). Novel INK4 proteins, p19 and p18, are specific inhibitors of the cyclin D-dependent kinases CDK4 and CDK6. Mol. Cell. Biol. 15:2672–2681.

Hirama, T., and Koeffler, P. H. (1995). Role of the cyclin-dependent kinase inhibitors in the development of cancer. Blood 86:841–854.

Hunter, T., and Pines J. (1994). Cyclins and Cancer II : Cyclin D and CDK inhibitors come of age. Cell 79:573–582.

Jeffrey, P. D., Russo, A. A., Polyak, K., Gibbs, E., Hurwitz, J., Massague, J., and Pavletich, N. P. (1995). Mechanism of CDK activation revealed by the structure of a cyclinA-CDK2. Nature 376:313–320.

Kamb, A., Gruis, N. A., Weaver-Feldhaus, J., Liu, Q., Harshman, K., Tavtigian, S. V., Stockert, E., Day, R. S., Johnson, B. E., and Skolnick, M. H. (1994). A cell cycle regulator potentially involved in genesis of may tumor types. Science 264:436–440.

Kato, J., Matsushime, H., Hiebert, S. W., Ewen, M. E., and Sherr, C. J. (1993). Direct binding of cyclin D to the retinoblastoma gene product (pRb) and pRb phosphorylation by the cyclin D-dependent kinase CDK4. Genes Dev. 7:331–342.

King, R. W., Jackson, P. K., and Kirschner, M. W. (1994). Mitosis in transition. Cell 79:563–571.

Koff, A., Ohtsuki, M., Polyak, K., Roberts, J. M., and Massague, J. (1993). Negative regulation of G1 in mammalian cells: inhibition of cyclin E-dependent kinase by TGF-beta. Science 260:536–539.

Koh, J., Enders, G. H., Dynlacht, B.D., and Harlow, E. (1995). Tumour-derived p16 alleles encoding proteins defective in cell-cycle inhibition. Nature 375:506–510.

Lee, J. M. and Bernstein, A. (1995) Apoptosis, cancer, and the p53 tumour suppressor gene. Cancer Metastasis Rev. 14:149–161.

Lew, D. J. and Reed, S. I. (1993). A proliferation of cyclins. Trends Cell Biol. 2:77–81.

Li, R., Waga, S., Hannon, G. J., Beach D., and Stillman, B. (1994). Differential effects by the p21 CDK inhibitor on PCNA-dependent DNA replication and repair. Nature 371:534–537.

Li, Y., Graham, C., Lacy, S., Duncan, A. M., and Whyte, P. (1993). The adenovirus E1A-associated 130-kD protein is encoded by a member of the retinoblastoma gene family and physically interacts with cyclins *A* and *E*. Genes Dev. 7L2366–2377.

Makela, T. P., Tassan, J. P., Nigg, E. A., Frutiger, S., Hughes, G. J., and Weinberg, R. A. (1994). A cyclin assiciated with the CDK activating kinase, MO15. Nature 371:254–257.

Matsuoka, S., Edwards, M. C., Bai, C., Parker, S., Zhang, P., Baldini, A., Harper, J. W. and Elledge, S. (1995). p57Kip2, a structurally distinct member of the p21Cip1 CDK inhibitor family is a candidate tumor suppressor gene. Genes Dev. 9:650–662.

Nobori, T., Miura, K., Wu, D. J., Lois, A., Takabayashi, K., and Carson, D. A. (1994). Deletions of the cyclin-dependent kinase-4 inhibitor gene in multiple human cancers. Nature 368:753–756.

Noda, A., Ning, Y., Venable, S. F., Pereira-Smith, O. M., and Smith J. R. (1994). Cloning of senescent cell-derived inhibitors of DNA synthesis using an expression screen. Exp. Cell Res. 211:90–98.

Parker, L. L., Sylvestre, P. J., Byrnes, M. J., Liu, F., and Piwnica-Worms, H. (1995). Identification of a 95-kDa WEE1-like tyrosine kinase in HeLa cells. Proc. Natl. Acad. Sci. USA 92:9638–9642.

Polyak, K., Kato, J. Y., Solomon, M. J., Sherr, C. J., Massague, J., Roberts, J. M., and Koff A. (1994). p27Kip1, a cyclin-Cdk inhibitor, links transforming growth factor-beta and contact inhibition to cell cycle arrest. Genes Dev. 8:9–22.

Poon, R. Y., and Hunter, T. (1995). Dephosphorylation of Cdk2 Thr160 by the cyclin-dependent kinase-interacting phosphatase KAP in the absence of cyclin. Science 270:90–93.

Ranade, K., Hussussian, C. J., Sikorski, R. S., Varmus, H. E., Goldstein, A. M., Tucker, M. A., Serrano, M., Hannon, G. J., Beach, D., and Dracopoli, N. C. (1995). Mutations associated with familial melanoma impair p16INK4 function. Nat. Genet. 10:114–116.

Reymond, A., and Brent, R. (1995) p16 proteins from melanoma-prone families are deficient in binding to Cdk4. Oncogene 11:1173–1178.

Sadhu, K., Reed, S. I., Richardson, H., and Russell, P. (1990). Human homolog of fission yeast cdc25 mitotic inducer is predominantly expressed in G2. Proc. Natl. Acad. Sci. USA 87:5139–5143.

Sardet, C., Vidal, M., Cobrinik, D., Geng, Y., Onufryk, C., Chen, A., and Weinberg, R. A. (1995). E2F-4 and E2F-5, two members of the E2F family, are expressed in the early phases of the cell cycle. Proc. Natl. Acad. Sci. USA 92:2403–2407.

Serrano, M., Hannon, G. J., and Beach, D. (1993). A new regulatory motif in cell-cycle control causing specific inhibition of cyclin D/CDK4. Nature 366:704–707.

Sherr, C. J. (1994). G1 phase progression: cycling on cue. Cell 79:551–555.

Sherr, C. J. (1993). Mammalian G1 cyclins. Cell 73:1059–1065.

Slingerland, J. M., Hengst, L., Pan, C. H., Alexander, D., Stampfer, M.R., and Reed, S. I. (1994). A novel inhibitor of cyclin-Cdk activity detected in transforming growth factor beta-arrested epithelial cells. Mol. Cell. Biol. 14:3683–3694.

Solomon, M. J. (1993). Activation of the various cyclin/cdc2 protein kinases. Current Opin. Cell Biol. 5:180–186.

Toyoshima, H., and Hunter, T. (1994). p27, a novel inhibitor of G1 cyclin-Cdk protein kinase activity, is related to p21. Cell 78:67–74.

Vairo, G., Livingston, D. M., and Ginsberg, D. (1995). Functional interaction between E2F-4 and p130: evidence for distinct mechanisms underlying growth suppression by different retinoblastoma protein family members. Genes Dev. 9:869–881.

Waga, S., Hannon, G. J., Beach, D., and Stillman, B. (1994). The p21 inhibitor of cyclin-dependent kinases controls DNA replication by interaction with PCNA. Nature 369:574–578.

Waldman, T, Kinzler, K. W., and Vogelstein, B. (1995). p21 is necessary for the p53-mediated G1 arrest in human cancer cells. Cancer Res. 55:5187–5190.

Weinberg, R. A. (1995). The retinoblastoma protein and cell cycle control. Cell 81:323–330.

Wick, S. T., Dubay, M. M., Imanil, I., and Brizuela, L. (1995). Biochemical and mutagenic analysis of the melanoma tumor suppressor gene product/p16. Oncogene 11:2013–2019.

Xiong, Y., Hannon, G. J., Zhang, H., Casso, D., Kobayashi, R., and Beach, D. (1993). p21 is a universal inhibitor of cyclin kinases. Nature 366:701–704.

Xiong, Y., Zhang, H., and Beach, D. (1993). Subunit rearrangement of the cyclin-dependent kinases is associated with cellular transformation. Genes Dev. 7:1572–1583.

Yang, R., Gombart, A. F., Serrano, M., and Koeffler, H. P. (1995). Mutational effects on the p16INK4a tumor suppressor protein. *Cancer Res.* 55:2503–2506.

Zhang, H., Hannon, G. J., and Beach, D. (1994). p21-containing cyclin kinases exist in both active and inactive states. Genes Dev. 8:1750–1758.

12

Two-Hybrid Analysis of Genetic Regulatory Networks

Russell L. Finley, Jr.
Roger Brent

There is a great need for general methods to characterize the proteins that contemporary biology makes available. The list of such proteins needing further characterization is growing and includes proteins already known to be important for specific cellular functions, mutant proteins identified in vivo or made in vitro, and very large numbers of proteins being identified by genome projects. Here we describe the extension of two-hybrid approaches to bear on this problem.

The success of two-hybrid systems is due to the fact that many cellular functions are carried out by proteins that touch one another. For example, the complex process of transcription initiation requires the ordered assembly of numerous interacting transcription factors with RNA polymerase and ancillary proteins into a protein machine that initiates transcription (Guarente 1996, Tjian and Maniatis 1994). This machine can be viewed as a network of interacting proteins, as can the machines that control other processes, such as DNA replication, protein translation, and the cell cycle. A full understanding of these processes will require knowledge not only of the proteins (parts) that make up each machine, but of the topological relationships (connections) that individual parts make with one another. Similarly, a full understanding of the function of any new protein will require knowledge of the interactions it makes with previously identified proteins. Currently, most new proteins are being identified by large-scale sequencing projects. For many of these new proteins, the sequence alone sheds little or no light on their function.

Two-hybrid systems have been used to probe the function of new proteins ever since they were developed (Fields and Song 1989; Chien et al. 1991). The first applications examined the interactions between proteins isolated by two-hybrid methods and relatively small numbers of test proteins (see, for example, Durfee et al. 1993; Gyuris et al. 1993; Harper et al. 1993; Zervos et al. 1993), but their use quickly spread to the analysis of many other proteins (Van Aelst et al. 1993; Yuan et al. 1993; Choi et al. 1994; Kranz et al. 1994; Marcus et al. 1994; Printen and Sprague, 1994). In anticipation of the utility of applying these methods to larger sets, we and others began devising ways to do so.

Larger scale two-hybrid approaches typically rely on interaction mating. In this method, the protein fused to the DNA-binding domain (the bait) and the protein fused to the activation domain are expressed in two different haploid yeast strains of opposite mating type (**a** and α), and the strains are mated to determine if the two proteins interact. Mating occurs when the two strains come into contact, and results in fusion to form a diploid yeast strain. Thus, an interaction can be determined by measuring activation of a two-hybrid reporter gene in the diploid strain.

As described later, interaction mating has been used to examine interactions between small sets of tens of proteins (Finley and Brent 1994; Finley and Brent 1995; Reymond and Brent 1995), larger sets of hundreds of proteins (R.L.F. and R.B., unpublished), to screen libraries (Bendixen et al. 1994), and to attempt to comprehensively map connections between proteins encoded by a small genome (Bartel et al. 1996). The primary advantage of this technique is that it reduces the number of yeast transformations needed to test individual interactions. For example, to test for interactions between a set of ten bait proteins and five prey proteins without interaction mating would require fifty transformations to create fifty strains that carry the pair-wise combinations of baits and preys. With mating, however, only fifteen transformations would be needed; ten for the different bait plasmids, and five for the different prey plasmids; and the resulting two sets of transformants would be mated to create the fifty combinations. The microbiology of the mating procedure (which is extremely simple) is detailed in the next section.

Interaction mating can be used to characterize small sets of proteins (as described in Protocol 1). In one example of this approach, we used interaction mating to characterize a set of seven *Drosophila* cyclin-dependent kinase (Cdk) interactors, or Cdis (Finley and Brent 1994). Strains expressing versions of the Cdis fused to an activation domain were mated with 74 different strains expressing different bait proteins, including Cdks from other species and four of the Cdis themselves. The results from this study illustrate the types of information that can be derived from such a characterization. First, the experiments showed that some of the Cdis interacted with different subgroups of seven highly related Cdk baits, suggesting that the Cdis recognize structural features shared by these Cdks but absent in the noninteracting Cdks; inspection of an alignment of the Cdk protein sequences suggested residues that may be important for specific interactions with certain Cdis. Second, Cdi3, *Drosophila* cyclin D, interacted much more strongly with human Cdk4 than with any of the other Cdks in the panel including the *Drosophila* Cdks, suggesting that there may be an as yet unidentified *Drosophila* Cdk4 homolog which is the true partner for cyclin D. Third, two of the Cdis interacted with two other Cdis, indicating in each instance that each Cdi has surfaces for binding to the Cdk and to another Cdi, and suggesting that these proteins form ternary or higher order complexes. Finally, the demonstration that two Cdis with no sequence similarity to previously identified proteins interacted with each other as well as with the Cdk, but not with a panel of over 60 other proteins, provided an additional clue to their functions, strongly supporting the idea that they function along with the Cdk in the network of proteins that regulates the cell cycle. These results demonstrate that examination of the interactions between even small numbers of proteins can provide a number of functional insights. Much larger sets of proteins can be characterized by scaling up these procedures as described below.

Interaction Mating

In this section, we present methods for performing interaction mating assays on small or large sets of proteins using the interaction trap, and in the next section we discuss use of interaction mating with other two-hybrid systems. The interaction trap (see chapter 4 and references therein) uses the *Escherichia coli* protein LexA as the DNA-binding domain and a protein encoded by a random *E. coli* sequence, the B42 "acid blob," as the transcription activation domain. Both proteins are expressed from multicopy (2μ) plasmids; the LexA fusion, or bait, is expressed from a plasmid containing the *HIS3* marker, and the activation domain-fused protein, or prey, is expressed from a plasmid containing the *TRP1* marker. In the most commonly used bait plasmid, pEG202, the bait is expressed from the constitutive yeast *ADH1* promoter. Related bait plasmids are available that express the bait fused to a nuclear localization signal (pNLex, see chapter 4), or that express the bait conditionally from the *GAL1* promoter (pGILDA, D. Shaywitz and C. Kaiser, personal communication). The most commonly used prey plasmid, pJG4-5, expresses proteins fused to the B42 activation domain, the SV40 nuclear localization signal, and an epitope tag derived from the influenza virus hemagglutinin, all driven by the yeast *GAL1* promoter, which is active only in yeast grown on galactose (Gyuris et al. 1993). Use of the *GAL1* promoter to express the prey allows toxic proteins to be expressed transiently and helps eliminate many false positives in interactor hunts.

The interaction trap uses two reporter genes that carry upstream LexA binding sites (operators): *LEU2* and *lacZ*. The *LEU2* reporters are integrated into the yeast genome and the *lacZ* reporters typically reside on 2μ plasmids bearing the *URA3* marker, though integrated versions are also available (R.L.F., R.B., S. Hanes, unpublished data). Several versions of the *LEU2* and *lacZ* reporters have been made that have a range of sensitivities based on the number of upstream LexA operators. In general, the *LEU2* reporters are more sensitive to a given interacting pair of proteins than the *lacZ* reporters (Estojak et al. 1995); recently, however, highly sensitive *lacZ* reporters have been used that contain several LexA operators and transcription terminator sequences downstream of the *lacZ* gene (S. Hanes, personal communication).

Several different combinations of strains, plasmids, and reporters can be used for mating. In one common version (Finley and Brent 1994), the strain expressing the bait (bait strain) is RFY206 (*MATa ura3-52 his3Δ200 leu2-3 lys2Δ201 trp1::hisG*) transformed with the *HIS3* bait plasmid and a *URA3 lacZ* reporter plasmid like pSH18-34. The strain expressing the activation domain-tagged protein (prey strain) is EGY48 (*MATα ura3 his3 leu2::3LexAop-LEU2 trp1 LYS2*) transformed with the *TRP1* prey plasmid. Patches of these two strains on agar plates are brought into contact by replica plating (see next section) and grown on a rich medium overnight. During this time, cells in the patches mate and fuse to form diploids. The cells are then transferred by replica-plating to plates on which only diploids can grow; these plates lack uracil, histidine, and tryptophan so that neither parental haploid can grow on them. In the protocols presented here, the *lacZ* reporter is measured using diploid selection indicator plates containing X-Gal, a chromogenic substrate for the *lacZ* gene product. However, it is worth mentioning that expression of the *LEU2* reporter can also be easily scored by putting the diploids on plates that lack leucine, and that in the future other reporters will likely be available. Furthermore, because both reporter genes exhibit a reduced sensitivity in diploid strains compared to haploid strains, the most sensitive versions of the *lacZ* or *LEU2* reporters are recommended for interaction mating assays.

Variants of this simple procedure are sometimes useful. In particular, because some baits activate transcription by themselves, it is often useful to conditionally express the activation domain hybrid so that one scores patches that show an *increase* in reporter gene expression in the presence of this hybrid. To do this, place the diploids on two different X-Gal plates, one that contains galactose, which results in expression of the prey, and one that contains glucose, which represses expression of the prey. Here, an interaction between the bait and prey will be detected when the diploid yeast containing them turn more blue on the galactose X-Gal plate than on the glucose X-Gal plate.

Interaction Mating—Small Scale

It is often informative to look for interactions between small sets of proteins, or between a given protein and a test set of ten to a hundred proteins. The test set, for example, might contain different allelic forms of the original bait, sets of structurally related proteins, sets of proteins known or suspected to be involved in some process, or unrelated proteins used to demonstrate the specificity of an interaction. Protocol 1 describes a convenient method to test small sets of proteins for interactions.

Protocol 1. Mating Assay—Small Scale for Tens of Different Bait or Prey Proteins

Materials

Bait strains: *S. cerevisiae* strain RFY206 (*MATa ura3-52 his3Δ200 leu2-3 lys2Δ201 trp1::hisG*) transformed with a *URA3* plasmid containing a *lacZ* reporter, such as pSH18-34, and various *HIS3* bait plasmids, such as derivatives of pEG202 that produce different LexA fusions. Each bait strain will contain a different bait plasmid.

Prey strains: *S. cerevisiae* strain EGY48 (*MATa ura3 his3 leu2*::3LexAop-*LEU2 trp1 LYS2*) transformed with *TRP1* prey plasmids, such as derivatives of pJG4-5 that produce different activation domain-tagged proteins or preys.

Sterile wooden applicator sticks (for example, FisherBrand 01-340).

Minimal glucose yeast plates lacking uracil and histidine (Glu/CM-Ura-His) (see chapter 4).

Minimal glucose plates lacking tryptophan (Glu/CM-Trp) (see chapter 4).

YEPD plates (see chapter 4).

Minimal X-Gal glucose plates lacking uracil, histidine, and tryptophan (Glu/CM-Ura-His-Trp X-Gal) (see chapter 4).

Minimal X-Gal galactose/raffinose plates lacking uracil, histidine, and tryptophan (Gal/Raff CM-His-Trp X-Gal) (see chapter 4).

Replica plater and sterile replica velvets.

Optional

Minimal glucose plates lacking uracil, histidine, tryptophan, and leucine (Glu/CM-Ura-His-Trp-Leu) (see chapter 4).

Minimal galactose/raffinose plates lacking uracil, histidine, tryptophan, and leucine (Gal/Raff CM-Ura-His-Trp) (see chapter 4).

Method

1. Streak different bait strains in horizontal parallel stripes on a Glu/CM-Ura-His plate. Streaks should be at least 3 mm wide and at least 5 mm apart, with the first streak starting about 15 mm from the edge of the plate. A 100 mm plate will hold eight different bait strains. Create a duplicate plate of bait strains for each different plate of prey strains to be used.
2. Likewise, streak different prey strains in vertical parallel stripes on a Glu/CM-Trp plate. As a control for baits that may activate transcription, include a prey strain that contains the prey vector pJG4-5 not encoding a fusion

protein (that is, encoding only the activation domain). Create a duplicate plate of prey strains for each plate of bait strains to be used.

3. Incubate plates at 30°C until there is heavy growth on the streaks. When taken from reasonably fresh cultures, for example plates that have been stored at 4°C for less than a month, streaked RFY206-derived bait strains take about 48 hours to grow and EGY48-derived prey strains take about 24 hours.
4. Press a plate of prey strains to a replica velvet, evenly and firmly so that yeast from all along each streak are left on the velvet. This plate may be reused if necessary. Press a plate of bait strains to the same replica velvet. This plate of bait strains cannot be reused as it is now contaminated with prey strains.
5. Lift the impression of the bait and prey strains from the velvet by pressing a YEPD plate on it. Incubate the YEPD plate for 24 hours at 30°C.
6. Replica YEPD plates to the following diploid selection indicator plates: Glu/CM-Ura-His-Trp X-Gal, Gal/Raff CM-Ura-His-Trp, and (optional: Glu/CM-Ura-His-Trp-Leu, and Gal/Raff CM-Ura-His-Trp-Leu). The YEPD plate should contain sufficient growth to enable a single impression on the velvet to be lifted by at least four indicator plates.
7. Patch control strains (see text of this chapter) onto the indicator plates and incubate at 30°C. Examine results daily. Diploids will grow and blue color will develop within 2 days.

Interaction Mating—Large Scale

With a few modifications, the procedure just described can be used to test for interactions between a single prey protein and hundreds of baits (Protocol 3, Figure 12-1). Large panels of bait strains can be collected and stored frozen indefinitely (Protocol 2) and then screened against any number of preys. One such set of bait strains contains over 700 different LexA-fusion proteins from our own work and from numerous other labs that use the interaction trap (R.L.F., R.B., A. Reymond, unpublished data). Screening a protein against such a panel enables one to quickly test its ability to interact with a large number of known proteins, most of which have been characterized to some extent, and have been chosen for study because of their known or suspected involvement in some biological process. Thus, the finding of an interaction between a tested protein and a member of the panel can often lead to immediate clues about the biological function of both proteins. While the number of proteins in the existing panel is far less than the number of proteins in a good library, this approach does offer the advantage of screening the test protein against a set of proteins enriched for those of current interest to the biological community. It is worth noting that these proteins come from many different organisms in which they are expressed in different tissues and at different developmental stages. Thus, it becomes possible to identify interacting partners that have not yet been isolated from the same species, or that are not expressed in tissues from which interaction libraries have been made.

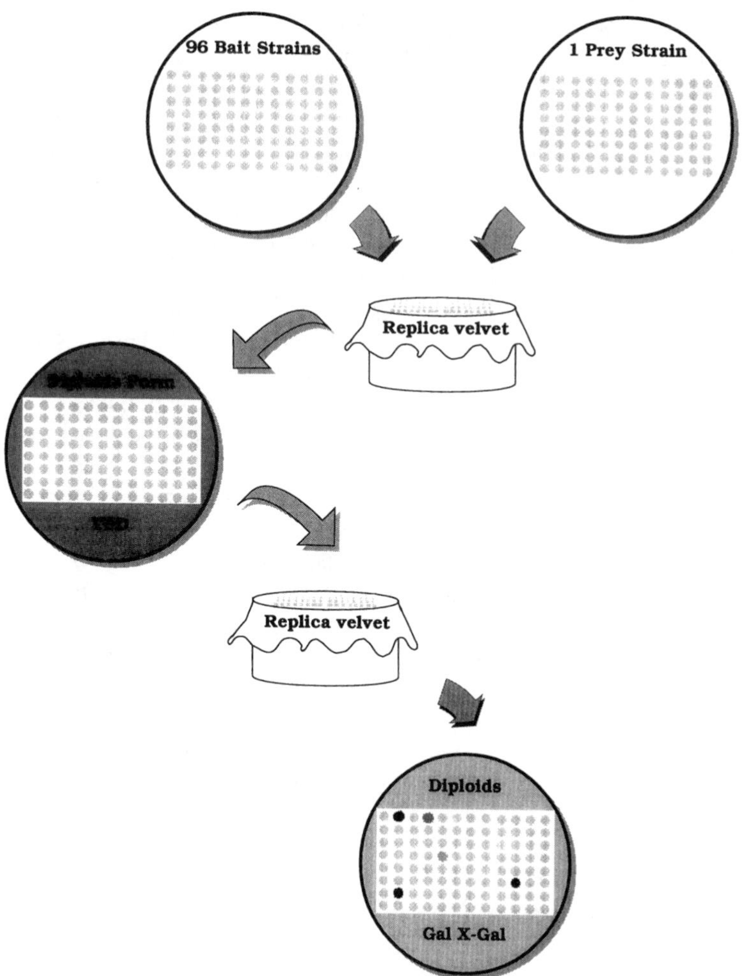

Fig. 12-1. (Top.) The plate on the left holds 96 different yeast strains in patches (or colonies) that each express a different bait protein. The plate on the right holds 96 patches, each of the same yeast strain (prey strain) that expresses a protein fused to an activation domain (prey). The plate of bait strains and the plate of prey strains are each pressed to the same replica velvet and the impression is lifted with a plate containing YPD medium. After one day of growth on the YPD plate, during which time the two strains mate to form diploids, the YPD plate is pressed to a new replica velvet and the impression is lifted with a plate containing diploid selection medium and an indicator like X-Gal. Blue patches (dark spots) on the X-Gal plate indicate that the *lacZ* reporter is transcribed, suggesting that the prey interacts with the bait at that location.

For some proteins, this approach offers additional advantages over screening a library using a traditional two-hybrid scheme. Proteins that activate transcription when fused to LexA or another DNA-binding domain can be difficult to use in conventional interactor hunts. Though methods are available to reduce the sensitivity of the reporter genes (Durfee et al. 1993; Estojak et al. 1995; chapters 2, 3, and 4), it is not always possible to reduce the reporter sensitivity below the threshold of activation for some baits. Moreover, reduction in reporter sensitivity carries with it the risk that the reporters will not detect weakly interacting proteins. Furthermore, spontaneously occurring yeast mutations, for example, those that increase the copy number of the bait plasmid, can increase the activating potential of weakly activating baits (R.L.F., R.B., A. Mendelsohn, unpublished data); strains carrying such mutations are typically scored as positive in the early stages of an interactor hunt, and they are not readily detected in schemes where the specificity test is performed by removing the bait plasmid from the strain containing the prey and mating the strain with other bait strains. Thus, an alternative for proteins that activate transcription as baits is to use them as preys to screen existing panels of baits, or even libraries of baits. Interaction mating approaches also have clear advantages for proteins that are somewhat toxic to yeast; the prey vector allows conditional expression of toxic proteins in the presence of a bait, and often the interaction can be observed as the reporters are activated even if the cells are inviable. An example of the use of interaction mating together with a large panel of bait strains to characterize a protein that both activates transcription and is toxic to yeast, *Drosophila* cyclin E (unpublished data), is discussed later in this chapter.

Protocol 2: Collecting Bait (and Prey) Strains

Materials

Freezing media: 1:1 solution of minimal glucose media lacking appropriate amino acids (for example, Glu/CM-Ura-His for bait strains) : sterile glycerol solution [65% (v/v) glycerol, 0.1 M $MgSO_4$, 25 mM Tris-HCl pH 7.4].

1.0 to 1.5 ml cryotubes.

Yeast strains freshly streaked to minimal glucose plates.

Sterile wooden applicator sticks.

Methods

1. Streak bait strains to Glu/CM-Ura-His plates, or prey strains to Glu/CM-Trp plates, and incubate at 30°C for 24 to 48 hours. Yeast should be taken from the plates and frozen no more than 4 days after being streaked.
2. With a sterile wooden applicator stick, take a dollop of yeast from the plates and inoculate 0.5 ml of freezing solution in a cryotube. Vortex lightly. This solution should have an OD_{600} over 3.0.
3. Alternatively, inoculate 0.5 ml of Glu/CM-Ura-His liquid media to an OD_{600}

less than 0.2, incubate at 30°C with shaking until OD_{600} = 1.5 to 2.0 (log phase), and add 0.25 ml of this culture to 0.25 ml of sterile glycerol solution in a cryotube.

4. Freeze by placing cryotubes in –80°C freezer. Most strains can be recovered after at least 2 years by scraping the surface of the ice and streaking to minimal glucose plates. Avoid allowing entire contents of cryotube to thaw.

Protocol 3: Mating Assay—Large Scale for Hundreds of Different Bait or Prey Strains.

Materials

Freshly streaked bait and prey strains (see Protocol 1).

One set of the following 150 × 15 mm plates for each test of interactions between an activation domain-tagged protein (in a prey strain) and 96 baits (bait strains): Glu/CM-Ura-His; Glu/CM-Trp; YEPD; Glu/CM-Ura-His-Trp X-Gal; Gal/Raff CM-Ura-His-Trp X-Gal.

Replica device and sterile velvets for 150 mm diameter plates. (A replica device can be fashioned from a box of 200 µl pipette tips by stretching a velvet over the top of the box.)

96-prong device (for example, DanKar MC-96) with 3 mm diameter flat-ended metal prongs in a 96-well configuration. Similar devices can be used in 48-well configurations for use with 100 mm plates.

0.5 to 4.0 ml sterilized tubes arranged in a 96-well configurations (for example, cluster tubes such as Costar #4411). Ideally these tubes can be capped and frozen at –80°C.

Glu/CM-Ura-His liquid media (see chapter 4).

Glu/CM-Trp liquid media (see chapter 4).

Sterile glycerol solution [65% (v/v) glycerol, 0.1 M $MgSO_4$, 25 mM Tris-HCl pH 7.4].

Methods

1. It is most convenient to place large numbers of bait strains in a 96-well configuration (Figure 12-1). This can be done by inoculating 2 ml of Glu/CM-Ura-His media in cluster tubes and growing to OD_{600} = 1.5 to 2.0. After making plates from these cultures (see step 2), add an equal volume of sterile glycerol solution, cap and freeze at –80°C.
2. Use the 96-prong device, sterilized in ethanol and flame, to transfer bait strains from the culture to the center of a 150 mm Glu/CM-Ura-His plate. Each plate can contain 96 different bait strains. Tens of identical plates can be made from one culture. Incubate the plates at 30°C for 48 hours or until all bait strains have grown to colonies 5 mm in diameter. These plates can be stored at 4°C for up to 2 months and used to inoculate another liquid culture when more plates are needed. Several positions on each plate should contain control strains with baits that activate various levels of transcription (see Recording the Results and Table 12-1).
3. Inoculate 50 ml of Glu/CM-Trp liquid media with a prey strain and grow at 30°C with shaking to OD_{600} = 1.5 to 3.0. Pour the culture into a sterile 150 mm plate, or into the sterile top from a box of 200 µl pipettes, and use the 96-

prong device, sterilized in ethanol and flame, to transfer the culture to Glu/CM-trp plates. On these plates, all 96 positions will contain the same prey strain.

4. Follow the replica-plating procedure from Protocol 1 to combine the bait and prey strains to a YEPD plate, and then after growth on the YEPD plate at 30°C for 24 hours, replica to X-Gal indicator, diploid selection plates (Glu/CM-Ura-His-Trp X-Gal and Gal/Raff CM-Ura-His-Trp X-Gal) (Figure 12-1).
5. Examine results after 2 days.

Interaction Mating Assay with Other Yeast Two-Hybrid Systems

In addition to the interaction trap, many other yeast two-hybrid systems have been developed (see Allen et al. 1995; Fields and Sternglanz 1994; Mendelsohn and Brent 1994, for reviews). All of these allow the analysis of individual protein-protein interactions and permit interactor hunts to isolate new proteins that interact with a bait. In some instances, plasmids or strains from one system can be used in another, but in other cases the components are incompatible because the yeast selectable markers on the different components differ. In addition, systems that use Gal4p as the DNA-binding domain cannot be used with yeast strains that have a wild-type *GAL4* gene, and therefore, since the Gal4 protein is required to activate the *GAL1* promoter, it cannot be used with systems that use the *GAL1* promoter to drive expression of the prey protein. Finally, use of interaction mating requires careful attention to the mating types of the strains and the selectable markers used to select the diploids.

Recording the Results

Interaction between bait and prey results in the interaction phenotypes: growth of the strain on medium lacking leucine, and transcriptional activation of the *lacZ* reporter and production of active β-galactosidase. On X-Gal plates, the β-galactosidase cleaves the X-Gal substrate, producing a product that turns the yeast colony blue. The amount of color provides a fast and simple method to approximate the level of *lacZ* expression in a strain. An interaction is scored when a diploid colony is more blue on the X-Gal plate containing galactose than on the X-Gal plate containing glucose.

Scoring these interactions benefits from inclusion of a number of controls. To control for common variations between the X-Gal plates, include control strains that contain baits that activate transcription to varying extents. Table 12-1 shows some baits with known activating abilities. Inclusion of such strains on every X-Gal plate enables one to normalize the amount of blue produced by an interaction. It is also useful to include a control strain to check that the plates contain the correct carbon sources, and to ensure that the *GAL1* promoter which drives the expression of the prey

Table 12-1 Activating and Nonactivating Baits.

Bait	Activation	Color on X-Gal[a]	Plasmid name	Reference
LexA-Gal4	strong	blue	pSH17-4	S. Hanes unpublished; Golemis and Brent 1992
LexA-Fos	strong	blue	VR1001	Lech et al. 1988; Golemis and Brent 1992
LexA-Myc	strong	blue	VR1004	Lech et al. 1988; Golemis and Brent 1992
LexA-Cdi2	moderate	light blue	pRF202-Cdi2	Finely and Brent 1994
LexA-HsCycC	moderate	light blue	pJG28-2	J. Gyuris and R. B. unpublished data
LexA-DmErk	moderate	light blue	pEG202-ERK	Zavitz, Zipursky, Finley, and Brent unpublished data
LexA-Mxi1	weak	very light blue	pLexA-Mxi1	Zervos et al. 1993
LexA-NF2 (N)	weak	very light blue	pAR-NF2(N-term)	A. Reymond and R. B. unpublished data
LexA-DmCdi11	weak	very light blue	pRF202-Cdi11	Finely and Brent 1994
LexA-BcdΔC	none	white	pRFHM1	Gyuris et al. 1993
LexA-Max	none	white	pLexA-Max	Zervos et al. 1993
LexA-HsCdc2	none	white	pJG14-1	Gyuris et al. 1993

[a] As determined on typical fresh X-Gal galactose/raffinose plates in a diploid strain containing pSH18-34 (S. Hanes unpublished data), the sensitive LexAop-*lacZ* reporter plasmid (R. Finley unpublished data).

protein is activated on the Gal/Raff plates and not on the Glu plates. An ideal control of this nature consists of a diploid strain, derived from a mating assay, that expresses an interacting pair of bait and prey proteins, such as any one of a number of well-characterized interacting pairs (Finley and Brent 1994; Gyuris et al. 1993; Zervos et al. 1993). An alternative to using X-Gal plates is to perform a filter lift assay for β-galactosidase activity in grown diploid colonies (chapters 2, 3, and 4). Finally, every bait should be tested to see if, and how much, it activates transcription in the absence of a prey, which can be simply accomplished by mating the bait strains to a strain containing the empty prey vector. Thus, a true interaction with a prey protein is scored when the amount of galactose-dependent activation of the *lacZ* reporter (for example, amount of blue) exceeds the amount produced in the absence of a prey.

Interpreting Interaction Data

Qualitative Interpretation

For large amounts of information flowing from interaction mating experiments, the problem of determining whether individual interactions are meaningful is multiplied. We consider a number of these separately.

True and False Positives Most interactions with affinity tighter than a K_D value of 10^{-6} will be detected. Although there may exist a weak positive correlation between apparent strength of binding and biological significance, many apparently weak interactions are real while some strong ones are not. The problem of determining which interactions have biological significance is therefore not trivial. At the moment, the most satisfying way to show biological significance is to verify the interaction by a different, biochemical technique, preferably coprecipitation from a cell in which both proteins are expressed. However, the interaction data alone can often point out probable true and false positives. For example, our experience indicates that highly specific interactions, such as between a protein that binds to one or a small set of highly related proteins and not to hundreds of unrelated proteins, are good candidates to pursue as biologically relevant. Conversely, we tend to give less weight to interactions between proteins that are so ubiquitous in the life of the cell (for example, members of the ubiquitin system or heat shock proteins) that the interactions might be meaningful but relatively uninformative.

True and False Negatives A problem less frequently considered is that of interactions that are not observed. Two observations suggest that some interactions that should be observed are not. One is that in library screens proteins that should be found occasionally are not. Failure to recover expected proteins in this instance might be due to trivial considerations, such as the absence of the protein from the library used. The second is the number of examples in which known interactions are either not observed, or are subject to directionality, being observed only when one of the two proteins is a bait and the other a prey (see, for example, Estojak et al. 1995). We are resigned that false negatives will arise, and we do not give the absence of interaction much weight in our data analysis. This doctrine may change as more sensitive detection methods are designed.

Multimeric Complexes Finally, it is worth noting that one can build up chains of individual binary interactions to suggest higher order complexes. This has worked well, for example, with proteins in signal transduction (Choi et al. 1994; Marcus et al. 1994; Printen and Sprague 1994; chapter 10), and the advent of mating techniques has made it even easier to build up such patterns (Finley and Brent 1994; C. Kaiser and D. Shaywitz, personal communication).

Inference of Function from Pattern of Interactions

One reason for developing interaction mating techniques was the hope that it would reveal contacts between test proteins and known proteins that would provide clues to the function of the test proteins. However, our first experiments revealed that clues to function might also be derived from the pattern of interactions a protein makes, without reference to the biochemical identity of the interacting proteins. A simple example, taken from our first experiments, illustrates this point. Cdi4 and Cdi11 both interact with *Drosophila* Cdc2c, and interaction mating experiments also revealed that Cdi4 interacts with Cdi11 (Finley and Brent 1994). From the pattern of interactions alone, these data are consistent with the idea that Cdi4, Cdi11, and Cdc2c could form a three-protein complex. It is possible that other such patterns of interactions, particularly conjoined with the crude affinity data, might signal other sorts of regulators. The algorithmic analysis of connectivity data for patterns of this type is an important area of future research.

Library Scale and Genome-Wide Characterization of Protein Networks

Interaction mating schemes can also be used on a larger scale, for screening libraries, and, eventually, to characterize complex genomes. One such scheme is to mate a pool of cells containing different activation domain-tagged proteins against a bait protein. Another is the converse of the original two-hybrid system. In this approach, a library of different proteins fused to a DNA-binding domain is used in an interactor hunt to find proteins that interact with an activation domain-tagged protein. Historically, the drawback to such approaches has been that libraries that express proteins fused to DNA-binding domains will contain a large number proteins that activate transcription when brought to DNA (Ma and Ptashne 1987), complicating the task of identifying yeast in which the reporters are active due to the presence of an interacting protein. One way to circumvent this difficulty would be to introduce the library into a yeast strain that contained a counter-selectable reporter gene (for example, *LexAop-LYS2* or *LexAop-URA3*), select against those yeast that contained activators, and then mate the "depleted" library with yeast of the opposite mating type that contain the test protein. Yet another way is to express the activation domain-tagged proteins from a conditional promoter like *GAL1* and compare reporter activation between replica plates on which they are and are not expressed, as described in Protocols 1 and 3, and in chapter 4).

Recently, Bartel et al.(1996) applied two-hybrid technology to characterize a small genome. They set out to identify all detectable binary interactions between proteins encoded by the bacteriophage T7 genome. They did this by making two libraries, one of DNA-binding domain hybrids and one of activation domain hybrids, which were expressed in yeast strains of opposite mating type. They then mated a pool of yeast that contained the entire library of activation domain hybrids with 30,000 of the strains expressing DNA-binding domain fusions, in groups of ten DNA-binding domain fusions so that they could readily single out those that activated transcription. They selected diploids in which the *HIS3* reporter was activated, and screened for activation of a second *lacZ* reporter using a filter assay. In this way, they identified 19 binary interactions between T7 encoded proteins. They further performed individual searches testing 34 specific DNA-binding hybrids against the entire activation domain library, and 11 specific activation domain hybrids against the entire DNA-binding domain hybrid library, again by interaction mating, and identified three additional interactions. Finally, they made a matrix of all of the yeast expressing DNA-binding domain hybrids involved in an interaction and mated them with yeast expressing all of the activation domain hybrids involved in an interaction to identify three more interactions.

By this means they detected a total of 25 interactions. Some of the interactions were previously known, while others confirmed interactions that had been suspected based on genetic or biochemical studies. Most importantly, 10 of the interactions detected in this two-hybrid tour de force identified connections between proteins not previously known to interact. This new information contains both clues to the function of individual proteins and clues as to how some may function together. An additional windfall from this approach, made possible by the fact that the two libraries were made from random fragments of the T7 genome, was the identification of a number of intramolecular interactions. The detection of these intramolecular interactions suggested possible contacts between domains of a polypeptide that might promote the formation of tertiary structure. The success of this genome-wide approach demonstrates that interaction mating techniques can be used to identify the networks of interacting proteins encoded by more complex genomes. The charting of such connections between proteins will provide insights into the functions of individual proteins and lead to a better understanding of how groups of proteins control biological processes.

Conclusions

The few years since the advent of two-hybrid systems have proven their utility in the study of defined protein interactions, in identification of new interacting proteins, and in the charting of genetic networks of proteins involved in processes from signal transduction to transcription regulation. These tremendous successes suggest that two-hybrid approaches like those discussed in this chapter may eventually be used to identify many, if not most, of the protein-protein contacts made in a cell or an organism.

Sequencing projects like the human genome initiative will soon provide us with the sequences of all of the expressed proteins. A good deal of insight into the function of these proteins can be derived from their sequences alone, but ultimately these data must be combined with other forms of information to understand the biology in detail. Information about contacts made by the proteins of a genome will complement and augment the sequence information. Such information will likely come from incremental scaling up of the methods described here, as well as from scaled-up versions of ideas such as those developed by Bartel et al. (1996). Connection data will also come from the thousands of labs using two-hybrid systems to identify and characterize specific proteins. Finally, it may also come from efforts to identify all of the proteins in the networks of interacting proteins in a cell using rapid sequential two-hybrid interactor hunts that use the proteins isolated in one hunt as starting points for further hunts, in a sort of "protein interaction walk" (R.L.F., unpublished data).

As discussed earlier, all two-hybrid approaches inevitably produce false positives, as well as interactions that do not occur in any biological setting. Thus, although they will be rich in information, connectivity maps derived from two-hybrid data will necessarily be imprecise. This need not be thought of as a significant drawback of genome-wide two-hybrid approaches, provided it is borne in mind that the information in a protein linkage map derives its utility in providing clues to important interactions which must be explored with further study, using other methods.

One example of an insight into protein function from a large scale two-hybrid approach is the identification of the *Drosophila* protein Roughex, Rux, as a protein that interacts strongly and specifically with *Drosophila* cyclin E (Thomas et al. 1997; Finley and Brent, unpublished data). Rux, a 335 amino acid protein whose sequence gives no clues to its function (Thomas et al. 1994), was in a panel of 600 bait proteins that we tested for interaction with a cyclin E prey. It was known that *rux* is required for normal eye development; loss of function *rux* mutants have rough eyes and aberrant cell cycle regulation in the eye imaginal disc from which the eye develops. Thomas et al. (1994) previously showed that a stripe of cells in the morphogenetic furrow of the developing eye disc must arrest transiently in the G1 phase of the cell cycle for proper development and this G1 arrest fails in *rux* mutant eye discs. Combined with this information, the finding that Rux interacts directly with cyclin E, a protein known to be required for progression through G1, immediately suggested that rux modulated cyclin activity, and inspired us to undertake specific genetic and biochemical experiments to test the hypothesis.

Scaled-up interaction mating assays are likely to be useful in the analysis of genetic diseases and other complex genetic traits. The first version of this idea, which has a long history, is that genes that modify the function of other genes may participate in the same process. A less obvious corollary of this idea became apparent several years ago: that, among the proteins that interact with a protein involved in a disease, those that interact differently with wild-type and disease state allelic forms of the protein are likely to be involved in the disease. Reymond and Brent (1995) undertook a test of this idea by studying the protein encoded by the *INK4* human tumor suppressor gene, encoding p16. Wild-type p16 interacts with two human cyclin-dependent kinases, Cdk4 and Cdk6, to inhibit their activity. Interaction mating showed that alleles of p16 found in cancer-prone families are deficient in their interaction with the kinases. Two unexpected conclusions arose from these experiments. One allele, p16-G101W, showed decreased interaction with Cdk4 but not with Cdk6, suggesting that its role in disease is unrelated to its action on Cdk6. Furthermore, another allele, p16-I49T, which is also found in the control population, is deficient in interaction with Cdk4, suggesting that this allele may also contribute to a tumor-prone phenotype. These findings underscore the fact that interaction mating with different alleles in a population will contribute to the analysis of complex polygenic traits.

The ability to conduct scaled-up two-hybrid analysis has come at a good time. The trickle of new genes and alleles has become a torrent. Robust and general approaches to the understanding of gene and pathway function will help us to the next step of biological understanding.

ACKNOWLEDGMENTS We thank L. Lok and members of the Brent laboratory, past and present, for helpful discussions, A. Mendelsohn for assistance in working out the interaction mating assay, and A. Reymond for help in collecting and maintaining the bait panel. We also thank P. Colas, E. Golemis, and C. Giroux for helpful comments on the manuscript. R.B. was supported by Hoescht AG and an American Cancer Society Faculty Research Award.

References

Allen, J. B., Walberg, M. W., Edwards, M. C., and Elledge, S. J. (1995). Finding prospective partners in the library: the two-hybrid system and phage display find a match. Trends Biochem. Sci. 20:511–516.

Bartel, P. L., Roecklein, J. A., SenGupta, D., and Fields, S. (1996). A protein linkage map of Escherichia coli bacteriophage T7. Nature Genetics 12:72–77.

Bendixen, C., Gangloff, S., and Rothstein, R. (1994). A yeast mating-selection scheme for detection of protein-protein interactions. Nuc. Acids Res. 22:1778–1779.

Brent, R., and Ptashne, M. (1984). A bacterial repressor protein or a yeast transcriptional terminator can block upstream activation of a yeast gene. Nature 312:612–615.

Chien, C.-T., Bartel, P. L., Sternglanz, R., and Fields, S. (1991). The two-hybrid system: A method to identify and clone genes for proteins that interact with a protein of interest. Proc. Natl. Acad. Sci. USA 88:9578–9582.

Choi, K. Y., Satterberg, B., Lyons, D. M., and Elion, E. A. (1994). Ste5 tethers multiple protein kinases in the MAP kinase cascade required for mating in *S. cerevisiae*. Cell 78:499–512.

Durfee, T., Becherer, K., Chen, P.-L., Yeh, S.-H., Yang, Y., Kilburn, A. E., Lee, W.-H., and Elledge, S. J. (1993). The retinoblastoma protein associates with the protein phophatase type 1 catalytic subunit. Genes Dev. 7:555–569.

Estojak, J., Brent, R., and Golemis, E. A. (1995). Correlation of two-hybrid affinity data with in vitro measurements. Mol. Cell. Biol. 15:5820–5829.

Fields, S., and Song, O. (1989). A novel genetic system to detect protein-protein interactions. Nature 340:245–246.

Fields, S., and Sternglanz, R. (1994). The two-hybrid system: an assay for protein-protein interactions. Trends Genet. 10:286–292.

Finley, R. L., Jr., and Brent, R. (1994). Interaction mating reveals binary and ternary connections between Drosophila cell cycle regulators. Proc. Natl. Acad. Sci. USA 91:12980–12984.

Finley, R. L., Jr., and Brent, R. (1995). Interaction trap cloning with yeast. In *DNA Cloning 2, Expression Systems: A Practical Approach*, B. D. Hames, and D. M. Glover, eds. Oxford, Oxford University Press. pp. 169–203.

Golemis, E.A., and Brent, R. (1992). Fused protein domains inhibit DNA binding by LexA. Mol. Cell. Biol. 12:3006–3014.

Guarente, L. (1996). Transcriptional coactivators in yeast and beyond. Trends Biochem. Sci. 20:517–521.

Guarente, L., and Ptashne, M. (1981). Fusion of *Eschericia coli* lacZ to the cytochrome c gene of *Saccharomyces cerevisiae*. Proc. Natl. Acad. Sci. USA 78:2199–2203.

Gyuris, J., Golemis, E., Chertkov, H., and Brent, R. (1993). Cdi1, a human G1 and S phase protein phosphatase that associates with Cdk2. Cell 75:791–803.

Harper, J. W., Adami, G. R., Wei, N., Keyomarsi, K., and Elledge, S. J. (1993). The p21 Cdk-interacting protein Cip1 is a potent inhibitor of g1 cyclin-dependent kinases. Cell 75:805–816.

Kranz, J. E., Satterberg, B., and Elion, E. A. (1994). The MAP kinase Fus3 associates with and phosphorylates the upstream signaling component Ste5. Genes Dev. 8:313–327.

Lech, K., Anderson, K., and Brent, R. (1988). DNA-bound Fos proteins activate transcription in yeast. Cell 52:179–184.

Ma, J., and Ptashne, M. (1987). A new class of transcriptional activators. Cell 51:113–119.

Marcus, S., Polverino, A., Barr, M., and Wigler, M. (1994). Complexes between STE5 and components of the pheromone-responsive mitogen-activated protein kinase module. Proc. Natl. Acad. Sci. USA 91:7762–7766.

Mendelsohn, A. R., and Brent, R. (1994). Applications of interaction traps/two-hybrid systems to biotechnology research. Curr. Op. Biotech. 5:482–486.

Printen, J. A., and Sprague, G. F., Jr. (1994). Protein-protein interactions in the yeast pheromone response pathway: Ste5p interacts with all members of the MAP kinase cascade. Genetics 138:609–619.

Reymond, A., and Brent, R. (1995). p16 proteins from melanoma-prone families are deficient in binding to Cdk4. Oncogene 11:1173–1178.

Rose, M., and Botstein, D. (1983). Construction and use of gene fusions to lacZ (beta-galactosidase) that are expressed in yeast. Methods Enzymol. 101:167–180.

Thomas, B. J., Gunning, D. A., Cho, J., and Zipursky, L. (1994). Cell cycle progression in the developing *Drosophila* eye: roughex encodes a novel protein required for the establishment of G1. Cell 77:1003–1014.

Thomas, B. J., Dong, X., Zavitz, K., Lane, M. E., Weigmann, K., Lehner, C. F., Finley, R. L., Brent, R., and Zipursky, S. L. (1997). Roughex regulates G1 arrest in the developing *Drosophila* eye by promoting cyclin A degradation. Genes Dev. *in press.*

Tjian, R., and Maniatis, T. (1994). Transcriptional activation: A complex puzzle with few easy pieces. Cell 77:5–8.

Van Aelst, L., Barr, M., Marcus, S., Polverino, A., and Wigler, M. (1993). Complex formation between RAS and RAF and other protein kinases. Proc. Natl. Acad. Sci., USA 90:6213–6217.

Yuan, Y. O., Stroke, I. L., and Fields, S. (1993). Coupling of cell identity to signal response in yeast: interaction between the α1 and STE12 proteins. Genes Dev. 7:1584–1597.

Zervos, A. S., Gyuris, J., and Brent, R. (1993). Mxi1, a protein that specifically interacts with Max to bind Myc-Max recognition sites. Cell 72:223–232.

Part III

VARIATIONS ON THE TWO-HYBRID THEME

With the standard two-hybrid approach, proteins are expressed as fusions in what might be to them a strange milieu and without other factors that might modify their function. Therefore, in some cases, interactions that occur in a protein's native environment do not occur in a standard two-hybrid assay. Several groups have devised ingenious ways to circumvent certain limitations of the assay. Tsan, Wang, Jin, Hwang, Bash, and Baer take the two-hybrid assay out of yeast cells into mammalian cells (chapter 13). Proteins of mammalian origin might then have an opportunity to interact in their normal environment and with their normal modifications. In chapter 14, Osborne, Lubinus, and Kochan describe a variation in which they coexpress an active protein kinase with the two-hybrid proteins, such that interactions that require phosphorylation of one of the proteins can be detected. In chapter 15, Berlin demonstrates that the two-hybrid system can be used to detect protein interactions that are mediated by the addition of a small molecule, in this case, FK506.

In chapter 16, Yang describes a two-hybrid variation in which peptides that bind to a protein are identified. With this strategy, he was able to identify peptides that bind to the retinoblastoma protein (Rb) and to correlate particular amino acids residues to binding affinity.

13

Mammalian Cells as Hosts for Two-Hybrid Studies of Protein-Protein Interaction

Julia Tsou Tsan
Zhuo Wei Wang
Ying Jin
Larn-Yuan Hwang
Robert O. Bash
Richard Baer

An important advantage of the two-hybrid system is that it allows protein-protein interactions to be studied in a relatively natural setting—that is, within the nuclei of eukaryotic cells. The two-hybrid assay was originally carried out in *Saccharyomyces cerevisiae* to evaluate the in vivo association of two yeast proteins (Fields and Song, 1989). The use of *S. cerevisiae* as host allowed this system to blossom into a powerful screening method to identify novel protein-protein interactions (Chien et al. 1991). Moreover, two-hybrid analysis in *S. cerevisiae* has proven applicable to the study of not only yeast proteins but also those of many other organisms. As such, the yeast two-hybrid system has already had a major impact on our understanding of mammalian cell biology (for example, Harper et al. 1993).

Although interactions involving mammalian proteins are often amenable to analysis in yeast, there are circumstances in which it would be preferable to study these interactions in their native environment. As with any physiological event, the formation of a protein complex may be subject to regulatory influences that determine where and when it will occur. However, the regulatory factors that control mammalian protein-protein interactions may not always be faithfully represented in yeast. In theory, this obstacle can be overcome by conducting two-hybrid assays in mammalian cells. There are also practical reasons why some investigators may wish to avoid using yeast as hosts for two-hybrid analysis, particularly if the appropriate equipment or expertise is not available.

The two-hybrid system was successfully adapted for use in mammalian cells by Dang et al. (1991). The mammalian assay relies on functional reconstitution of GAL4-VP16, an artificial transcription factor containing the DNA-binding domain of the yeast Gal4 protein (Ma and Ptashne 1987) fused to an acidic transactivation domain of the herpes simplex virus VP16 protein (Treizenberg et al. 1988). In HeLa cells, the GAL4-VP16 hybrid readily stimulates transcription of reporter plasmids bearing the UAS_G sequence, a recognition site for the DNA-binding domain of Gal4p (Sadowski et al. 1988). Therefore, Dang et al. (1991) constructed mammalian expression vectors encoding hybrid polypeptides in which the Gal4p DNA-binding domain was fused to one protein moiety (GAL4-X) and the VP16 transactivation domain was fused to another (VP16-Y). These vectors were cotransfected into Chinese hamster ovary (CHO) cells along with G5E1b-CAT, a Gal4p-responsive reporter plasmid with five UAS_G sites located upstream of coding sequences for bacterial chloramphenicol acetyltransferase (Lillie and Green 1989). According to the paradigm established with the yeast two-hybrid system (Fields and Song 1989), if proteins X and Y interact stably in vivo, then coexpression of both hybrid polypeptides should induce transcription of the G5E1bCAT reporter gene (Figure 13-1A). To establish the feasibility of the mammalian assay, Dang et al. (1991) evaluated dimerization of the well-characterized leucine zipper motifs of c-Fos and c-Jun. Thus, separate vectors that encode either the Gal4p DNA-binding domain fused to the leucine zipper of c-Fos (GAL4-Fos) or the VP16 transactivation domain fused to the leucine zipper of c-Jun (VP16-Jun) were transfected into CHO cells along with G5E1bCAT. As anticipated, transcription of the reporter gene was not stimulated by either vector alone, since the hybrid proteins encoded by these vectors lack either a transactivation domain (GAL4-Fos) or an appropriate DNA-binding domain (VP16-Jun). In contrast, co-expression of both proteins induced a large increase in CAT activity to levels more than 25-fold higher than those found with GAL4-Fos or VP16-Jun alone.

The mammalian two-hybrid system has grown in popularity since its introduction, in large part because of its reproducibility and the technical ease of the assay. In this chapter, we provide a simple protocol for mammalian two-hybrid analysis, and we describe expression vectors and reporter constructs that are available for use in the assay. Finally, we will weigh the relative strengths and weaknesses of the mammalian assay, and its use as an adjunct to the yeast two-hybrid screening method will be considered.

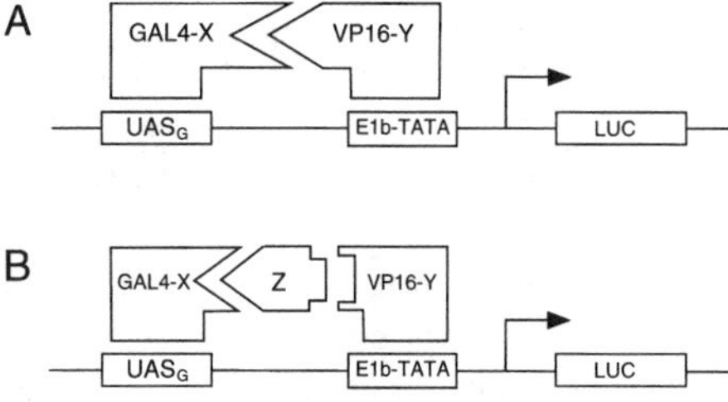

Fig. 13-1. Two-hybrid analysis in mammalian cells. (A) The conventional two-hybrid assay for detection of a protein-protein interaction. The Gal4p-responsive reporter gene is illustrated; it contains a UAS_G recognition site for the DNA-binding domain of Gal4p, a TATA element from the adenoviral E1b gene, and an open reading frame that encodes luciferase. Cells are cotransfected with the reporter gene and expression vectors encoding the GAL4-X and VP16-Y fusion proteins. In vivo association between the X and Y moieties of these proteins induces transcription of the reporter gene. (B) The bridge two-hybrid assay to detect formation of a multicomponent protein complex. Cells are cotransfected with the reporter gene, the expression vectors for GAL4-X and VP16-X, and an expression vector encoding a wild-type protein (Z). The reporter gene is induced by formation of a stable tenary complex involving sequences from X, Y, and Z.

Description of Methods

The Reporter Plasmids

In the mammalian version of the two-hybrid system, transcriptional activation of a Gal4p-responsive reporter gene provides a quantitative measure of in vivo interaction between the two hybrid proteins (GAL4-X and VP16-Y). For this purpose Dang et al. (1991) originally used G5E1bCAT, a reporter plasmid that contains the TATA element of the adenoviral Elb gene and five tandem copies of UAS_G (5'-CGGAGTACTGTCCTGCG-3') positioned immediately upstream of coding sequences for bacterial chloramphenicol acetyltransferase (Lillie and Green 1989). Gal4p-responsive reporter plasmids based on the firefly luciferase gene are also available; these were generated by transferring the UAS_G and TATA sequences from

G5E1bCAT into the pGL2-basic or pGL3-basic plasmids (Promega) (Hsu et al. 1994). The resultant plasmids (G5E1bLUC and G5LUC, respectively) provide the advantages commonly attributed to luciferase reporters, including high sensitivity and a broad range of linear response (de Wet et al. 1987).

In addition to the Gal4p-responsive reporter, it is also advantageous to include a control reporter to normalize each sample for transfection efficiency (Hollon and Yoshimura 1989). Expression vectors that encode bacterial β-galactosidase, such as pSV-β-galactosidase (Promega) or pCMVβ (CLONTECH), serve as convenient control plasmids.

The Gal4p-Fusion Vectors

The pSG424 plasmid was designed for mammalian expression of fusion proteins that contain the DNA-binding domain of the yeast Gal4 protein (Sadowski and Ptashne 1989). This vector contains a polylinker of common restriction enzyme sites preceded by coding sequences for the amino-terminal 147 residues of Gal4p. Thus, a hybrid reading frame that encodes a Gal4p-fusion protein (GAL4-X) can be generated by inserting foreign cDNA sequences into the polylinker region of pSG424. Transcripts of the hybrid reading frame are initiated from the SV40 early promoter and their processing is facilitated by the SV40 polyadenylation signal.

pSG424 and related Gal4p-fusion vectors (pGALO and pLR60) have been used extensively to identify transcriptional transactivation domains (Sadowski and Ptashne 1989; Dang et al. 1991; Raycroft and Lozano, 1992); in this assay, foreign amino acid sequences that imbue the Gal4p DNA-binding domain with the ability to activate transcription of a Gal4p-responsive reporter gene are operationally defined as a transactivation domain (Ptashne 1988). These same vectors can also be used to generate Gal4p-fusion polypeptides for two-hybrid analysis of protein-protein interaction. However, a Gal4p-fusion protein that contains a transactivation domain within its foreign moiety is not suitable for two-hybrid analysis since it will activate transcription of the Gal4p-responsive reporter independent of its interaction with the VP16-fusion polypeptide. Therefore, in mammalian two-hybrid assays of protein-protein interaction, sequences that harbor a measurable transactivation domain should be expressed within the context of the VP16-fusion protein.

Sadowski et al. (1992) subsequently produced a series of improved Gal4p-fusion vectors (pM1, pM2, and pM3). These have a more versatile array of polylinker restriction sites, and they provide significantly greater yields of plasmid DNA when propagated in *Escherichia coli.*

The VP16 Fusion Vectors

The pNLVP16 plasmid was constructed for the expression of VP16 fusion proteins in mammalian cells (Dang et al. 1991). The vector has a polylinker of unique restriction sites preceded by an artificial reading frame that includes the eleven amino-terminal residues of Gal4p (MKLLSSIEQAC), a nuclear localization signal from the SV40 large T antigen (PKKKRKVD), and the acidic transactivation domain of the herpes simplex virus VP16 protein (VP16 residues 411-456). pNLVP16 has been used successfully to express numerous VP16-fusion proteins for two-hybrid analysis in mammalian cells (Dang et al. 1991; Fearon et al. 1992; Kato et al. 1992; Finkel et al. 1993; Hsu et al. 1994; Wadman et al. 1994; Osada et al. 1995).

We constructed a series of VP16-fusion vectors with additional features that can also be used in mammalian two-hybrid experiments. Thus, a 0.3 kb *Bgl*II/*Sal*I fragment spanning the artificial VP16 reading frame designed by Dang et al. (1991) was excised from pNLVP16 and ligated into the 3.1 kb *Bgl*II/*Sal*I backbone fragment of pM1. The pM1 backbone allows significantly greater yields of plasmid DNA from *E. coli*. Desirable restriction endonuclease sites were then added to the polylinker and removed from the backbone in order to generate the pVP-Nco vector. Nucleotide sequences that encompass the VP16 reading frame and polylinker of pVP-Nco are illustrated in Figure 13-2; the complete sequence of pVP-Nco has been deposited in the Genbank database (accession number U55939). The pVP-FLAG expression vector was then constructed by inserting an oligonucleotide that contains additional restriction sites, as well as coding sequences for the FLAG epitope (DYKDDDDK, Kodak/IBI) (Hopp et al. 1988), into the polylinker of pVP-Nco (Figure 13-3). Similarly, sequences encoding the influenza hemagglutinin epitope (YPYDVPDYASL) recognized by the 12CA5 monoclonal antibody (Field et al. 1988) were added to pVP-Nco in order to generate the pVP-HA1, pVP-HA2, and pVP-HA3 vectors (Figure 13-4).

The Host Cells for Mammalian Two-Hybrid Assays

Mammalian two-hybrid analysis is usually conducted by transient transfection of adherent cells. A variety of adherent cell lines have been used, including CHO cells, C3H10T1/2 fibroblasts, and HeLa cells (Dang et al. 1991; Chakraborty et al. 1992; Hsu et al. 1994). The human 293 line of transformed embryonal kidney cells is especially convenient for mammalian two-hybrid studies (see Protocol), primarily due to its high efficiency of DNA transfection (Graham and Van der Eb 1973; Graham et al. 1977).

Two-hybrid studies can also be conducted in hematopoietic lines that grow in suspension. For example, protein-protein interactions involving

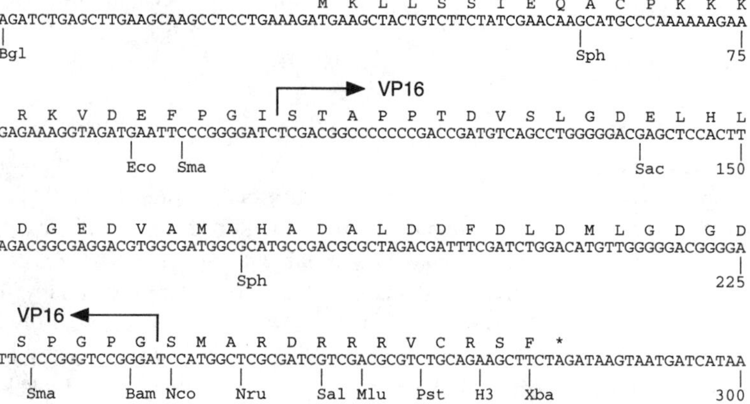

Fig. 13-2. The pVP-Nco expression vector. Nucleotide sequences encompassing the artificial VP16 reading frame of pVP-Nco are illustrated. The reading frame encodes the eleven amino-terminal residues of Gal4p (MKLLSSIEQAC), a nuclear localization signal from SV40 large T antigen (PKKKRKVD) and the acidic transactivation domain of VP16 (Dang et al. 1991). The numbering of the nucleotide sequences begins with a unique *Bgl*II site located upstream of the VP16 reading frame. The polylinker sequence includes unique recognition sites for *Bam*HI, *Nco*I, *Nru*I, *Sal*I, *Mlu*I, *Pst*I, *Hind*III, and *Xba* I. The complete nucleotide sequence of pVP-Nco (3,230 basepairs) is available from the Genbank database (accession number U55939).

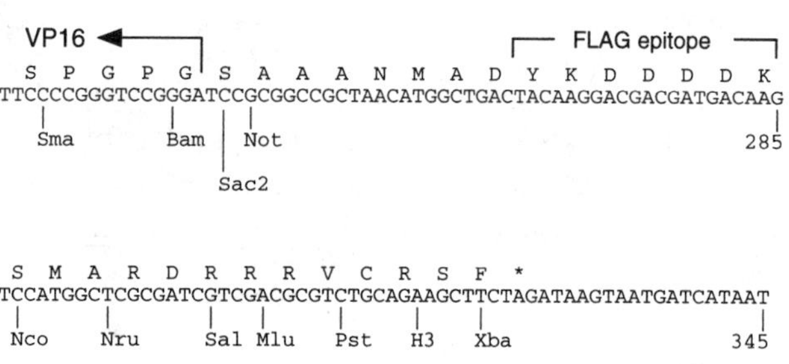

Fig. 13-3. The pVP-FLAG expression vector. Nucleotide sequences encompassing the coding sequence for the FLAG epitope and the polylinker are illustrated (the remaining sequences of pVP-FLAG are identical to those of pVP-Nco). Sequence numbering begins with the unique *Bgl*II site (see Fig. 13-2). All of the illustrated restriction sites are unique except *Sma*I.

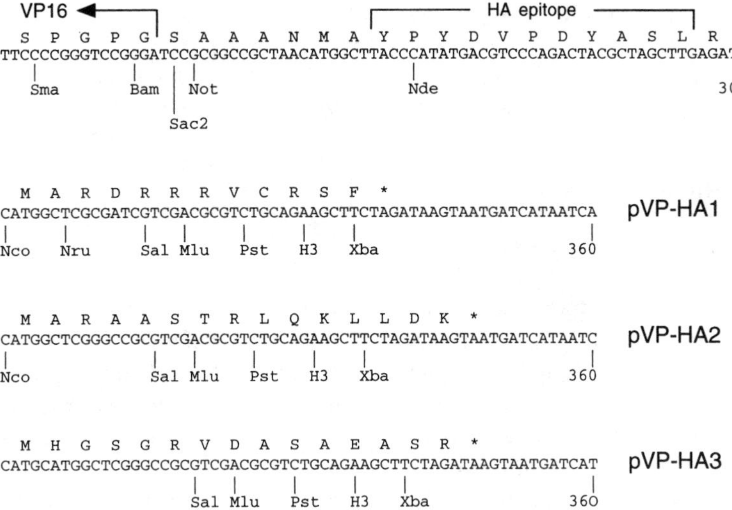

Fig. 13-4. The pVP-HA series of expression vectors. The nucleotide sequence encoding the influenza hemmagglutinin (HA) epitope is illustrated, along with the distinct polylinker regions of pVP-HA1, pVP-HA2, and pVP-HA3 (the remaining sequences of these vectors are identical to those of pVP-Nco). Sequence numbering begins with the unique *Bgl*II site (see Fig. 13-2). All of the illustrated restriction sites are unique except *Sma*I and *Nde*I. The downstream restriction sites of pVP-HA1, pVP-HA2, and pVP-HA3 (*Sal*I, *Mlu*I, *Pst*I, *Hind*III, *Xba*I) are each shifted by one residue with respect to the VP16 reading frame.

factors implicated in human T cell leukemia (e.g., the viral TAX and cellular TAL1 polypeptides) have been characterized by two-hybrid analysis in Jurkat, a cell line originally derived from a patient with T cell acute leukemia (Hsu et al. 1994; Wadman et al. 1994; Yin et al. 1995). In general, it should be possible to perform two-hybrid experiments in any stable mammalian line that is reasonably susceptible to DNA transfection. Nevertheless, some consideration should be given to the unique properties of each potential host. For example, two-hybrid assays in which the retinoblastoma susceptibility gene product is expressed as a GAL4-Rb fusion protein cannot be performed in 293 cells; in this setting, the GAL4-Rb polypeptide can activate transcription of a Gal4p-responsive reporter gene independent of Rb-mediated interaction with the VP16-X fusion protein (data not shown). The 293 line was originally established by transformation with DNA fragments of the Adenovirus Type 5 genome (Graham et al. 1977). Thus, the self-activating behavior of GAL4-Rb in 293 cells may represent the interaction of its Rb sequences with the endogenously-expressed adenoviral E1A protein, which avidly binds the Rb pocket (Whyte et al. 1988) and has a potent transactivation domain of its own (Lillie and Green 1989). In contrast, the GAL4-Rb chimera does not exhibit self-activating behavior in other cell lines; for example, GAL4-Rb has been used successfully in CHO cells to detect in vivo interaction between Rb and the transforming E7 protein encoded by human papillomavirus (Hoang et al. 1995).

Examples

Standard Two-Hybrid Assays

Figure 13-5 illustrates the results of a two-hybrid experiment conducted in 293 cells using the procedure described in the Protocol. E47 and TAL1 are mammalian transcription factors that harbor the basic helix-loop-helix (bHLH) motif, a conserved domain of 50 to 60 amino acids that mediates protein dimerization and sequence-specific DNA recognition (Murre et al. 1989; Chen et al. 1990). Protein-protein interaction between the bHLH domains of E47 and TAL1 allows formation of a stable heterodimer (E47/TAL1) with DNA-binding activity (Hsu et al. 1991). Thus, as shown in Figure 13-5, co-expression of the hybrid GAL4-E47 and VP16-TAL1 polypeptides induced luciferase activity to levels at least 100-fold higher than either polypeptide alone (compare lane 5 with lanes 2 and 6). The in vivo association of E47 and TAL1 can also be demonstrated in the reciprocal two-hybrid assay using the GAL4-TAL1 and VP16-E47 fusion proteins (Hsu et al. 1994).

Bridge Two-Hybrid Assays

The mammalian two-hybrid system has also been used to unravel interactions involved in the assembly of multicomponent protein complexes. For example, the bHLH sequences of TAL1 associate avidly with LM02 (formerly called RBTN2 or TTG2) (Valge-Archer et al. 1994; Wadman et al. 1994), a nuclear oncoprotein that contains the cysteine-rich LIM motif (Sanchez-Garcia and Rabbitts 1994). In view of the ability of TAL1 to interact with both E47 and LM02, it was important to establish whether the bHLH domain of TAL1 binds these proteins in a mutually exclusive manner or whether it mediates the assembly of a ternary complex by simultaneous association with both E47 and LM02. These possibilities can be distinguished in a "bridge" two-hybrid assay (see Figure 13-1B). As shown in Figure 13-5, coexpression of the GAL4-E47 and VP16-LM02 chimeras does not yield a significant increase of luciferase activity in the mammalian two-hybrid assay (lane 3), indicating that E47 and LM02 do not interact directly with one another in vivo. However, a large induction of luciferase activity occurs when these polypeptides are co-expressed in the presence of wildtype TAL1 (lane 9). This result implies that TAL1 can form a bridge between the hybrid GAL4-E47 and VP16-LM02 polypeptides by interact-

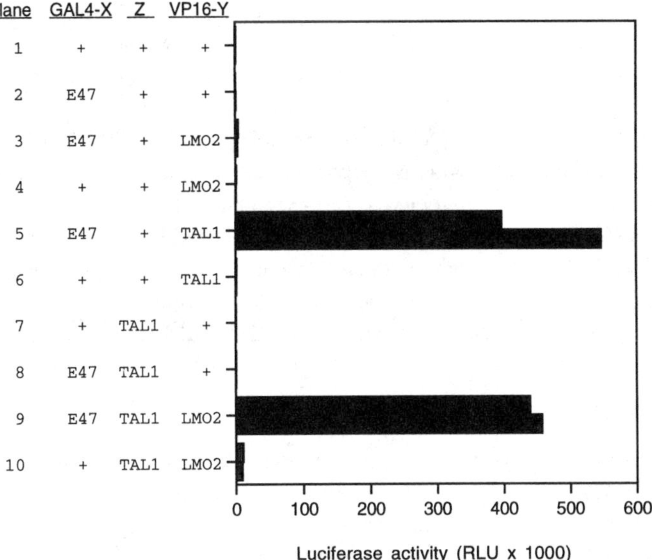

Fig. 13-5. Mammalian two-hybrid analysis of protein-protein interactions involving the E47, TAL1, and LM02 polypeptides. Each dish of 293 cells was transiently cotransfected with the G5E1bLUC reporter plasmid (1.0 µg), the pSV-ß galactosidase control plasmid (1.5 µg), and the three indicated expression plasmids. The GAL4-X expression plasmid (3 µg) encoded either the parental Gal4p DNA-binding domain (pM1; denoted by "+" in the GAL4-X column) or the GAL4-E47 hybrid polypeptide. The VP16-Y plasmid encoded either the parental VP16 transactivation domain (pVP-Nco; denoted by "+" in the VP16-X column) or the indicated hybrid polypeptide (VP16-LM02 or VP16-TAL1). Cells were also cotransfected with 1.5 µg of an empty expression vector (pCMV4; indicated by "+" in the Z column) (Andersson et al. 1989) or a vector encoding the wild-type TAL1 polypeptide (TAL1/pCMV4). Duplicate transfections were conducted for each combination of expression plasmids, and the normalized luciferase activities obtained from each transfection are illustrated.

ing simultaneously with sequences from both E47 and LM02 (Wadman et al. 1994).

Osada et al. (1995) have shown that LM02 can also interact with GATA-1, a tissue-specific transcription factor required for normal erythroid development (Pevny et al. 1991). Although TAL1 and GATA-1 do not directly interact with one another, a bridge two-hybrid experiment revealed that the GAL4-TAL1 and VP16-GATA-1 chimeras activate transcription of a Gal4p-responsive reporter gene when coexpressed in the presence of wild-type LM02 (Osada et al. 1995), implying that LM02 can interact simultaneously with both TAL1 and GATA-1 to form a stable ternary complex (TAL1/LM02/GATA-1). These data, together with previous evidence for assembly of the E47/TAL1/LM02 heterotrimer (Wadman et al. 1994), suggested that a quaternary complex composed of E47, TAL1, LM02, and GATA-1 would also be stable in vivo. Indeed, Osada et al. (1995) observed induction of a Gal4p-responsive reporter gene when the GAL4-E47 and VP16-GATA-1 hybrids were coexpressed with wild-type versions of both TAL1 and LM02 together. Hence, the mammalian two-hybrid system was sufficiently sensitive to detect the formation of a multicomponent complex of four proteins interacting in series (E47/TAL1/LM02/GATA-1).

Discussion

The mammalian two-hybrid system has proven to be a reliable method to study in vivo interactions between two or more known proteins. The range of proteins tested in the mammalian system is expanding; most are nuclear factors such as bHLH proteins, leucine zipper proteins, GATA factors, LIM proteins, the viral TAX and E7 gene products, proteins with the Rb-pocket motif (Rb and p107), and subunits of the $TF_{II}D$ and $TF_{II}F$ transcription initiation factors (Dang et al. 1991; Aso et al. 1992; Chakraborty et al. 1992; Kato et al. 1992; Finkel et al. 1993; Hsu et al. 1994; Wadman et al. 1994; Xia et al. 1994; Hoang et al. 1995; Osada et al. 1995). However, the mammalian two-hybrid assay has also been used to detect interactions involving proteins that normally reside in the cytoplasm (Fearon et al. 1992; Takacs et al. 1993). This result implies that the corresponding hybrid proteins are translocated to the nucleus, where reporter gene transcription occurs, by virtue of their Gal4p or VP16 moieties. The DNA-binding domain of Gal4p is sufficient for nuclear localization of at least some GAL4-X hybrid polypeptides (e.g., GAL4-β-galactosidase) in yeast (Silver et al. 1984), and it may perform a similar function in mammalian cells. VP16-fusion proteins encoded by the pNLVP vector (and its derivatives) possess the nuclear localization signal from SV40 large T antigen, which presumably facilitates their translocation into the nuclei of mammalian cells (Dang et al. 1991).

In their original report, Dang et al. (1991) observed heterodimerization of the Fos and Jun leucine zipper proteins but not Jun/Jun homodimerization. These observations raised the possibility that some low-affinity interactions, though physiologically significant, may nonetheless escape detection in mammalian two-hybrid assays. Furthermore, some interactions are only observed in one of the two possible orientations of the assay. For example, in vivo association between LCK and the carboxy-terminal cytoplasmic domain of CD4 registers in the mammalian two-hybrid system when LCK sequences are expressed as a GAL4-LCK fusion but not when expressed as a VP16-LCK fusion (Fearon et al. 1992). Similarly, heterodimerization between the bHLH domains of E47 and TAL2 is detectable when mammalian two-hybrid assays are conducted with GAL4-TAL2 and VP16-E47 hybrids, but not with GAL4-E47 and VP16-TAL2 hybrids (Xia et al. 1994). Several mechanisms can be invoked to explain the failure of mammalian two-hybrid experiments to detect certain known protein-protein interactions, including conformational incompatibility, inappropriate subcellular localization, or instability of one or both of the hybrid proteins. Despite these examples of false-negative results from the mammalian two-hybrid assay, false-positive results have not as yet been reported.

The mammalian two-hybrid system may serve as a useful secondary assay to evaluate novel protein-protein interactions identified from two-hybrid screening in yeast (Wu et al. 1996). The proportion of false-positive clones that are generated during a yeast two-hybrid screen has been reduced considerably by recent improvements in the screening strategy (Bartel et al. 1993; Durfee et al. 1993; see chapters 2, 3, and 4). Nevertheless, it remains essential to distinguish interactions that are physiologically meaningful from those that occur fortuitously when the proteins in question are overexpressed ectopically in yeast. The need for supporting evidence is especially pressing when the yeast two-hybrid screen uncovers an interacting protein that is novel or an unanticipated association between known proteins. In the absence of functional data, most investigators seek additional evidence from coimmunoprecipitation analysis of mammalian cell lysates and from in vitro assays of protein-protein interaction, such as affinity chromatography with glutathione S-transferase fusion proteins (discussed by Luban and Goff 1995). However, another tier of supporting data can be obtained by mammalian two-hybrid analysis (Wu et al. 1996). For example, cDNAs that emerge from a yeast two-hybrid screen can be recloned into a VP16 fusion vector, and the encoded chimera (VP16-prey) tested for interaction in mammalian cells with the GAL4-bait hybrid. The pVP-FLAG and pVP-HA vectors were designed to facilitate co-immunoprecipitation analysis of cDNAs that perform favorably in the mammalian two-hybrid system. Thus, by using the appropriate restriction enzymes (for example, *Bam*HI, *Sac*II, *Not*I), the prey cDNA can be excised from pVP-FLAG or pVP-HA along with a Kozak translation initiator and sequences encoding an amino-terminal epitope tag (Figures 13-3 and 13-4). After transfer of the tagged cDNA into another expression vector (such as pCMV4; Andersson et al. 1989), coimmunoprecipitation analysis can be performed using antibodies specific for the bait (which are often available) and monoclonal reagents that recognize the FLAG or HA epitopes of the prey (Hopp et al. 1988; Field et al. 1988).

Future Directions

To date, the mammalian two-hybrid system has been used primarily to examine interactions involving known proteins. Within this limit, the system has been successfully exploited to uncover novel protein-protein interactions, to localize the protein motifs responsible for a given interaction, and to identify the interacting subunits of a multi-component protein complex. A future objective will be to develop two-hybrid screening procedures that can be used in mammalian cells as effectively as those currently used in yeast. Ideally, a mammalian version of the yeast two-hybrid screening method would identify interactions that are physiologically meaningful but might not occur when the relevant proteins are expressed ectopically in yeast. Important steps in this direction have been made by Fearon et al. (1992), who designed Gal4p-responsive reporter genes that allow selection of mammalian cells in a two-hybrid assay. In one example, CHO cells were stably transformed with a reporter gene that encodes hygromycin B resistance. The resultant cell line displayed high sensitivity to hygromycin and a low frequency of spontaneous resistance. Moreover, hygromycin resistance was readily and specifically induced by transient transfection with expression plasmids that encode two interacting hybrid proteins (GAL4-Fos and VP16-Jun) (Fearon et al. 1992). These studies provide a foundation for the development of two-hybrid screening methods that can be performed in mammalian cells.

Finally, the mammalian two-hybrid system can potentially serve as an assay to evaluate how a given protein-protein interaction is influenced by regulatory factors such as extracellular stimuli and intracellular signal transduction pathways (Chakraborty et al. 1992). Ideally, these studies would be conducted with cells stably transformed with the two-hybrid components, including a reporter gene and expression constructs that encode the interacting hybrid proteins. Once the appropriate cell line was established, a range of different regulatory signals could be tested for their effect on the protein interaction of interest.

Protocol

Two-Hybrid Analysis in 293 Cells

1. The 293 human embryonal kidney cell line can be obtained from the American Type Culture Collection. Maintain the cells in a 75 cm^2 flask with 20 ml of low glucose DMEM (supplemented with 2 mM glutamine, 100 μg/ml penicillin G, 100 μg/ml streptomycin, and 10% fetal calf serum) at 37°C in 5% CO_2. When the cells are confluent, split 1:10 every 3 days or 1:50 every 5 days. Since 293 cells are somewhat sensitive to temperature, we always use warm media and avoid leaving the cells outside the incubator for too long.
2. One day before transfection:
   a. Harvest the cells from a confluent 75 cm^2 flask by removing the media and adding 2.5 ml of 0.25% trypsin, 1 mM EDTA. Incubate at 37°C for no more than 3 minutes. Dilute the trypsinized cells with 7.5 ml of media and spin the cell suspension at 1,500 rpm for 5 minutes. Resuspend the cell pellet in 5 ml of media and gently vortex the cell suspension. Count the cells by staining with trypan blue and dilute them to 6×10^5 cells/ml in culture medium. In general, we obtain 18 to 20×10^6 cells from each 75 cm^2 flask of confluent 293 cells.
   b. We usually prepare duplicate culture dishes for each transfection. Place 2

ml of media into each 35 mm dish and add 0.5 ml of the 293 cell suspension. Therefore, each 35 mm dish will be seeded with 3×10^5 cells.

3. On the day of transfection:
   a. Dilute 10 μg of DNA (typically, 2 μg of the G5E1bLUC reporter plasmid, 3 μg of the GAL4-X expression plasmid, 3 μg of the VP16-Y expression plasmid, and 2μg of the pSVßgal control plasmid) into 440 μl of water in a 1.5 ml centrifuge tube.
   b. Add 60 μl of 2M $CaCl_2$ (kept at 4°C) to the tube drop-wise and mix gently.
   c. To a separate 1.5 ml tube, add 500 μl of 2× HBS, (50 mM Hepes pH 7, 280 mM NaCl, 1.5 mM NaH_2PO_4). Use a 1 ml pipette to gently bubble air through the 2× HBS solution. Add 500 μl of the DNA/$CaCl_2$ mixture slowly (drop-wise) to the gently mixing solution of 2× HBS.
   d. After mixing, allow the solution to precipitate at room temperature for 20 minutes.
   e. Vortex the precipitating DNA solution and use a 1 ml pipetteman to sprinkle 0.3 ml onto each 35 mm dish of 293 cells. Since 293 cells are weakly adherent, avoid swirling the dish during transfection. Incubate the transfected cells at 37°C in a 5% CO_2 incubator for ~48 hours.
4. Two days after transfection:
   a. To harvest the cells, aspirate the media and add 2 ml of TEN buffer (40 mM Tris/HCl pH 7.5, 1 mM EDTA, and 150 mM NaCl) to each dish. Incubate at room temperature for 5 minutes.
   b. Transfer the cell suspension to a 15 ml conical centrifuge tube (on ice). Pellet the cells by centrifugation at 1500 rpm for 5 minutes (4°C). If the cell lysates are to be analyzed by immunoblotting, then wash the cells twice with 2 ml of cold PBS to eliminate residual serum. Otherwise, proceed directly to the cell lysis procedure.
   c. Lyse the transfected cells of each dish by vigorous vortexing in 90 μl of luciferase lysis buffer (prechilled at 4°C). Transfer each lysate to a chilled 1.5 ml centrifuge tube. Clear the lysate of cell debris by centrifugation in a microfuge for 10 minutes at 4°C, and transfer the supernatant to a fresh tube. Luciferase lysis buffer is prepared by mixing 7 volumes of buffer A (100 mM K_2HPO_4 pH 7.8, 3 mM $MgCl_2$, 1 mM dithiothreitol), 2.5 volumes of buffer B (220 mM TrisHCl pH 7.8, 1 mg/ml soybean trypsin inhibitor, 20 μg/ml aprotinin), and 0.5 volumes of 2% Nonidet P-40.
   d. Use approximately 5 μl of each lysate for the luciferase assay and the β-galactosidase assay.
5. Normalizing the data:
   a. Measure the luciferase activity (LUC_S) and β-galactosidase activity (β-GAL_S) of each sample.
   b. Calculate the mean β-galactosidase activity of all samples in the entire experiment (β–GAL_E).
   c. To correct for variation in transfection efficiency between samples, determine the normalized luciferase activity of each sample (LUC_N) using the following formula:

$$LUC_N = (LUC_S)\ (\beta\text{-}GAL_E)\ /\ (\beta\text{-}GAL_S)$$

ACKOWLEDGMENTS We thank Marilyn Gardner for preparing the manuscript. We are also especially grateful to Chi Dang, Toren Finkel, Michael Green, Kathy Martin, Ivan Sadowski, and Lillian Xie Xu for generously providing reagents and advice. This work was supported by grants CA46593 (R. Baer) and CA63544 (R. Bash) from the National Cancer Institute. R. Baer is the recipient of a Faculty Research Award from the American Cancer Society (FRA-421). L.-Y. Hwang is supported in part by a fellowship from the Leukemia Association of North Central Texas.

References

Andersson, S., Davis, D. N., Dahlback, H., Jornvall, H., and Russell, D. W. (1989). Cloning, structure, and expression of the mitochondrial cytochrome P-450 sterol 26-hydroxylase, a bile acid biosynthetic enzyme. J. Biol. Chem. 264:8222–8229.

Aso, T., Vasavada, H. A., Kawaguchi, T., Germino, F. J., Ganguly, S., Kitajima, S., Weissman, S. M., and Yasukochi, Y. (1992). Characterization of cDNA for the large subunit of the transcription initiation factor TFIIF. Nature 355:461–464.

Bartel, P., Chien, C. T., Sternglanz, R., and Fields, S. (1993). Elimination of false positives that arise in using the two-hybrid system. BioTechniques 14:920–924.

Chakraborty, T., Martin, J. F., and Olson, E. N. (1992). Analysis of the oligomerization of myogenin and E2A products in vivo using a two-hybrid assay system. J. Biol. Chem. 267:17498–17501.

Chen, Q., Cheng, J.-T., Tsai, L.-H., Schneider, N., Buchanan, G., Carroll, A., Crist, W., Ozanne, B., Siciliano, M. J., and Baer, R. (1990). The *tal* gene undergoes chromosome translocation in T cell leukemia and potentially encodes a helix-loop-helix protein. EMBO J. 9:415–424.

Chien, C.-T., Bartel, P. L., Sternglanz, R., and Fields, S. (1991). The two-hybrid system: A method to identify and clone genes for proteins that interact with a protein of interest. Proc. Natl. Acad. Sci. USA 88: 9578–9582.

Dang, C. V., Barrett, J., Villa-Garcia, M., Resar, L. M. S., Kato, G. J., and Fearon, E. R. (1991). Intracellular leucine zipper interactions suggest c-Myc hetero-oligomerization. Mol. Cell. Biol. 11:954–962.

de Wet, J. R., Wood, K. V., DeLuca, M., Helinski, D. R., and Subramani, S. (1987). Firefly luciferase gene: structure and expression in mammalian cells. Mol. Cell. Biol. 7:725–737.

Durfee, T., Becherer, K., Chen, P.-L., Yeh, S.-H., Yang, Y., Kilburn, A. E., Lee, W.-H., and Elledge, S. J. (1993). The retinoblastoma protein associates with the protein phosphatase type 1 catalytic subunit. Genes Dev. 7:555–569.

Fearon, E. R., Finkel, T., Gillison, M. L., Kennedy, S. P., Casella, J. F., Tomaselli, G. F., Morrow, J. S., and Dang, C. V. (1992). Karyoplasmic interaction selection strategy: A general strategy to detect protein-protein interactions in mammalian cells. Proc. Natl. Acad. Sci. USA 89:7958–7962.

Field, J., Nikawa, J.-I., Broek, D., MacDonald, B., Rodgers, L., Wilson, I. A., Lerner, R. A., and Wigler, M. (1988). Purification of a RAS-responsive adenylyl cyclase complex from *Saccharomyces cerevisiae* by use of an epitope addition method. Mol. Cell. Biol. 8:2159–2165.

Fields, S., and Song, O.-K. (1989). A novel genetic system to detect protein-protein interactions. Nature 340:245–246.

Finkel, T., Duc, J., Fearon, E. R., Dang, C. V., and Tomaselli, G. F. (1993). Detection and modulation in vivo of helix-loop-helix protein-protein interactions. J. Biol. Chem. 268:5–8.

Graham, F. L., Smiley, J., Russell, W. C., and Nairn, R. (1977). Characteristics of a human cell line transformed by DNA from human adenovirus type 5. J. Gen. Virol. 36:59–72.

Graham, F. L., and Van der Eb, A. J. (1973). Transformation of rat cells by DNA of human adenovirus 5. Virol. 54:536–539.

Hoang, A. T., Lutterbach, B., Lewis, B. C., Yano, T., Chou, T.-Y., Barrett, J. F., Raffeld, M., Hann, S. R., and Dang, C. V. (1995). A link between increased transforming activity of lymphoma-derived MYC mutant alleles, their defective regulation by p107, and altered phosphorylation of the c-Myc transactivation domain. Mol. Cell. Biol. 15:4031–4042.

Hollon, T., and Yoshimura, F. K. (1989). Variation in enzymatic transient gene expression assays. Anal. Biochem. 182:411–418.

Hopp, T. P., Prickett, K. S., Price, V. L., Libby, R. T., March, C. J., Cerretti, D. P., Urdal, D. L., and Conlon, P. J. (1988). A short polypeptide marker sequence useful for recombinant protein identification and purification. Bio/Technology 6:1204–1210.

Hsu, H.-L., Cheng, J.-T., Chen, Q., and Baer, R. (1991). Enhancer-binding activity of the tal-1 oncoprotein in association with the E47/E12 helix-loop-helix proteins. Mol. Cell. Biol. 11:3037–3042.

Hsu, H.-L., Wadman, I., and Baer, R. (1994). Formation of in vivo complexes between the TAL1 and E2A polypeptides of leukemic T cells. Proc. Natl. Acad. Sci. USA 91:3181–3185.

Kato, G. J., Lee, W. M. F., Chen, L., and Dang, C. V. (1992). Max: functional domains and interaction with c-Myc. Genes Dev. 6:81–92.

Lillie, J. W., and Green, M. R. (1989). Transcription activation by the adenovirus E1a protein. Nature 338:39–44.

Ma, J., and Ptashne, M. (1987). Deletion analysis of GAL4 defines two transcriptional activating segments. Cell 48:847–853.

Murre, C., McCaw, P. S., and Baltimore, D. (1989). A new DNA binding and dimerization motif in immunoglobulin enhancer binding, *daughterless*, *MyoD*, and myc proteins. Cell 56:777–783.

Osada, H., Grutz, G., Axelson, H., Forster, A., and Rabbitts, T. H. (1995). Association of erythroid transcription factors: complexes involving the LIM protein RBTN2 and the zinc-finger protein GATA-1. Proc. Natl. Acad. Sci. USA 92:9585–9589.

Pevny, L., Simon, M. C., Robertson, E., Klein, W. H., Tsai, S.-F., D'Agati, V., Orkin, S. H., and Costantini, F. (1991). Erythroid differentiation in chimeric mice blocked by a targeted mutation in the gene for transcription factor GATA-1. Nature 349:257–260.

Ptashne, M. (1988). How eukaryotic transcriptional activators work. Nature 335:683–689.

Raycroft, L., and Lozano, G. (1992). A convenient cloning vector containing the *GAL4* DNA-binding domain. Gene 118:143–144.

Sadowski, I., Ma, J., Treizenberg, S., and Ptashne, M. (1988). GAL4-VP16 is an unusually potent transcriptional activator. Nature 335:559–560.

Sadowski, I., and Ptashne, M. (1989). A vector for expressing GAL4(1-147) fusions in mammalian cells. Nucl. Acids Res. 17:7539.

Sadowski, I., Bell, B., Broad, P., and Hollis, M. (1992) GAL4 fusion vectors for expression in yeast or mammalian cells. Gene 118:137–141.

Sanchez-Garcia, I., and Rabbitts, T. H. (1994). The LIM domain: a new structural motif found in zinc-finger-like proteins. Trends Genet. 10:315–320.

Silver, P. A., Keegan, L. P., and Ptashne, M. (1984). Amino terminus of the yeast GAL4 gene product is sufficient for nuclear localization. Proc. Natl. Acad. Sci. USA 81:5951–5955.

Takacs, A. M., Das, T., and Banerjee, A. K. (1993). Mapping of interacting domains between the nucleocapsid protein and the phosphoprotein of vesicular stomatitis virus by using a two-hybrid system. Proc. Natl. Acad. Sci. USA 90:10375–10379.

Treizenberg, S.J., Kingsbury, R.C., and McKnight, S.L. (1988) Functional dissection of VP16, the trans-activator of herpes simplex virus immediate early gene expression. Genes Dev. 2:718–729.

Valge-Archer, V. E., Osada, H., Warren, A., Forster, A., Li, J., Baer, R., and Rabbitts, T. H. (1994). Two oncogenes RBTN2 and TAL1 involved in T cell acute leukaemias produce proteins which complex with each other in erythroid cells. Proc. Natl. Acad. Sci. USA 91:8617–8621.

Wadman, I., Li, J., Bash, R. O., Forster, A., Osada, H., Rabbitts, T. H., and Baer, R. (1994). Specific in vivo association between the bHLH and LIM proteins implicated in human T cell leukemia. EMBO J. 13:4831–4839.

Whyte, P., Buchkovich, K. J., Horowitz, J. M., Friend, S. H., Raybuck, M., Weinberg, R. A., and Harlow, E. (1988). Association between an oncogene and an antioncogene: the adenovirus E1A proteins bind to the retinoblastoma gene product. Nature 334:124–129.

Wu, L.C., Wang, Z.W., Tsan, J.T., Spillman, M.A., Phung, A., Xu, X.L., Yang, M.-C. W., Hwang, L.-Y., Bowcock, A.M., and Baer, R. (1996). Identification of a RING protein that can interact in vivo with the BRCA1 gene product. Nature Genet. 14:430–440.

Xia, Y., Hwang, L.-Y., Cobb, M. H., and Baer, R. (1994). Products of the TAL2 oncogene in leukemia T cells: bHLH phosphoproteins with DNA-binding activity. Oncogene 9:1437–1446.

Yin, M.-J., Paulssen, E. J., Seeler, J.-S., and Gaynor, R. B. (1995). Protein domains involved in both in vivo and in vitro interactions between human T-cell leukemia virus type I Tax and CREB. J. Virol. 69:3420–3432.

14

Detection of Protein-Protein Interactions Dependent on Post-Translational Modifications

Mark A. Osborne
Manuel Lubinus
Jarema P. Kochan

The two-hybrid system has shown tremendous utility in the characterization of protein-protein interactions that occur in mammalian and other cells. However, a feature common to many cell signaling systems is the post-translational modification of proteins which initiates a signaling cascade based on the interactions that occur subsequently (Pawson 1995). The best studied of these modifications is specific tyrosine phosphorylation of receptors and other proteins following the activation of cell-surface receptors (Hunter 1995; Heldin 1995). One of the limitations of the original form of the two-hybrid system has been its limited utility to detect interactions that are dependent on post-translational modifications (such as tyrosine phosphorylation) that do not commonly occur in *Saccharomyces cerevisiae*; only those processes that occur within the yeast cytosol or nucleus will be able to modify the two fusion partners. Higher eukaryotic cells use tyrosine phosphorylation as a major modification in signal transduction pathways that respond to extracellular stimuli. *S. cerevisiae* does posess dual-specificity kinases (Gartner et al. 1992; Lim et al. 1993), but to date no mono-specific tyrosine kinases have been described. Although the two-hybrid system has worked well with autophosphorylating tyrosine kinases as baits (Pandey et al. 1994; O'Neill et al. 1995; Gustafson et al. 1995; Pandey et al. 1995; Xing et al. 1994), can it be used for proteins that are tyrosine kinase substrates but not kinases themselves? Mammalian two-

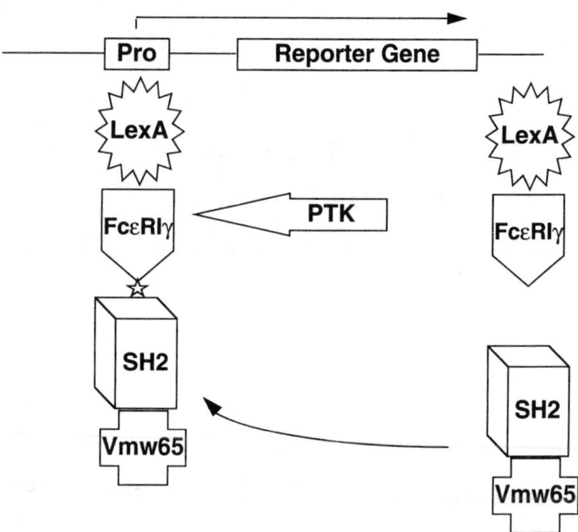

Fig. 14-1. Schematic diagram of the yeast tribrid system. The LexA-FcεRIγ CT fusion protein is phosphorylated by the tyrosine kinase Lck (PTK), the phosphotyrosine residue represented by a star. Bound to its cognate operator (Pro), it is able to interact with an SH2-transcriptional activation domain fusion protein (SH2-Vmw65). This bimolecular complex then activates transcription of the reporter gene (*lacZ*, encoding β-galactosidase). The absence of tyrosine phosphorylation results in no interaction and, therefore, no β-galactosidase is produced.

hybrid systems have been described (Vasavada et al. 1991; Fearon et al. 1992; chapter 13), but these are impractical for screening large numbers of transfectants, which is necessary to identify rare interacting clones. We and others (Lioubin et al 1996; Keegan and Cooper 1996) have solved this problem by the introduction of a third component, a cytosolic tyrosine kinase, which can *trans*-phosphorylate substrates in the yeast cell (the tribrid system, Figure 14-1). This now permits the investigation of phosphotyrosine-dependent signal transduction pathways, which are often critical in higher eukaryotic cell function. Although we have only utilized the three-component approach to investigate tyrosine phosphorylation, it should be possible to use a similar approach to analyze other translational or allosteric regulation.

Novel Reagents

Plasmids

The tribrid system uses three plasmids that direct the expression of the two typical fusion proteins and a third protein, the tyrosine kinase (see Figure 14-2). The DNA-binding domain used is that of the bacterial protein LexA (see chapters 3 and 4), and the transcriptional activation moiety is amino acids 410–490 from the Vmw65 protein of herpes simplex virus type I (Dalrymple et al. 1985); similar to the HSV2 protein VP16). The LexA fusion vector is p4402, containing the *TRP1* selectable marker and the pRS314 backbone (Sikorski and Hieter 1989). The Vmw65 fusion vector is p4064, a derivative of pSD10a (Dalton and Treisman 1992), containing the *URA3* selectable marker, and is derived from the pRS316 backbone (Sikorski and Hieter 1989). The tyrosine kinase vector bears the *LEU2* selectable marker, and is derived from pRS415 (Stratagene). All three plasmids contain a strong inducible *GAL10/CYC1* hybrid promoter (Johnston 1987; Dalton and Treisman 1992), a multiple cloning site, and the 3' transcriptional termination signal of *CYC1* (Dalton and Treisman 1992; Russo and Sherman 1989). The multiple cloning sites of all three plasmids are similar, with common sites for cloning cDNAs into both the Vmw65 and LexA plasmids. The details of the construction of all plasmids have been published previously (Osborne et al. 1995; Dalton and Treisman 1992). Both the LexA-fusion plasmid and the Vmw65-fusion plasmid contain the nuclear localization signal of SV40 T antigen (Kalderon et al. 1984), and the Vmw65-fusion vector also contains the 9E10 c-myc epitope for fusion protein detection. The tyrosine kinase plasmid does not have a nuclear localization sequence.

There are a number of significant advantages to the tribrid system. These include the inducible promoters, which make it possible to conditionally express all three of the proteins and thereby eliminate any growth disadvantages that might be conferred on the host cell by overexpression of certain proteins when using constitutive promoters. The Vmw65-fusion protein vector, in which cDNA libraries are generated, is amenable to the use of *Bst*XI linkers, which greatly reduce the background of non-insert containing library members (Aruffo and Seed 1987). The plasmids are all CEN-based, so copy number fluctuations should be minimal within the host cell.

Microbial Strains Used

The *S. cerevisiae* reporter strain used in our studies is S-260 (*MATα ura3::Col E1 operator (x6)-lacZ leu2-3,112 trp1-1 ade2-1 can1-100 ho*) and for mating assays, W303-1a (*MATa leu2-3,112 trp1-1 his3-11,15 ura3-*

A. LexA fusion("bait"); p4402

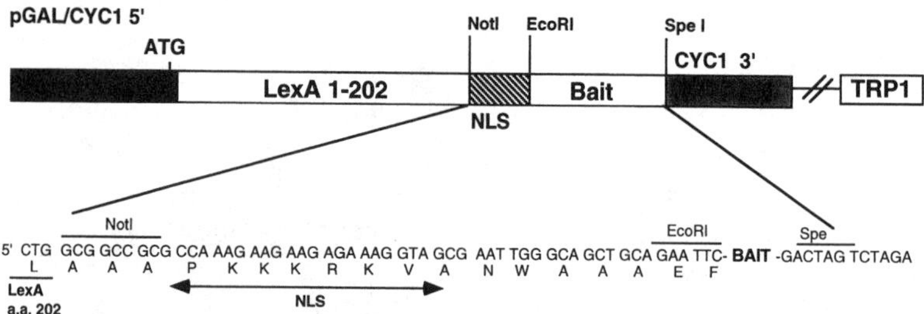

B. Vmw65-cDNA fusion ("prey"); p4064

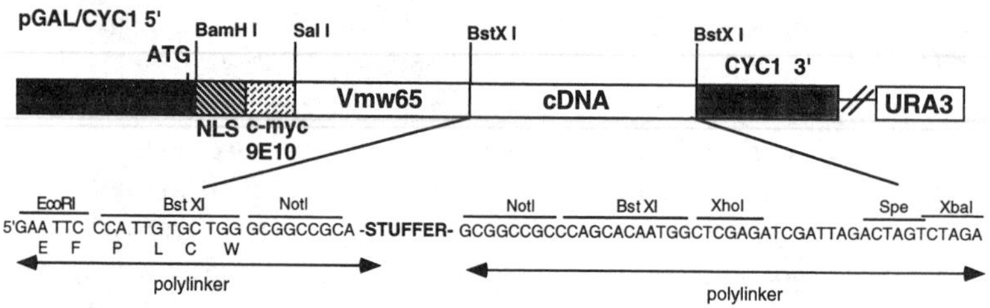

C. Kinase; p4140

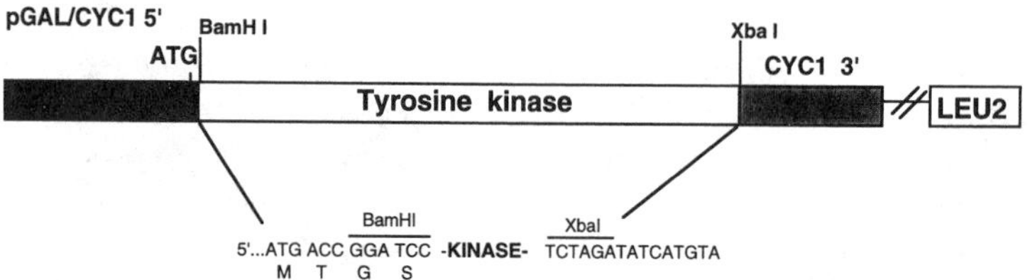

Fig. 14-2. Plasmids in this study: all plasmids contain a galactose-inducible promoter, and transcription initiation and termination sequences from the yeast *CYC1* gene. Both the LexA(p4402) and the Vmw65(p4064) plasmids contain a nuclear localization signal (NLS) from SV40 T antigen; the Vmw65 plasmid also contains the 9E10 c-myc-epitope tag.

52 can1-100 ho) . Rescue of library plasmids from S-260 was accomplished by electroporation into *Escherichia coli* strain KC8 (*pyrF::Tn5 hsdR leuB600 trpc9830 lacΔ74 strA galK hisB436*).

PCR and Sequencing Primers

Primers suitable for sequencing and PCR of DNAs subcloned into the three plasmids that we have found to work well are:

1. LexA (anneals ~75 bp 5' to the polylinker)
   5' TCG TTG ACC TTC GTC AGC AGA GCT TCA 3'
2. Vmw65 (anneals ~50 bp 5' to the polylinker)
   5' TCG AGT TTG AGC AGA TGT TTA CCG ATG 3'
3. *GAL10/CYC1* 5' (for kinase inserts; anneals 30 bp 5' to *Bam*HI site)
   5' TTA CTA TAC TTC TAT AGA CAC GCA 3'
4. *CYC1* 3' UTR (will sequence from the 3' end of the polylinker of all three plasmids; anneals 25 bp 3' to the *Xba* I site)
   5' GAG GGC GTG AAT GTA AGC GTG AC 3'

Commercial Sources for Reagents Reinforced nitrocellulose must be used for the ß-galactosidase filter assays, since unsupported membranes shatter easily. We have used Schleicher and Schuell (grade BA-S), Micron Separations (NitroPure), and Sartorius (reinforced cellulose nitrate) with equal success, and others may be acceptable as well. Large, 245 × 245 mm bioassay dishes are manufactured by Nunc (#166508) and are distributed by a number of suppliers.

Media Recipes YEPD (rich medium) is 10 g yeast extract, 20 g bacto peptone, 20 g glucose, 30 mg adenine sulfate per liter. For plates, add 15 g agar.

Plates for KC8 uracil selection are prepared by mixing 6 g Na_2HPO_4, 3 g KH_2PO_4, 0.5 g NaCl, 1 g NH_4Cl, 2 g casamino acids, and 15 g agar in 1 liter of water. Autoclave 20 minutes, and add 1 ml 1M $MgSO_4$, 10 ml 20% (w/v) glucose, 0.1 ml 1M $CaCl_2$, and 5 ml 4 mg/ml tryptophan (all filter-sterilized). Ampicillin is added to 100 μg/ml.

Yeast selective plates are prepared by standard methods (Guthrie and Fink 1991), with the addition of 0.2 ml 10N NaOH per liter to raise the pH to ~6. This greatly assists the hardening of agar.

Methods

We will describe only steps in our method that differ from those typically used in two-hybrid screening which have been described in chapters 2, 3, and 4.

Determination of Bait Suitability for Tribrid Screening

First, the bait cDNA must be subcloned into the LexA-fusion vector (p4402) to make an in-frame fusion protein. It is best to confirm the sequence of the insert, particularly if it was generated by PCR, in order to ensure that the two reading frames mesh correctly. Second, the bait must be tested to verify that the protein joined to LexA does not confer the ability to activate transcription of the reporter gene. If the interactor hunt that is being considered makes use of the trans-phosphorylation of the bait by a tyrosine kinase, it is critical to ensure that the self-activation test is performed with the kinase present. We have found several proteins that, when fused to LexA, activate transcription only when they are coexpressed with a tyrosine kinase. Thus, it is important to verify that the LexA fusion protein does not activate transcription on its own both in the presence and absence of kinase. If transcription is not activated, it is important to confirm that the LexA-fusion protein and the tyrosine kinase are being expressed. Protocol 1 accomplishes these two steps in a single round of experiments.

Protocol 1

1. Transform strain S-260 with the LexA-fusion protein expression vector along with the plasmid directing the expression of the tyrosine kinase. Select colonies on Trp^- Leu^- glucose plates.
2. Pick a colony to inoculate 5 ml Trp^- Leu^- glucose liquid medium and grow overnight at 30°C.
3. Overlay the transformation plate with a nitrocellulose filter. Allow the filter to become completely wetted from beneath.
4. Remove the filter and place, colony side up, on a Trp^- Leu^- galactose plate. Incubate the plate at 30°C overnight (18 hours).
5. The next morning, perform a β-galactosidase filter assay (see chapter 2, 3, or 4) on the nitrocellulose filter. If the colonies do not turn blue within ~30 minutes, proceed to step 6.
6. Measure the OD_{600} of the overnight culture started in step 2. Centrifuge the remaining cells for 5 minutes at 3,000 × g.
7. Wash the cell pellet in 5 ml H_2O, and recentrifuge. Resuspend the pellet in 1 ml Trp^- Leu^- liquid medium lacking carbon source.
8. Prepare two tubes for each transformant to be analyzed with 4.5 ml Trp^- Leu^- liquid medium in each. To one tube, add 0.5 ml 20% (w/v) glucose and to the other, add 0.5 ml 20% (w/v) galactose.
9. Add a portion of the washed cell culture to each of the tubes containing 5 ml of medium to obtain a final OD_{600} of 0.2. Shake the cultures at 30°C for 4 to 6 hours.
10. Centrifuge the cell cultures 5 minutes at 3,000 × g to pellet the cells. Wash the pellet in 1 ml sterile H_2O and transfer the culture to a microcentrifuge tube. Spin again (15 seconds, maximum speed) to pellet the cells and remove the supernatant.
11. Add 100 to 300 μl of 1× Laemmli sample buffer to each pellet, resuspend by pipetting up and down, and boil for 5 minutes.
12. Centrifuge the lysates 2 to 5 minutes at maximum speed to precipitate unlysed cells and debris. Load an SDS-PAGE gel and analyze the products for LexA-fusion protein and tyrosine kinase expression.

Notes on Protocol 1 It is necessary to wash the yeast cells in step 7 to remove all residual glucose, since glucose will interfere with induction of the *GAL 10* promoter by galactose through the catabolite repression pathway (Johnston 1987). Although other protocols for yeast cell lysis may call for the use of glass beads to increase the number of lysed cells, we have found that this dramatically increases the proteolysis observed as well. Controls to include in the expression test include using a transformant containing LexA protein alone (relative mobility of 25,000 daltons) to verify the reduced mobility of a LexA-fusion protein (when using anti-LexA antibodies). Running lysates from both the glucose and galactose-grown cultures will provide a relative level of protein expression induction. One should easily detect an anti-LexA immunoreactive species in the galactose-grown cultures but not in those grown in glucose.

A second immunoblot with anti-phosphotyrosine antibodies will confirm the expression of the tyrosine kinase, as most protein tyrosine kinases will phosphorylate endogenous yeast proteins. We have observed that different tyrosine kinases give different patterns of phosphorylation of yeast proteins. For example, the src-family protein tyrosine kinases Lck and Lyn phosphorylate a large number of endogenous yeast proteins (Osborne et al. 1995). Jak3, in contrast, gives rise to only a few phosphotyrosine containing bands (S. Bowen and J.P.K., unpublished results). Others have observed similar results with c-src and v-src protein kinases (Okada et al. 1993; Murphy et al. 1993). Although v-src expression is lethal in *S. cerevisiae* (Boschelli et al. 1993), we have not observed in our experiments any growth impairment due to expression of any tyrosine kinases in yeast. The control extract of uninduced cells grown in parallel (glucose as a carbon source) eliminates confusion that may be due to yeast proteins that immunoreact with the anti-LexA and/or anti-phosphotyrosine antibodies.

Verification of *Trans*-Phosphorylation of LexA-Fusion Proteins

If the LexA-fusion protein is expressed (and does not activate transcription on its own) and endogenous yeast proteins are phosphorylated by the tyrosine kinase, it is also important to verify that the tyrosine kinase is phosphorylating the LexA-fusion protein. Immunoprecipitation (IP) with anti-LexA antibodies followed by immunoblotting with anti-phosphotyrosine antibodies will clearly show whether *trans*-phosphorylation of the LexA-fusion protein has occurred. Control strains grown with LexA alone (which contains no tyrosine residues) and without the kinase plasmid should also be assayed. Alternatively, phosphorylation of the activation domain hybrid may be important for the interaction. If this is the case, skip to Protocol 3.

Protocol 2

1. Inoculate 5 ml Trp$^-$ Leu$^-$ glucose liquid medium with a yeast colony bearing the appropriate LexA-fusion plasmid and kinase plasmid. Grow overnight at 30°C.
2. The next morning, inoculate 10 ml of Trp$^-$ Leu$^-$ galactose liquid medium with an appropriate volume of washed yeast cells (see Protocol 1, steps 6 through 10) to give an OD_{600} of 0.2. Grow at 30°C for 4 to 6 hours.

3. Centrifuge cells, wash the pellet with sterile H_2O, respin, wash with 1 ml sterile H_2O, spin down in preweighed sterile microcentrifuge tubes.
4. Remove the supernatant and weigh the tubes, recording the mass of the wet cell pellet.
5. Resuspend the pellet in 2 ml/g wet weight of 0.1M Tris-Cl (pH 9.4), 10 mM DTT. Incubate at 30°C for 15 minutes.
6. Centrifuge for 2 minutes at 6,000 rpm and resuspend the pellet in 5 ml/g wet weight of: 1.2 M sorbitol, 20 mM HEPES pH 7.4 containing 0.25 mg yeast lytic enzyme (ICN # 360942; also known as Zymolyase) and 15 μl of glusulase (NEN # NEE154) per ml. Incubate at 30°C with occasional rocking for 15 to 45 minutes.
7. Check for spheroplasting by microscopic inspection—mix 10 μl of cell suspension with 10 μl of H_2O and look to see if there are any cells remaining. Membrane ghosts should be plentiful and unlysed cells should be few. If not, continue incubation for up to 1 hour.
8. Spin down spheroplasts for 5 minutes at 6,000 rpm in a microcentrifuge.
9. Resuspend pellet in PBS containing 1% Triton X-100 (protease inhibitors may help as well as 1 mM $NaVO_4$, a phosphatase inhibitor), pipette up and down to fully resuspend spheroplasts. Incubate on ice for 5 minutes. Suspension should be clear, not cloudy like a suspension of cells. Check for lysis by microscopic examination if desired.
10. Perform an immunoprecipitation using standard methods. Analyze immunoprecipitates by SDS-PAGE followed by immunoblotting.

Notes on Protocol 2 We have investigated the tyrosine phosphorylation of LexA-fusion proteins by immunoprecipitating with anti-LexA antibodies and then immunoblotting with anti-phosphotyrosine antibodies. It should also be possible to use bait-specific antibodies instead of anti-LexA antibodies, or to reverse the order of the reagents—IP with anti-phosphotyrosine first, followed by anti-bait antibodies. We have found that preparing spheroplasts, although time consuming, is worth the effort since proteolysis is reduced relative to glass bead lysis. It is important to include several controls: glucose induction to control for endogenous yeast proteins immunoreacting with the antibody used; a no-kinase control (transforming the LexA-fusion plasmid along with the kinase vector, pRS415, allows growth in Trp^- Leu^- medium), and a control using the LexA-fusion vector without an insert. LexA has no tyrosine residues, so it should not be modified by a tyrosine kinase.

The tyrosine phosphorylation of proteins is often dependent on the particular kinase used. It may be necessary to test several different kinases in order to ascertain the best for phosphorylation of a specific "bait." Once you are satisfied that phosphorylation of the LexA-bait fusion protein is occurring, the interactor screen can be performed as described below.

cDNA Library Transformation

Techniques for cDNA library construction have been described in chapter 5. (Specific details regarding construction of libraries in the Vmw65-fusion vector can be found in Osborne et al. 1995.) Although there are several libraries constructed in the Vmw65 fusion protein vector [HeLa (Dalton and Treisman 1992), rat mast cell (Osborne et al. 1995), and human peripheral blood lymphocytes (M. L. and J. P. Kochan, unpublished)], it is important to select a library from which it is reasonable to expect to identify interactors. One way to examine the suitability of a specific library is to determine whether the bait cDNA is present by PCR analysis. Optimistically, one can assume that if the bait is expressed in the cells from which the library is derived, its binding partners will be also. It is certainly less work to make use of existing libraries than to construct one's own; however, there is no substitute for the construction of a high-quality library from the appropriate cell type in order to obtain interactors of relevance to the system under study.

A pilot transformation of the cDNA library into S-260 carrying the LexA-fusion and kinase plasmids will give an estimate of how many colonies to expect per mg of library DNA. This number can then be used to scale up the transformation for the screen. We attempt to plate 20,000 colonies per large plate and find it cumbersome to deal with more than 20 plates at one time. If more than 400,000 colonies are to be screened, transformations performed on successive days may be less problematic.

Protocol 3: Large-Scale Library Transformation

1. Inoculate a culture of 25 ml Trp$^-$ Leu$^-$ glucose liquid with a transformant containing the LexA-bait fusion and the tyrosine kinase plasmid (for example, a colony from the plate used in Protocol 1, step 1). Grow the culture at 30°C with shaking overnight.
2. Dilute the overnight culture 1:10 in a final volume of 200 ml Trp$^-$ Leu$^-$ glucose liquid medium and grow overnight at 30°C with shaking.
3. The next morning, count the cells in the overnight culture from step 2 and inoculate 3 separate 2 liter flasks containing 300 ml Trp$^-$ Leu$^-$ glucose liquid with the overnight culture to a final cell density of 2×10^6 cells/ml. Incubate at 30°C for 5 hours with shaking.
4. Centrifuge cells and process for LiOAc transformation as usual (see chapters 2, 3, and 4). Transform with an appropriate amount of cDNA library to generate 20,000 colonies per plate $\times$ 15 plates = 300,000 colony-forming units.
5. After heat shock, plate cells on 245×245 mm (Nunc #166508) Trp$^-$ Leu$^-$ Ura$^-$ glucose plates. In addition, plate a small (0.0002%) amount of the transformation mixture on small (10 cm) Trp$^-$ Leu$^-$ Ura$^-$ glucose plates, in order to estimate the total number of colonies screened. Incubate the plates at 30°C for 24 to 36 hours.
6. When colonies are clearly visible but not touching, overlay with reinforced nitrocellulose (22×22 cm) sheets. Allow to completely wet from beneath, then transfer (colony side up) to Trp$^-$ Leu$^-$ Ura$^-$ galactose plates for 18 hours at 30°C.

7. Perform β-galactosidase filter assay.
8. Pick blue colonies as they appear with a sterile toothpick onto a small Trp⁻ Leu⁻ Ura⁻ glucose plate, and incubate this plate for 2 to 3 days at 30°C.

Notes on Protocol 3 Do not autoclave the nitrocellulose before placing it on the transformation plates. Autoclaving causes wrinkling which interferes with the ability of the nitrocellulose to lie flat on the agar surface. It is helpful to ensure that the Trp⁻ Leu⁻ Ura⁻ glucose plates are not wet when plating the transformation mixture. A volume of cell suspension of 3 to 4 ml per plate has worked well in our hands. It is important to spread the transformation mixture carefully to ensure even distribution of colonies, as a highly dense area of cells will give rise to very small colonies among which a blue colony may be difficult to detect. The liquid nitrogen permeabilization step is best accomplished in a large, glass baking dish, followed by allowing the filters to equilibrate to room temperature while sitting on a piece of dry Whatman 3 mm paper. This allows the colonies, which tend to "sweat" as they warm up, to dry a bit before becoming wet with Z buffer in the assay. This step minimizes the possibility of colonies smearing and running into each other during the color development step. We cut pieces of Whatman 3 mm paper to fit in the lid of the Nunc bioassay dishes and then add ~25 ml of Z buffer to saturate. Be sure that the Whatman 3 mm paper is perfectly flattened in the lid, to ensure even contact with the nitrocellulose filter. Once the filters are placed on the Whatman paper saturated with Z buffer, additional Z buffer may be needed in order to maintain an appropriate level of moisture. Too much buffer results in runny colonies; too little results in poor color development. Keep the filters covered with the bottom part of the plates for two reasons: (1) so that the filters do not dry out; and (2) to avoid incurring the wrath of other laboratory members offended by the pungent aroma of β-mercaptoethanol in the Z buffer.

Considerations for Library Transformations

The transformation mixture may be plated in several different ways. One is to plate it directly on Trp⁻ Leu⁻ Ura⁻ glucose plates, allow the transformants to grow for 24 to 36 hours, and then transfer them to reinforced nitrocellulose filters, which are then placed, colony side up, on Trp⁻ Ura⁻ Leu⁻ galactose plates for 18 hours. The filter is then processed for a β-galactosidase assay as usual. Blue colonies may be scraped directly from the nitrocellulose filter, or, alternatively, identified and picked from the glucose "master" plate. Picking from the "master" plate can be tricky, as the colonies on the glucose plate are often small and easily smeared by the process of lifting to nitrocellulose and the positives identified on the filter after assay may be difficult to match against the "master" glucose plate later. The ease of the former method (scraping directly from the nitrocellulose) is tempered by the fact that the cells that constitute the colony being picked from the nitrocellulose have been grown on galactose. Thus, if one of the plasmids encodes a toxic protein, this may make the rescue of living cells from a colony exposed to galactose difficult. We have not observed this phenomenon, but it bears keeping in mind. Picking from the "master" plate as well as from the nitrocellulose filter is probably the safest alternative. Another method to consider is to plate the transformaton mixture directly onto nylon membranes (such as Hybond from Amersham) placed on the Trp$^-$ Ura$^-$ Leu$^-$ glucose plates. The colonies will then grow directly on the filters, which are then transferred to Trp$^-$ Ura$^-$ Leu$^-$ galactose plates after 24 to 36 hours. The β-galactosidase assay is then performed. Although this method involves the fewest manipulations, we find that the overall transformation efficiency for plating directly on nylon filters is reduced to one fifth that of plating onto agar plates directly.

Recovery of Blue Colonies

There are several technical considerations to consider when performing large-scale screenings of yeast colonies. It is important to keep the filters moist with Z buffer during the time it takes for blue colonies to appear (typically 15 to 90 minutes from the start of the assay), and to ensure that the filter paper placed under the nitrocellulose is flat so that the nitrocellulose can make contact with it uniformly. If bubbles form between the nitrocellulose filter and the filter paper, the X-gal solution will not reach the colonies, and positives may not be identified in this region. Blue colonies are then picked by either scraping the filter (or the corresponding "master" plate) and patching them onto a Trp$^-$ Ura$^-$ Leu$^-$ glucose plate (some cells that have been dipped in liquid nitrogen do live). Typically, the patches will show growth in 2 to 3 days at 30°C.

Toward the end of the assay, it becomes increasingly difficult to discern "true" positives from the emerging background of blue colonies. The intensity of the blue background is dependent on the particular "bait," as some seem to display higher incidences of background than others. It is often helpful to note the time elapsed since the start of the assay when identifying blue colonies, so that they can be ranked according to intensity during subsequent analysis. We have not found any "true" positives that take longer than 90 minutes to turn blue during an initial screen.

Purification of positive colonies Since it is not generally feasible to scrape the blue colony from the filter or "master" plate without touching neighboring colonies, it is important to obtain a pure population before proceeding with analysis. Each patch contained on the plate generated in step 8 of Protocol 3 is therefore a mixed population of both the desired "positive" cells and contaminating non-interactors. The following protocol directs the isolation of a "pure" positives.

Protocol 4: Colony Purification

1. After the patches of cells from step 8 of protocol 3 have grown, scrape the cells with a toothpick into 1 ml of Trp^- Ura^- Leu^- media lacking carbon source. If there are a number of single, isolated colonies in the patch, scrape them all together.
2. Dilute 100 μl of the cell suspension in 900 μl of H_2O and determine the OD_{600}. Assume that 1 $OD_{600} = 1.5 \times 10^6$ colony forming units/ml, and calculate the volume needed to deliver 200 cells.
3. Dilute the cell suspension to 200 colony forming units/100 μl, and plate onto a 10 cm Trp^- Ura^- Leu^- glucose plate. Incubate at 30°C for 1 to 2 days.
4. Overlay each plate with a nitrocellulose filter, and transfer to a Trp^- Ura^- Leu^- galactose plate for 18 hours. Perform a β-galactosidase assay.
5. Identify a well-isolated blue colony on each filter, and pick the corresponding colony from the glucose plate. Patch the colony onto a fresh Trp^- Ura^- Leu^- glucose plate. Incubate at 30°C for 1 to 2 days.
6. Scrape a portion of the patch into 1 ml of 50% (v/v) glycerol and freeze at −70°C as a backup in case of disaster.

Notes on Protocol 4 The conversion factor for OD_{600} to colony forming units must be determined empirically, and may vary depending on the bait or other considerations. Once pure populations of each positive have been obtained, they can now be used to ask two important questions: (1) Does the Vmw65 fusion protein ("prey") interact specifically with the LexA fusion protein used?; (2) Is the interaction dependent on protein tyrosine kinase activity?

Testing for Specificity and Kinase-Dependence

Once a pure population of colonies bearing interacting cDNAs exists, many researchers may be tempted to get sequence information as quickly as possible. Although this may be the most immediately satisfying approach, it is not always the most time- and cost-effective. The rescue of cDNA library plasmids from yeast via *E. coli* is quite time consuming, and, given the likelihood that some of the plasmids may prove uninteresting to the investigator (false positives and nonspecific interactors, for example), it is more efficient to eliminate these classes of clones quickly.

The two questions posed at the end of the previous section require two different transformations. The first, determining if the interaction is specific, can be addressed by curing the host strain of the LexA fusion plasmid (p4402; *TRP1* marker) and then mating the resulting strain (still carrying the Vmw65-cDNA fusion and tyrosine kinase plamids) to W303-1a (*MATa*) transformed with a number of different LexA fusion protein transformants (see chapters 2, 3, and 4). The resulting diploids can then be tested individually by nitrocellulose filter β-galactosidase assay following galactose induction. For the second question, determining if the interactions observed require the tyrosine kinase, the host strain should be screened for transformants that have lost the tyrosine kinase plasmid (*LEU2* marker); these cells will be Trp+ Ura+ Leu-. These double transformants can then be screened by filter assay to determine which, if any, transformants retain β-galactosidase activity. Those that are still active do not require the kinase for interaction.

The two different transformant profiles required for these two determinations can be identified in the same experiment. To do this, triple-transformant yeast (which are blue in the β-galactosidase assay) are grown under non-selective conditions (in rich media such as YEPD) for several generations to allow for the loss of plasmids, which will occur randomly since they are no longer being subjected to nutritional selection (see Figure 14-3). In our experience, the following protocol allows for identification of all classes of transformants.

Protocol 5: Plasmid Segregation

1. For each positive interactor, inoculate a 2 ml culture in YEPD medium. Shake at 30°C for 24 to 48 hours.
2. Dilute the culture 1:3000 and plate 75 μl on a fresh YEPD agar plate. Incubate at 30°C overnight.
3. The next day, replica plate each YEPD plate to two plates—one Trp- Ura- glucose (plate #1) and one Ura- Leu- glucose (plate #2). Incubate at 30°C overnight.
4. Compare the two replicas. Identify two colonies on each plate that do not grow on the other replica. Patch these colonies to a fresh agar plate (supplemented with the appropriate amino acids) for further analysis.

Notes of Protocol 5 Comparison of the two replica plates will reveal the two types of colonies sought. If a colony grows on plate #1 (LexA+ Vmw65-cDNA+) but does not grow on plate #2 (Vmw65-cDNA+ kinase+), then it must have lost the kinase plasmid, since it will not grow on the plate lacking leucine (plate #2). This colony may be picked for analysis of kinase dependence, and should be patched or streaked onto a Trp- Ura- glucose plate. It is easiest and most convincing to patch several colonies onto one plate along with positive and negative controls and then assay for β-galactosidase activity. Trp+ Ura+ Leu- colonies that remain blue after β-galactosidase assay do not require the tyrosine kinase for an interaction. This also permits evaluation of the relative intensities of the interactors with each other. If a colony grows on plate #2 but not on plate #1, then it has lost the LexA fusion plasmid, since it will not grow on a plate lacking tryptophan. Patch this colony onto a Ura- Leu- glucose plate for use in mating assays to determine the specificity of its interaction.

Mating Assay to Test for Specificity

We find it most convenient to maintain glycerol stocks of several LexA fusion protein bait plasmid transformants in the strain W303-1a. These can

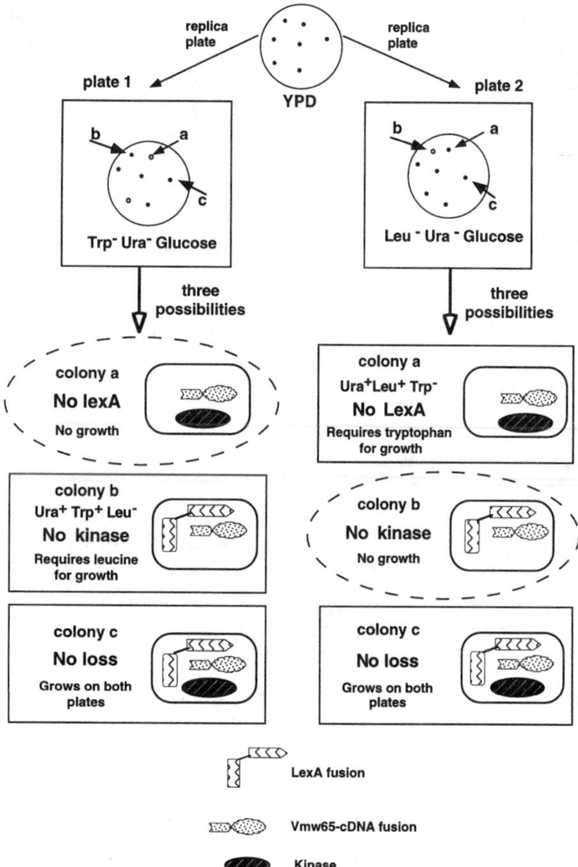

Fig. 14-3. Plasmid segregation (Protocol 5): Positive isolates are grown for 48 hours in YEPD liquid media, plated on YEPD agar and replica plated onto Trp- Ura- and Leu- selective media containing 2% glucose. After 24 hours, individual colonies are compared on each plate; those lacking the kinase (colony b) or the LexA-fusion protein (colony a) are patched on selective media for additional testing.

then be individually mated to the Ura- Leu- interactors to test for specificity. Mating is accomplished by mixing the LexA fusion bait in strain W303-1a with the library/kinase transformant (Ura^+ Leu^+ Trp^-) in S-260 on a YEPD plate. After incubating for several hours to overnight at 30°C, replica plate to a Trp- Leu- Ura- glucose plate and incubate at 30°C overnight. Only the diploids will grow. This plate is then overlaid with nitrocellulose and the filter induced on a Trp- Leu- Ura- galactose plate overnight, followed by β-galactosidase filter assay. If blue colonies are observed only for the combination of the bait of interest and the library/kinase, but not with the extraneous baits and the library/kinase, it is now appropriate to rescue the library plasmid in *E. coli* for sequencing and further manipulation.

At the end of the above analysis, it will be known how specific the interaction is for a particular LexA fusion, in addition to whether the tyrosine kinase is required for the interaction to be detected. The Vmw65-cDNA fusion plasmid can now be rescued from these cells into *E. coli* by standard methods and analyzed by restriction endonuclease digestion, DNA sequencing, and retransformation back into S-260 to verify the interaction. This last step is particularly important, as we have observed blue colonies whose β-galactosidase activity is not linked to the plasmid (and is therefore an artifact). β-galactosidase units may also be measured to quantify the interaction. We commonly observe 1000 units of activity for src-homology 2 (SH2)-phosphotyrosine interactions with a background of less than 10 units (Osborne et al. 1995). In the next several sections, we introduce techniques for additional analysis that are not specific to the tribrid system and can be adapted to any two-hybrid system.

Structure-Function Analysis

More often than not, two interacting proteins under study have not been crystallized and therefore detailed structural information about their binding surfaces are not available. Given the time-consuming and expensive nature of X-ray crystallography, alternative methods of identifying points of contact between two protein molecules have a wide use in facilitating an understanding of protein interactions. The two-hybrid system is an excellent tool to address this issue, since it permits the rapid identification of interactor pairs. For example, this system has been used to identify critical residues for peptide binding to the retinoblastoma protein. Another practical use of two-hybrid technology is the saturation mutagenesis of protein domains in order to investigate residues important for protein function. For example, residues within tyrosine kinases responsible for conferring substrate specificity could be identified by screening for mutant kinases that gain the ability to phosphorylate a previously unrecognized substrate. Restrictions on the amino acid sequence of the substrate molecule could be addressed similarly, by mutagenizing a nonsubstrate and screening for those that are able to be phosphorylated by the kinase under study.

There are a number of points to consider when contemplating a structure-function type of interactor screen. The first is that although the two-hybrid system allows one to identify interacting molecules easily, it is less straightforward to identify meaningful mutants that eliminate an interaction. Any of a number of possible changes in system components could result in lack of transcription of the reporter gene. For example, mutations that alter the length of the protein (by introducing a termination codon) would be seen as potential mutants in the screen, as would any mutations in the host strain or the plasmids that alter expression levels of the critical components. Therefore, it is simpler to assay for gain-of-function interactor mutants than to look for the elimination of an interaction. This method also has its drawbacks. Mutagenesis of fusion protein genes can lead to unforseen consequences related to the transcriptional nature of the two-hybrid assay. For example, mutations within the LexA-fusion protein moeity that confer transcriptional activation function on the bait itself would be falsely scored as positive interactors, and mutations that reduce the effectiveness of the transcriptional activation function of Vmw65 would, conversely, reduce their detectability and be missed. At present, we have only attempted to identify gain of function mutants within the context of the Vmw65 fusion protein.

Methods for Generation of Mutants

Although there are several chemical methods for mutagenesis of DNA, we have found that the polymerase chain reaction is the easiest . The error rate for typical PCR (using Taq polymerase) is generally estimated at 10^{-4} per nucleotide. This rate is suitable for introducing mutations into large (>1 kb) target sequences, but may be too low for shorter ones. Inclusion of Mn^{2+} or lowering the dNTP concentration of one nucleotide also has been reported to increase nucleotide misincorporation (Leung et al. 1989). Whether or not to use these error-enhancing methods is up to the individual researcher.

Protocol 6: PCR Mutagenesis

1. Subclone the target for mutagenesis into the Vmw65 activation domain vector. Verify expression and the fact that it does not interact with the "bait" that the mutants will interact with.
2. Perform 30 cycles of PCR using primers that anneal 200 to 300 bp 5' and 3' to the target. Reamplify if necessary to generate 1 to 2 μg of PCR product.
3. Subclone the PCR product back into the Vmw65 "prey vector" and prepare DNA from *E. coli* transformants. This is the mutant cDNA library.
4. Transform S-260 bearing LexA-bait and kinase (if needed) plasmids with the library generated in step 3. Select transformants on Trp^- Leu^- Ura^- glucose plates. Do β-galactosidase assay after galactose induction to identify mutants.

Notes on Protocol 6 To obtain the maximum amount of information possible from a mutant screen, it is advisable to perform several PCR reactions in parallel to eliminate the possibility of an early-cycle mutation "jackpot" polluting the PCR pool. The PCR products are then subcloned by standard methods into the two-hybrid vector of choice, followed by amplification and transformation into S-260 containing the appropriate bait and kinase. The number of library members to generate is up to the individual investigator; in the example described below, 20,000 was sufficient. To increase the mutation rate of the PCR, two changes have been reported to have the greatest impact (Leung et al. 1989). The first is reducing the relative amount of dATP in the reaction to 1/5 of that normally used, while keeping the other three nucleotides the same. The second is to include 0.5 mM $MnCl_2$ in the reaction. Making both alterations in the reaction conditions is reported to raise the overall mutation rate to as high as 2%, and to make equally likely both transition and transversion mutations (Leung et al. 1989). The disadvantages with highly mutagenic PCR is that the yield of product is reduced and that the overall mutation rate may be so high that multiple mutations may be found within a DNA, complicating assignment of a function to a single mutated amino acid.

Gap-Repair Alternative to Subcloning

One alternative to subcloning the PCR products into a vector (which is time consuming and occasionally problematic) is to generate the desired products in vivo via homologous recombination (see Figure 14-4) (Ma et al. 1987). This requires the use of a gapped plasmid (the vector cleaved with restriction endonucleases that leave noncomplementary ends) and a PCR product with ends that are complementary to those of the gapped plasmid. We have found it easy to start with the target DNA in the Vmw65-fusion vector, and to amplify it with one primer that anneals 5' to the Vmw65 activation domain (primer 3 above) and another primer that anneals 3' to the *CYC1* 3' UTR (the pBluescript T3 primer from Stratagene works well). This gives the target DNA overhangs of about 300 bp each, which will facilitate homologous recombination with the gapped vector. The result is the formation of a circular plasmid containing two regions where the recombination occurred, and DNA encoding the mutant interacting protein.

Protocol 7: Gap Repair

1. Digest the Vmw65 fusion protein vector with restriction endonucleases to generate noncomplementary ends. *Eco*RI and *Xba*I work well. Purify linear DNA.
2. Mix 0.1 μg of vector generated in step 1 with 1 μg of PCR product from Protocol 6. Transform S-260 by standard methods. Select transformants on appropriate selective medium.
3. Assay transformants for β-galactosidase activity following galactose induction with filter assay.
4. Pick positive colonies, rescue the plasmid in *E. coli,* and analyze the insert by DNA sequencing and retransformation.

Notes on Protocol 7 The Vmw65 fusion vector must be completely digested with restriction endonucleases that cleave in the polylinker and that generate noncomplementary ends. This last point is important, since complementary ends would facilitate reclosure of the vector, which would give rise to a Ura^+ colony without an insert, thereby increasing the numbers of

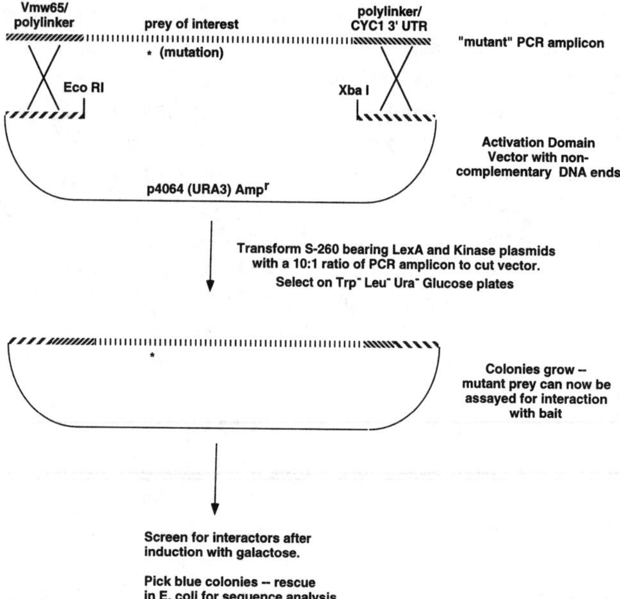

Fig. 14-4. Gap repair method for transformant recovery. As an alternative to subcloning mutant cDNAs into p4064, mutant plasmids can be recovered through in vivo recombination. After amplification of the insert by PCR using primers that include ~300 bp of flanking sequence, the PCR product is mixed with p4064 cleaved with *Eco*RI and *Xba*I (top). This mixture is used to transform S-260 carrying the appropriate LexA and kinase plasmids. The resulting transformants are selected and screened immediately for interactors. The yeast gap-repair enzymes close the gap in p4064 by utilizing sequence from the PCR product. This eliminates the need for high-efficiency cloning and *E. coli* amplification.

colonies that must be screened in order to identify a mutant. Remember to do a control transformation with the gapped vector alone to gauge the relative background from vector reclosure. In order to generate the large amount of PCR product needed for this method, a single round of mutagenic PCR may be performed and then the reaction split to several tubes and reamplified under standard (less mutagenic) conditions.

For example, to determine the amino acid residues of an SH2 domain that are critical for binding to a phosphotyrosine-containing bait, a wide-spectrum mutagenesis procedure can be followed. First, a library of mutant SH2 domains joined to Vmw65 is generated. This library is then screened for members that are able to bind to the bait and activate transcription of *lacZ*. The resulting plasmids are rescued in *E. coli* and sequenced. The location of the altered amino acids will allow the investigator to identify residues important for binding to the particular substrate under study. It would be of interest to determine the specificity of the mutant SH2 domains for the substrate under study, in order to determine whether the amino acids changed confer binding to a small or wide class of substrates. This can be readily accomplished by the mating assay described previously, where a number of tyrosine-phosyphorylated baits are mated to the various mutant SH2 domains. Mutagenesis of the kinase or the bait plasmids can be pursued as well, in order to define specific amino acid residues that enable an interaction to occur. Negative strategies can also be envisioned, if the investigator can easily eliminate false negatives (as discussed previously).

Examples

The primary use of the two-hybrid system is to identify cDNAs that encode interactors with a given protein "bait." The tribrid system adds the possibility of identifying interactors that require tyrosine phosphorylation of either the LexA-fusion protein, the cDNA-fusion protein, or both. We have used the system to clone many novel genes, most of which contain domains (SH2 and PTB) (Pawson 1995; van der Geer and Pawson 1996) that interact with phosphotyrosine-containing proteins encoded within various cDNA libraries (Osborne et al. 1995; Osborne et al 1996; M. Lubinus and J. Kochan, unpublished).

Another use is the characterization of specific interactor pairs. The interaction between an SH2 domain and its phosphotyrosine substrate makes major physical contacts (as determined by crystallographic study) with the phosphotyrosine and the residue located 3-amino acids distal to it (the Y+3 position) (Songyang et al. 1995). Src-family kinases prefer to phosphorylate proteins containing the sequence YXXL (single amino acid code, X is any amino acid). Syk, a tyrosine kinase found in B lymphocytes and mast cells, contains two SH2 domains, and has been demonstrated through biochemical methods (Kihara and Siraganian 1994; Shiue et al. 1995) to associate with the IgE receptor β and γ subunits (both of which contain dual YXXL motifs phosphorylated by src-family kinases). We cloned the cDNAs for the IgE receptor subunits and the Syk SH2 domains into the LexA and Vmw65 fusion vectors, respectively. In addition, mutants of the IgE receptor subunits were generated that converted either the critical tyrosine or leucine residue (within the YXXL motif) to phenylalanine or alanine, respectively. All fusions were tested for expression by immunoblot. The wild-type subunits and the mutants were then tested for interaction with both Syk SH2 domains, or either SH2 domain alone, joined to Vmw65. The interactions observed (Osborne et al. 1995) (see Table 14-1) were all consistent with the biochemical observations made by

Table 14-1 Interactions characterized between FcERIβ and Syk SH2 domains using the tribrid system

LexA fusion	Syk N SH2	Syk C SH2	Syk N+C SH2
FcERIβ	–	+	+
FcERIβFY	–	–	–
FcERIβAL	–	–	–
FcERIβYF	–	+	+
FcERIβLA	–	+	+
FcERIβFF	–	–	–
FcERIβAA	–	–	–

Note: + Indicates β-galactosidase activity, – indicates no activity.

others, but provided even more detail about the interactions—which SH2 domain contacted which phosphotyrosine, for example. The tribrid system can therefore provide detailed information regarding the nature of the contacts between two proteins modified by specific tyrosine phosphorylation.

A third example is the use of the tribrid system to perform structure-function analysis on either of the two components of the pair. For example, the Vmw65-Syk N-terminal SH2-domain fusion does not interact with any LexA-ITAM-fusion proteins when coexpressed with the Lck tyrosine kinase (Osborne et al. 1995). We sought to identify residues that, when mutant, would confer the ability of the SH2 domain to interact with the β subunit of IgE receptor. The SH2 domain was amplified by PCR and subcloned into the Vmw65 fusion plasmid to generate a library. This library was then transformed into S-260 carrying the LexA-β and Lck plasmids. Twenty thousand colonies were screened, and several hundred blue colonies were obtained. Plasmids from several of these were recovered and analyzed for mutations by DNA sequencing. The ability of these mutant SH2 domains to interact with various substrates is now under investigation (M. Osborne and J. Kochan, unpublished). The reverse experiment, to identify residues within the phosphotyrosine-containing protein that enable the SH2 domain to bind, can also be performed.

Problems

The *GAL10* promoter allows recovery of transformants without the expression of their encoded proteins, a significant advantage for use with proteins that might be toxic or confer a growth disadvantage to the host cell. This advantage comes at a price, however. Since the induction of transcription of all three components occurs after the transformants are selected and grown on plates, this necessitates the use of large numbers of plates in order to screen the colonies. The only reporter gene available in the host strain we use (S-260) is *lacZ*, so a screen is all we have attempted. A selection could be arranged similar to those used in typical two-hybrid screens (see chapters 2, 3, and 4); however, this would require the induction of transcription for a much longer period of time in order to allow the host strain to express the selectable reporter protein to sufficient levels to allow growth. This would eliminate the advantage that a short induction of gene expression confers, since several days of growth are necessary to identify interactors based on a growth selection.

The finding that a particular interaction requires tyrosine kinase activity does not ensure that the interaction detected is physiologically significant. To bolster confidence, it is important to construct mutant Lex-A or Vmw65-fusions in which the appropriate tyrosine residues have been mutated or the region surrounding them has been altered such that a bona fide interaction would not be expected to take place with such a protein. The final proof that the observed interaction occurs in vivo must be demonstrated in the cell.

Potential Applications

In addition to the examples described above, several additional uses for the tribrid system are possible. These include:

1. *Identification of novel tyrosine-containing proteins that interact with specific SH2 domain proteins.* In the manner opposite to that used for cloning of SH2 domain-containing proteins that interact with a phosphotyrosine-containing bait, an SH2 domain of interest is cloned into the LexA-fusion vector, and a cDNA library is screened with the tyrosine kinase plasmid present. Interactors are then characterized in the usual manner. One should first verify that the SH2 domain to be used as a bait is able to bind to phosphotyrosine-containing proteins. One way that this system can be verified is to transform a known protein partner that requires phosphorylation of tyrosine residue for the interaction to occur. These experiments verify the correct expression of the SH2 domain-containing protein, and allow for the screening of tyrosine-containing proteins that might interact with the SH2 domain-containing protein. In the event that no known tyrosine-containing proteins are observed to interact with the SH2 domain, an alternative approach is available. Immunoprecipitates from yeast extracts expressing the LexA-SH2 domain can be incubated with extracts from cells containing tyrosine phosphoproteins that would be expected to bind the SH2 domain in question, followed by anti-phosphotyrosine immunoblotting. If tyrosine-phosphorylated proteins can associate with the LexA-SH2 domain, then the bait plasmid can be used for a screen.

2. *Identification of novel tyrosine kinases that specifically phosphorylate a tyrosine-containing protein and thereby permit interaction with an SH2 or PTB domain.* A cDNA library could be constructed in the kinase plasmid, and members of this library would be screened for protein tyrosine kinases

that are able to phosphorylate a tyrosine-containing protein. The tyrosine-containing substrate of the kinase could be a LexA- or Vmw65-fusion protein. In order to detect the phosphorylation of the tyrosine, a binding partner of the phosphotyrosine-containing protein must be known. The cDNA encoding this protein is then subcloned into the other fusion protein (Vmw65 or LexA) plasmid such that the phosphorylation of the tyrosine-containing protein will facilitate interaction with its binding partner. There are two technical considerations to consider with this approach. The first is that a well-characterized binding partner to a tyrosine-phosphorylated protein may not be available. If this is the case, any SH2 domain that has a demonstrated ability to bind to the phosphotyrosine-containing region could be used. The second is the possibility that if the tyrosine-containing protein is used as a bait while joined to LexA, it may become able to activate transcription on its own or when tyrosine phosphorylated. This problem can be eliminated if the tyrosine-containing kinase substrate is fused to Vmw65, but this may not always be possible (if, for example, the SH2 domain used to detect the interaction is able to activate transcription when joined to LexA). If another kinase is able to phosphorylate the tyrosine-containing protein, then this problem can be addressed before attempting a screen.

3. *Identification of proteins that abrogate protein-protein interactions.* In biological systems, there are a number of proteins that function to prevent protein-protein interactions through allosteric modifications, such as phosphorylation, or through steric hinderance (Rotin et al. 1992). The tribrid system can be used to study such interactions. For example, if a protein-protein interaction can be detected by the two-hybrid system, a library could be constructed in a third plasmid (such as that used to express tyrosine kinases in example 2) to disrupt the interaction. In this case, like that discussed above with regard to structure-function analysis, one would be screening for lack of an interaction, which presents certain technical difficulties. The reverse two-hybrid system (Vidal et al 1996a, 1996b; chapter 7) provides one possible solution to this problem.

4. *Identification and generation of dominant negative proteins.* The tribrid system could be adapted for use in identification and characterization of proteins that have increased affinity for a binding partner. For example, if an SH2 domain and a tyrosine-containing protein interact in the tribrid system, a fourth plasmid could be introduced to express mutant forms of either of the binding partners. This protein would not need to be expressed as a fusion protein, and if its expression were able to reduce or disrupt the binding of the two fusion proteins, it would be a candidate as a dominant negative mutant. Screening for such dominant negative mutants would be more complicated than typical two-hybrid screening, but also potentially more rewarding. The identification of a particular mutation that confers dominant binding to a specific protein permits further experimentation to reveal the function of the protein in cells where it is expressed.

5. *Screening for therapeutically important compounds.* The yeast two-

hybrid system provides a mechanism by which two proteins are separated from their normal host cell environment. This provides an opportunity to identify molecules that may disrupt the interaction in a manner independent of any other proteins, which may interfere or otherwise complicate a biomolecular assay. These molecules would be candidate drugs for the abrogation of a specific protein-protein interaction in vivo, and would then be subjected to further testing. Specificity could easily be determined using different bait/prey combinations to further refine the analysis. The inexpensive and rapid nature of the yeast tribrid-based screen would enable many candidate drugs to be tested at very low cost. Although the yeast cell wall may present a barrier to certain small molecular weight compounds, many will be able to penetrate the cell to potentially disrupt the interaction. This disruption would be readily detectable, since liquid β-galactosidase assays are very sensitive. The two-hybrid system is also amenable to high-throughput assay in 96-well microtiter plates. The use of an in vivo assay to identify disruptors of protein-protein interactions also presents challenges. Any molecule that reduces or eliminates β-galactosidase activity may do so through a number of mechanisms, including interfering with transcription, translation, or β-galactosidase itself. These problems can be partially overcome by doing another assay in parallel, which detects protein-protein interaction unrelated to the drug target, using the same compound libraries.

6. *Alternatives to β-galactosidase as a reporter protein.* Although β-galactosidase is easily detected in permeabilized cells, it has limitations for quantitatively determining the relative strength of a pairwise interaction. One alternative reporter protein is green fluorescent protein (Cubitt et al. 1995), which functions in yeast, is quantitative, and does not necessitate the lysis of cells to detect. The use of GFP as a reporter also raises the possibility of using cell sorting to enrich for interactors in screening a cDNA library. Instead of plating large numbers of colonies on large numbers of individual plates, the transformation mixture could be grown under noninducing conditions to increase the numbers of cells, followed by galactose induction and flow-activated cell sorting (FACS) The fluorescent cells could then be grown on a relatively small number of plates and screened for β-galactosidase acitvity. This would allow vastly greater numbers of transformants to be screened with little additional effort.

Conclusions

The tribrid system, a modification of the two-hybrid system, is an excellent method for the detection of protein-protein interactions that require post-translational modifications such as tyrosine phosphorylation. Other modification activities, such as serine/threonine or dual-specificity kinases, phosphatases, acylases, ubiquitinating enzymes, glycosidases, and sulfatases, may also be considered for introduction into the plasmid used for kinase expression as described. Expression of these proteins may negatively impact yeast cell growth, so modification of induction conditions may be necessary for optimal use.

Although we have used the tribrid system to characterize novel phosphotyrosine-containing protein-SH2 domain interactions as well as other pairwise interactions, further manipulations of the system should dramatically increase its potential utility. As discussed previously, the inducible promoters used in all of the plasmids allow for the recovery of cDNAs that encode proteins that may be toxic to yeast when expressed at high levels.

ACKNOWLEDGMENTS The authors thank the members of the Kochan laboratory for stimulating discussions during the conception and development of the tribrid system. Stephen Dalton constructed all of the early versions of the plasmids used in the tribrid system.

References

Aruffo, A., and Seed, B. (1987). Molecular cloning of a CD28 cDNA by a high-efficiency COS cell expression system. Proc. Natl. Acad. Sci. USA 84:8573–8577.

Boschelli, F., Uptain, S.M., and Lightbody, J.J. (1993). The lethality of $p60^{v\text{-}src}$ in *Saccharomyces cerevisiae* and the aactivation of $p34^{CDC28}$ kinases are dependant on the integrity of the SH2 domain. J. Cell Sci. 105:519–528.

Cubitt, A.B., Heim, R., Adams, S.R., Boyd, A.E., Gross, L.A., and Tsein, R.Y. (1995). Understanding, improving and using green fluorescent proteins. Trends Biochem. Sci. 20:448–455.

Dalrymple, M.A., Mcgeoch, D.J., Davison, A.J., and Preston, C.M. (1985). DNA sequence of the herpes simplex virus type 1 gene whose product is responsible for transcriptional activation of immediate early promoters. Nucl. Acids Res. 13:7865–7879.

Dalton, S., and Treisman, R. (1992). Characterization of SAP-1, a protein recruited by serum response factor to the c-*fos* serum response element. Cell 68:597–612.

Fearon, E.R., Finkel, T., Gillison, M.L., Kennedy, S.P., Casella, J.F., Tomaselli, G.F., Morrow, J.S., and Dang, C.V. (1992). Karyoplasmic interaction selection strategy: A general strategy to detect protein-protein interactions in mammalian cells. Proc. Natl. Acad. Sci. USA 89:7958–7962.

Fields, S. and Sternglanz, R. (1994). The two-hybrid system: an assay for protein-protein interactions. Trends Genet. 10:286–292.

Gartner, A., Nsamyth, K., and Ammerer, G. (1992). Signal transduction in *Saccharomyces cerevisiae* requires tyrosine and threonine phosphorylation of FUS3 and KSS1. Genes. Dev. 6:1280–1292.

Gustafson, T.A., He, W., Craparo, A., Schaub, C.D., and O'Neill, T.J. (1995). Phosphotyrosine-Dependent Interaction of SHC and Unsulin Receptor Substrate 1 with the NPEY Motif of the Insulin Receptor via a Novel Non-SH2 Domain. Mol. Cell. Biol. 15:2500–2508.

Guthrie, C. and Fink, G.R. (1991). *Guide to Yeast Genetics and Molecular Biology.* Meth. Enz. 194:1–933.

Heldin, C-H. (1995). Dimerization of Cell Surface Receptors in Signal Transduction. Cell 80:213–223.

Hunter, T. (1995). Protein Kinases and Phsophatases: The Yin and Yang of Protein Phosphorylation and Signaling. Cell 80:225–236.

Johnston, M. (1987). A model fungal gene regulatory mechanism: the *GAL* genes of *Saccharomyces cerevisiae*. Microbiol. Rev. 51, 4:458–476.

Kalderon, D., Roberts, B.L., Richardson, W.D., and Smith, A.E. (1984). A short amino acid sequence able to specify nuclear locaton. Cell 39:499–509.

Keegan, K. and Cooper, J.A. (1996) Use of the two hybrid system to detect the association of the protein-tyrosine-phosphatase, SHPTP2, with another SH2-containing protein, Grb7. Oncogene 12:1537–1544.

Kihara, H., and Siraganian, R.P. (1994). Src homology 2 domains of Syk and Lyn bind to tyrosine-phosphorylated subunits of the high affinity IgE receptor. J. Biol. Chem. 269:22427–22432.

Leung, D.W., Chen, E., and Goeddel, D.V. (1989). A Method for Random Mutagenesis of a Defined DNA Segment Using a Modified Polymerase Chain Reaction. Technique 1:11–15.

Lim, M-Y., Dailey, D., Martin, G.S., and Thorner, J. (1993). Yeast *MCK1* Protein Kinase Autophosphorylates at Tyrosine and Serine but Phosphorylates Exogenous Substrates at Serine and Threonine. J. Biol. Chem. 268:21155–21164.

Lioubin, M.N., Algate, P.A., Tsai, S., Carlberg, K., Aebersold, R., and Rohrschneider, L.R. (1996) p150^{ship}, a signal transduction molecule with inositol polyphosphate-5-phosphatase activity. Genes Dev. 10:1084–1095.

Ma, H., Kunes, S., Schatz, P.J., and Botstein, D. (1987). Plasmid construction by homologous recombination in yeast. Gene 58:201–216.

Murphy, S.M., Bergman, M., and Morgan, D.O. (1993). Suppression of c-Src activity by C-terminal Src kinase involves the c-Src SH2 and SH3 domains: analysis with Saccharomyces Cerevisiae. Mol. Cell. Biol. 13, 9:5290–5300.

O'Neill, T.J., Craparo, A., and Gustafson, T.A. (1995). Characterization of an Interaction between Insulin Receptor Substrate 1 and the Insulin Receptor by Using the Two-Hybrid System. Mol. Cell. Biol. 14:6433–6442.

Okada, M., Howell, B.W., Broome, M.A., and Cooper, J.A. (1993). Deletion of the SH3 domain of Src interferes with regulation by the phosphorylated carboxyl-terminal tyrosine. J. Biol. Chem. 268:18070–18075.

Osborne, M.A., Dalton, S., and Kochan, J.P. (1995). The Yeast Tribrid System—Genetic Detection of *trans*-phosphorylated ITAM-SH2-Interactions. Bio/Technology 13:1474–1478.

Osborne, M.A., Zenner, G., Lubinus, M., Zhang, X., Songyang, Z., Cantley, L.C., Majerus, P., Burn, P., and Kochan, J.P. (1996). The Inositol S-Phosphatase SHIP binds to immuno receptor signaling motifs and responds to high-affinity IgE receptor aggregation. J. Biol. Chem. 271:29271–29278.

Pandey, A., Lazar, D.F., Saltiel, A.R., and Dixit, V.M. (1994). Activation of the Eck receptor protein tyrosine kinase stimulates phosphatidylinositol 3-kinase activity. J. Biol. Chem. 269:30154–30157.

Pandey, A., Duan, H., Di Fiore, P.P., and Dixit, V.M. (1995). The Ret Receptor Protein Tyrosine Kinase Associates with the SH2-containing Adapter Protein Grb10. J. Biol. Chem. 270:21461–21463.

Pawson, T. (1995). Protein modules and signalling networks. Nature 373:573–580.

Rotin, D., Margolis, B., Mohammadi, M., Daly, R.J., Daum, G., Li, N., Fischer, E.H., Burgess, W.H., Ullrich, A., and Schlessinger, J. (1992). SH2 domains prevent tyrosine dephosphorylation of the EGF receptor: identification of Tyr992 as the high-affinity binding site for SH2 domains of phospholipase Cγ. EMBO J. 10:559–567.

Russo, P., and Sherman, F. (1989). Transcription terminates near the poly(A) site in the CYC1 gene of the yeast Saccharomyces cerevisiae. Proc. Natl. Acad. Sci. USA 86:8348–8352.

Shiue, L., Green, J., Karas, J.L., Morgenstern, J.P., Ram, M.K., Taylor, M.K., Zoller, M.J., Zydowsky, L.D., Bolen, J.B., and Brugge, J.S. (1995). Interaction of $p72^{syk}$ with the γ and β subunits of the high- affinity receptor for immunoglobulin E FcεRI. Mol. Cell. Biol. 15, 1:272–281.

Sikorski, R.S., and Hieter, P. (1989). A system of shuttle vectors and yeast host strains designed for efficient manipulation of DNA in *Saccharomyces cerevisiae*. Genetics 122:19–27.

Songyang, Z., Carraway, III, K.L., Eck, M.J., Harrison, S.C., Felman, R.A., Mohammadi, M., Schelessinger, J., Hubbard, S.R., Smith, D.P., Eng, C., Lorenzo, M.J., Ponder, B.A.J., Mayer, B.J., and Cantley, L.C. (1995). Catalytic specificity of protein tyrosine kinases is critical for selective signaling. Nature 373:536–539.

van der Geer, P., and Pawson, T. (1996). The PTB domain: a new protein module implicated in signal transduction. Trends Biochem. Sci. 20:277–280.

Vasavada, H.A., Ganguly, S., Germino, F.J., Wang, Z.X., and Weissman, S.M. (1991). A contingent replication assay for the detection of protein-protein interactions in animal cells. Proc. Natl. Acad. Sci. USA 88:10686–10690.

Vidal, M., Brachmann, R.K., Fattaey, A., Harlow, E., and Boeke, J.D. (1996). Reverse two-hybrid and one-hybrid systems to detect dissociation of protein-protein interactions. Proc. Natl. Acad. Sci. USA 93:10315–10320.

Vidal, M., Braun, P., Chen, E., Boeke, J.D., and Harlow, E. (1996). Genetic characterization of a mammalian protein-protein interaction domain by using a yeast reverse two-hybrid system. Proc. Natl. Acad. Schi. USA 93:10321–10326.

Xing, Z., Chen, H-C., Nowlen, J.K., Taylor, S.J., Shalloway, D., and Guan, J-L. (1994). Direct Interaction of v-Src with the Focal Adhesion Kinase Mediated by the Src SH2 Domain. Molec. Biol. Cell 5:413–421.

Yang, M., Wu, Z., and Fields, S. (1995). Protein-peptide interactions analyzed with the yeast two-hybrid system. Nucl. Acids Res. 23:1152–1156.

15

Protein Interactions Mediated by Small Molecule Ligands

Vivian Berlin

Drug development efforts entail an iterative process of isolating or synthesizing small molecules with desired biological or biochemical properties, defining their mechanism of action, and refining structure to acheive more specific or potent activities. Until recently, most drug development efforts have focused on designing molecules that inhibit enzymatic activities. As information accumulates about the role of protein-protein interactions in regulating fundamental cellular processes, interest in the development of small molecules that modulate protein-protein interactions is growing.

CyclosporinA (CsA), FK506, and rapamycin constitute a class of small molecule ligands that mediate protein-protein interactions. These small molecules are potent immunsuppressants that bind with high affinity to proteins known collectively as immunophilins. CsA binds to cyclophilin (Handshumacher et al. 1984); FK506 and rapamycin bind to FKBP12 (Harding et al. 1989; Siekierka et al. 1989). These protein-drug complexes, in turn, bind to specific target proteins and inhibit their activity. The CsA- and FK506- immunophilin complexes bind to and inhibit calcineurin (Liu et al. 1991; Liu et al. 1992; Clipstone and Crabtree 1992; Fruman et al. 1992), a calcium-dependent phosphatase required for lymphokine gene expression (McCaffrey et al. 1993; Frantz et al. 1994). By contrast, rapamycin, complexed with FKBP12 binds to a distinct target, Rapt1 (also called RAFT1 and FRAP) (Chiu et al. 1994; Sabatini et al. 1994; Brown et al. 1994), required for subsequent lymphokine-induced cell division. Thus by promoting protein-protein interactions, CsA, FK506, and rapamycin inhibit intracellular signaling pathways in lymphocytes (Tocci et al. 1989; Bierer et al. 1990; Mattila et al. 1990; Flanagan et al. 1991).

This chapter describes the use of the two-hybrid system to identify targets of the FKBP12/rapamycin complex. The objective was to identify proteins that interact with FKBP12 in the presence, but not in the absence, of rapamycin (Chiu et al. 1994). Preliminary experiments were conducted to test the feasibility of the approach. Of concern was whether FKBP12, fused to a DNA-binding domain, would be in the proper conformation to bind a ligand and interact with a protein. Reconstruction experiments conducted to analyze the interaction of FKBP12 and calcineurin as a function of FK506 were critical for establishing the optimal conditions for conducting two-hybrid screens to identify rapamycin-dependent interactors (Berlin and Chiu 1995). This example demonstrates the feasibility of the approach and can serve as a general guide to define, characterize, and assay protein interactions mediated by small molecules ligands such as hormones, peptides, or synthetic organic molecules.

Principle and Applications

The two-hybrid system detects protein–protein interaction in yeast as a function of the transcriptional activation of reporter genes (for example, *HIS3* and *lacZ*). Normally, transactivation occurs via interaction of proteins fused to the DNA-binding domain (bait) and transactivation domain. However, if protein interaction is mediated by a small molecule ligand, transactivation will occur only in the presence of the ligand (Figure 15-1A). The ligand in question may complex directly with one or both members of the protein pair, mediating a tripartite interaction. No transactivation of reporter genes will occur if a ligand does not bind to one of the proteins (Figure 15-1B).

Protein-protein interactions mediated by small molecule ligands are amenable to characterization using the two-hybrid system. Protein targets of a drug or hormone receptor can be identified by conducting two-hybrid library screens on media containing the drug or hormone of interest (Chiu et al. 1994; Lee et al. 1995). Proteins that interact with the receptor in a ligand-dependent fashion are candidate targets. Subsequent analysis using the two-hybrid system can be conducted to define protein domains or identify the amino acid residues required for tripartite interaction.

The two-hybrid system can be adapted to a microtiter assay format for performing high-throughput screens for small molecule ligands that mediate protein-protein interaction. Screens can be designed using panels of related receptors and targets to identify small molecule ligands that mediate a unique set of interactions (Figure 15-2). If a ligand of interest has a low degree of specificity and interacts with multiple receptors or targets, deriva-

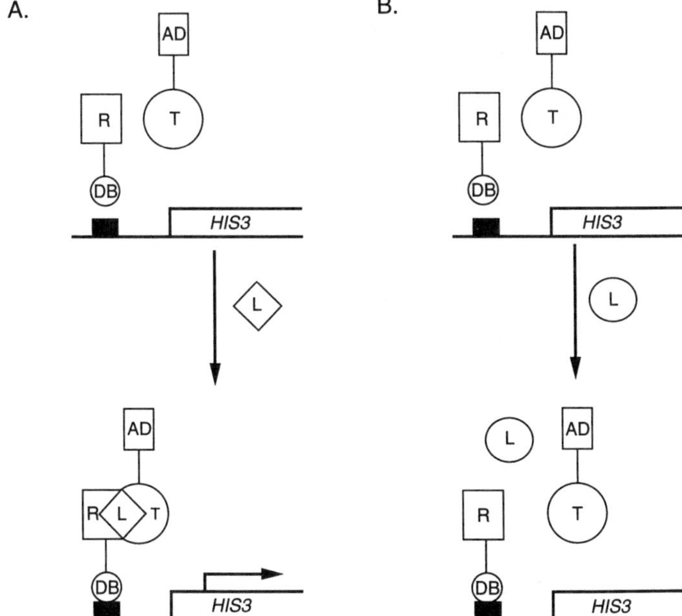

Fig. 15-1. (A) Ligand-dependent transactivation in the two-hybrid system. Ligand (L) mediates interaction of a ligand receptor (R) fused to a DNA-binding domain (DB) and the target protein (T) fused to an activation domain (AD). The tripartite complex activates transcription (arrow) of the reporter genes, *HIS3* and *lacZ*, under the control of upstream activation sequences (black box). (B) When a ligand fails to bind to the receptor or protein target, no transcriptional activation of reporter genes occurs.

tives of the parent compound can be synthesized and retested in the two-hybrid system to identify those that have increased specificity. One concern in undertaking a drug screening effort using the two-hybrid system is that yeast may have poor uptake or efficient efflux of certain classes of compounds, yielding a high frequency of false negatives.

Methods

System Validation

Using the two-hybrid system to detect a protein-protein interaction mediated by a small molecule ligand is feasible if the ligand receptor, when fused to a DNA-binding domain, retains its ability to bind ligand and exert its biological effect. Experiments can be conducted to address these con-

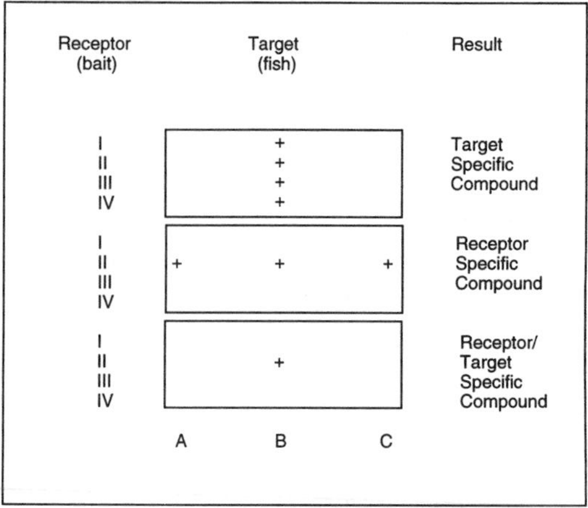

Fig. 15-2. Use of the two-hybrid system for drug screening. Shown is a hypothetical scheme for conducting a screen for compounds that mediate interaction between a receptor (I,II,III, or IV) and target (A,B, or C). Yeast two-hybrid strains containing different receptor-target pairs are tested against a library of small molecule analogs using a microtiter format. Using this scheme compounds can be identified that are receptor-specific, target-specific or that mediate interaction between a specific receptor-target pair.

cerns. For example, prior to conducting screens to identify the target of the FKBP12-rapamycin complex, we tested the ability of FKBP12 to interact with calcineurin as a function of FK506. A variety of baits were constructed in which FKBP12 was fused to the LexA or Gal4p DNA-binding domain, containing or lacking a polyglycine linker at the fusion junction. A mutant of FKBP12 that increases the sensitivity of yeast to rapamycin was included in the analysis. The baits were assayed for their ability to interact with calcineurin fused either to VP16 or to the Gal4p activation domain.

Wild-type FKBP12 interacted with calcineurin in an FK506 dose-dependent fashion in both the Gal4p (Hannon et al. 1993) and LexA/VP16 (Vojtek et al. 1993) two-hybrid systems. Introduction of the polyglycine linker in the baits did not have a significant effect on drug-mediated interaction. The FKBP12 mutant showed enhanced FK506-dependent interaction with calcineurin when assayed in the Gal4p system. However, in the LexA/VP16 system, the FKBP12 mutant paired with VP16 alone showed significant background transactivation of the *HIS3* and *lacZ* reporters and, therefore, was not used in subsequent screens (Table 15-1).

The issue of whether a ligand receptor fused to a DNA-binding domain retains its biological activity is also a concern. This concern can be addressed if the ligand receptor, when introduced into the appropriate cell type, produces an assayable phenotype, such as hormone-dependent transcription or drug sensitivity. For example, hormone receptor baits can be assayed in cells lacking the receptor and containing a reporter under the control of a hormone-inducible promoter. The ability of a hormone receptor bait to confer hormone-inducible expression of the reporter gene provides a measure of the bait's biological activity. If yeast contains a homolog of the ligand receptor of interest, the ability of receptor baits to complement a deletion of the yeast gene also can be assessed. For example, we tested the ability of FKBP12 baits to complement a deletion of the yeast gene for FKBP12 (*FKB1*) which renders cells resistant to rapamycin. Human FKBP12 baits were introduced into an *fkb1* deletion strain which was then assayed for growth on media containing different concentrations of rapamycin. Native human FKBP12 confers sensitivity to low levels of rapamycin, completely complementing the deletion of yeast *FKB1*. However, human FKBP12 fused to either the LexA or Gal4p DNA-binding domain partially complements the *FKB1* deletion. Drug sensitivity is observed, but only at relatively high concentrations of rapamycin (Table 15-2). This analysis was critical for choosing the FKBP12 baits and rapamycin concentrations for two-hybrid screens.

Two-hybrid System: Components and Conditions

The original two-hybrid system (Fields and Song 1989) has undergone many modifications. The current systems use either the LexA or the Gal4p DNA-binding domain and a variety of activation domains such as VP16 (Vojtek et al. 1993), B42 (Gyuris et al. 1993), or the Gal4p activation domain (Bartel et al. 1993; Chevray and Nathans 1992). Components of the two-hybrid system are described below with respect to possible alterations that would facilitate the detection and analysis of ligand-mediated protein interactions.

Yeast Strains Commonly used yeast two-hybrid strains contain the *HIS3* and *lacZ* reporter genes integrated in the chromosome, under the control of Gal4p or LexA binding sites. Modification of two-hybrid strains may be desirable if yeast contains a homolog of the ligand receptor. For example, disruption of *FKB1*, the gene for FKBP12, was critical for performing two-hybrid screens with rapamycin. The *fkb1* deletion strain containing a human FKBP12 bait was partially resistant to rapamycin, permitting screens to be performed using a concentration of rapamycin that had no effect on cell growth.

Gene disruptions can be performed using standard techniques (Rothstein 1991). Briefly, the coding sequences of the gene to be disrupted, contained on a linear fragment, are replaced with a selectable marker. Since

Table 15-1 FK506 mediates interaction of human FKBP12 and calcineurin in the two-hybrid system

Plasmids		Growth on no Histidine			Color on X-gal	
		FK506 (μg/ml)				
Binding Domain Fusion	Activation Domain Fusion	0	0.02	0.2	NO FK506	FK506, 0.2 μg/ml
pGAL4-FKBP12	PGAL4-hCNA	-	+	++	White	Blue
pGAL4-FKBP12*	pGAL4 - hCNA	-	++	+++	White	Blue
pGAL4-SNF1	PGAL4 - hCNA	-	-	-	White	White
pGAL4-FKBP12	pGAL4	-	-	-	White	White
pLexA-FKBP12	pVP16	-	ND	-	White	White
pLexA-FKBP12	pVP16-CNA	-	ND	+++	White	Blue
pLexA-G_6-FKBP12	pVP16-CNA	-	ND	+++	White	Blue
pLexA-FKBP12*	pVP16-CNA	++	ND	+++	Blue	Blue

The yeast strain Hf7C (gift of G. Hannon) (top) or L40 (bottom) was transformed with the indicated pairs of plasmids. The DNA binding-domain fusions contain the Gal4p DNA-binding domain or LexA-fused in-frame to wild-type or mutant (*) human FKBP12 or to the yeast Snf1p. The FKBP12 mutant increases the sensitivity of yeast to rapamycin (unpublished results). pLexA-G6-FKBP12 contains a spacer of six glycine residues. The Gal4p or VP16 activation domain is unfused or fused to human calcineurin (hCNA). Expression of the *HIS3* and *lacZ* reporters constitutes evidence of interaction which is detected as growth in the absence of histidine and β-galactosidase activity (blue color on media containing X-Gal), respectively. "-", no growth; "+", "++", "+++", poor, moderate, and good growth, respectively, assessed by the rate of appearance and size of colonies.

Table 15-2 FKBP12 fusions mediate rapamycin toxicity in yeast

DNA-Binding Domain-Fusion Plasmid	Growth on Media Containing Rapamycin (ng/ml)					
	15.6	31.2	62.5	125	250	500
pLexA-hFKBP12	+	+	+	+	+	+/–
pLexA-G_6-hFKBP12	+	+	+	+/–	+/–	–
pLexA-da	+	+	+	+	+	+
pGAL4-hFKBP12	+	+	+	+	+	+
pGAL4-hFKBP12*	+	+	+	+/–	+/–	+/–
pGAL4-G6-hFKBP12	+	+	+	+	+	+
pGAL4-G6-hFKBP12*	+	+	+	+/–	+/–	+/–
PGAL4	+	+	+	+	+	+
HFKBP12	+/–	–	–	–	–	–

Yeast strain VBY567 (top) or (bottom) containing a deletion of *FKB1* was transformed with plasmids encoding the indicated LexA or Gal4p fusions or Gal4p domains alone. Human FKBP12 sequences were wild type or mutant (*) as described in the legend to Table 15-1. One LexA-fusion contains an unrelated protein (da). Certain constructs contain a spacer of six glycine residues (G6) at the fusion junction. Human FKBP12 (hFKBP12), not fused to a DNA-binding domain, but expressed from the yeast *GAL1* promoter is shown. Transformants were plated on media containing various concentrations of rapamycin. "-", no growth; "+", growth; "+/-, slow growth, and reduced colony size.

many of the two-hybrid yeast strains contain an *ade2* allele, *ADE2* can be used as the selectable marker. A DNA fragment containing ligand receptor coding sequences replaced by *ADE2*, flanked by noncoding sequences, is used to transform a two-hybrid yeast strain to adenine prototrophy. Replacement of the wild-type allele with the disruption is verified by Southern analysis and an assay of the expected mutant phenotype (for example, drug resistance).

Bait Plasmids Depending on the two-hybrid system used, coding sequences of the ligand receptor are fused in-frame to either the LexA or to the Gal4p DNA-binding domain. Fusions can be constructed that contain or lack a polyglycine linker at the fusion junction. Introduction of a polyglycine linker to separate the DNA-binding domain from the ligand receptor may facilitate proper folding of each domain. If possible, the functionality of the baits should be tested, as described in the preceding section for the FKBP12 baits.

cDNA Libraries cDNA libraries, with inserts cloned 3' of the activation domain in the activation domain vector, can be generated from random- or oligo(dT)-primed polyA$^+$ RNA. Size selection of the cDNAs to bias the library toward long or short inserts is an option. The advantage of short, random-primed libraries is that peptide domains sufficient for interaction with the receptor-ligand complex can be identified rapidly. Using such a library, we were able to identify a region of 133 amino acids within the 289 kD Rapt1 protein that was sufficient to interact with the FKBP12-rapamycin complex (Chiu et al. 1994). However, the disadvantage of using short random libaries is that the minimum region required for interaction may be larger than that encoded by the inserts. Also, short clones may have interaction properties absent in the parent native protein (Vojtek et al. 1993).

Basal Transcription of Reporter Genes Before proceeding with the library screens, it is advisable to determine the background level of transcriptional activation of the reporter genes in strains containing a bait paired with the plasmid encoding the activation domain alone. Background may be a problem if the ligand receptor mediates transcriptional activation or if the ligand affects basal expression of the reporter genes. In the former case, fragments of the ligand receptor can be tested to identify those that no longer activate transcription but that still bind to ligand. In the latter situation, different two-hybrid strains can be tested for the effect of ligand on reporter gene expression. For example rapamycin significantly increased basal expression of *HIS3* in the strains used for the Gal4p system, but not in those used for the LexA/VP16 system .

Growth of yeast due to basal transcription of *HIS3* can be controlled using 3-aminotriazole (3-AT), an inhibitor of imidazole glycerol phosphate dehydratase, the *HIS3* gene product. However, high levels of 3-AT (>20 mM) increase the length of time for positives to appear in two-hybrid screens. These stringent selection conditions may increase the frequency of chromosomal mutations that permit growth in the absence of histidine, thereby increasing the frequency of false positives appearing in two-hybrid screens. Therefore, it is advisable to chose a bait and two-hybrid yeast strain that exhibits a low, or preferably undetectable, level of expression of the reporter genes on media containing ligand prior to conducting a screen for interacting proteins.

Library Screens for Ligand-Dependent Protein Interactors

Pooling Yeast Transformants

In screens to identify ligand-dependent interactors, it is advantageous to grow transformants representative of the cDNA library prior to applying selection for transformants expressing the reporter genes. For both the Gal4p (Hannon et al. 1993) and LexA/VP16 (Vojtek et al. 1993) systems, transformants are plated on synthetic media lacking tryptophan and leucine (-Trp and -Leu) to select for the bait and activation domain plasmids, respectively. The optimal number of transformants is several times the number of individual recombinants in the cDNA library to ensure complete representation of the library. Transformants are then pooled and stored for subsequent screening using the following protocol:

1. Follow the standard protocol for transforming yeast harboring the bait plasmid with the cDNA library in the activation domain vector (see chapters 2, 3, and 4). Recommended conditions for one liter of cells grown to A_{600} of 0.5 OD units (1 OD unit is approxmately 2×10^7 cells/ml) are to resuspend cells in a final volume of 25 ml and transform with 125 μg of the cDNA library plus 500 μg of carrier DNA.
2. Following the heat shock, the cells are pelleted, resuspended in 25 ml of H_2O, and plated on -Trp and -Leu media at a densitiy of 0.5 ml of cell suspension per 15 cm plate. Aliquots of 1 and 5 μl of the transformation mix are diluted into 100 μl of H_2O and plated on -Trp and -Leu media to monitor the transformation frequency. Typical frequencies range from $5-10 \times 10^4$ transformants/μg of DNA.
3. Cells are grown for 3 days at 30°C.
4. Cells are scraped from all 50 plates, pooled, washed 3 x with 200 ml H_20, resuspended in 50% glycerol, and stored at −80°C.

Selection for Ligand-Dependent Protein-Protein Interactors.

The following protocol is used to screen library transformants for ligand-dependent interactors (Table 15-3).

1. In aliquot of the pooled yeast transformants is outgrown in YEPD (Sherman 1991) at 30°C, for one doubling. The initial inoculum should not be higher than A_{600} of 0.4 OD units.
2. Cells are plated onto selective media, with the aim of one million per 15 cm plate. The selection media is synthetic media (Sherman, 1991) lacking

Table 15-3 Protocol to identify ligand-mediated protein-protein interactors

Test	Phenotype	Retain	Discard
Growth on selective media containing ligand (1–3)	His^+ $LacZ^+$	√	
	His^+ $LacZ^-$		√
Retest growth of His^+ $LacZ^+$ transformants on selective media plus or minus ligand (4)	Ligand-dependent His^+ $LacZ^+$	√	
	Ligand-independent His^+ $LacZ^+$		√
Plasmid loss to test linkage of His^+ $LacZ^+$ with activation domain and bait plasmids (5)	Cosegregation	√	
	Independent segregation		√
Recovery of activation domain plasmid and retransformation into strains with the original and an unrelated bait (6–7)	Ligand-dependent His^+ $LacZ^+$ with original bait	√	
	Ligand-dependent His^+ $LacZ^+$ with original and unrelated bait		√

This protocol is for two-hybrid systems using the *HIS3* and *lacZ* reporter genes. Numbers refer to the protocol described in the text.

leucine, tryptophan, and histidine and containing the ligand of interest. This media selects for the bait and library plasmids and for expression of the *HIS3* reporter gene, respectively. In some cases, the inclusion of 3-AT in the selective media may be necessary (see Discussion). To monitor the number of transformants plated, a series of dilutions is plated onto -Trp and -Leu plates. Cells are incubated at 30°C until His^+ colonies appear. This occurs within 3 to 10 days, depending on the selective conditions used for the screen.

3. His^+ transformants are screened for expression of *lacZ*. This can be accomplished in one of two ways. Step 2 can be performed using phosphate-buffered (pH 7.0) selective plates containing 5-bromo-4-chloro-3-indolyl-ß-D-galactoside (X-gal) so that the His^+ and $lacZ^+$ phenotypes are monitored simultaneously. A more sensitive method of detection is to lift His^+ transformants onto nitrocellulose filters (Schleicher and Schuell). The cells are then permeabilized in liquid nitrogen and placed on Whatman paper (#1) prewet with Z buffer (60 mM $Na_2HPO_4{\cdot}7H_20$, 60 mM $NaH_2PO_4{\cdot}H_20$, 10 mM KCl, 1 mM $MgSO_4{\cdot}7H_20$) containing 1 mg/ml X-gal. Blue color can be detected in a few minutes in some cases, or only after overnight incubation in other cases. Viable cells can be recovered from the filters using this protocol.

4. $His^{+,}$ $lacZ^+$ transformants are retested for growth on media lacking histidine (-His) and containing or lacking ligand. Candidates that grow in the presence, but not in the absence of ligand are chosen for the next step.

5. Plasmid loss can be performed to determine whether the His^+, $lacZ^+$ phenotype is plasmid-dependent. Each candidate is grown in nonselective medium (YEPD) to permit plasmid loss. Derivatives of each candidate that have lost either the bait plasmid, library plasmid, both plasmids, or neither plasmid (Trp^-

Leu+; Trp+ Leu-; Trp- Leu-, or Trp+ Leu+, respectively) are tested again for growth on -His plates containing ligand. Candidates for which only the Trp+ Leu+ derivative grows on media containing ligand and lacking histidine are characterized further.

6. DNA is prepared from candidates showing a plasmid- and ligand-dependent His+ phenotype (Hoffman and Winston 1987). The DNA is used to transform *E. coli* strain MC1066 [*ΔlacI-A x 74 galU galK strepA^r trpC9830 leuB6 pyrF::tn5(kan^r) Rec+ Ara+*] to leucine prototrophy to recover the library plasmids encoding the putative ligand target.

7. Plasmid DNA recovered in *E. coli* is used to transform yeast strains containing the original bait or an unrelated bait. Clones specific for the original bait that retain a ligand-dependent His+, lacZ+ phenotype are characterized further, while those that no longer show ligand dependence or are nonspecific, that is, exhibit a His+, lacZ+ phenotype when paired with the unrelated bait, are eliminated from further analysis.

8. Candidate clones that fulfill all the above criteria are analyzed by DNA sequencing on both strands, starting with oligonucleotides from vector sequences as primers.

High-Throughput Screens

To develop high-throughput screens based on the two-hybrid system, we devised a procedure to quantitate protein-protein interactions mediated by a small molecule. The assay is designed to measure expression of the *lacZ* reporter gene using a substrate of ß-galactosidase that produces a chemiluminescent signal when cleaved. The assay is performed in microtiter plates, allowing 1000 to 2000 compounds to be screened per week. The assay is as follows:

1. Inoculate yeast cells from a single colony into 50 ml of synthetic medium lacking leucine and tryptophan (Sherman 1991). Incubate the flask overnight at 30°C with shaking (~200 rpm).
2. Dilute the overnight culture to a final A_{600} of 0.5 in growth medium. Dispense 135 μl aliquots of the cell suspension into wells of a round bottom microtiter plate pre-loaded with 15 μl/well of the compound to be tested at various concentrations. (The compounds are dissolved in 5% dimethyl sulfoxide (DMSO) so that the final DMSO concentration added to cells is 0.5%, which does not affect yeast cell growth.) Cover microtiter plates and incubate at 30°C for 4 hours with shaking at 300 rpm.
3. Centrifuge the microtiter plate for 10 minutes at 2000 rpm. Remove the supernatant and wash the cells with 225 μl phosphate buffered saline.
4. Dispense 100 μl of lysis buffer (100 mM K_2HPO_4, pH 7.8; 0.2% Triton X-100; 1.0 mM dithiothriotol) into each well, cover, and incubate for 30 minutes at room temperature with shaking at 300 rpm.
5. Dispense into each well of a Microfluor plate (Dynatech Laboratories, Chantilly, Virginia), 50 ml of the chemiluminescent substrate, Galacton Plus™ (Tropix, Inc., Bedford, Massachusetts) in diluent (100 mM Na_2HPO_4, 1 mM $MgCl_2$, pH 8.0). To these wells, transfer 20 μl of cell lysate and incubate in the dark for 60 minutes at room temperature.
6. Add 75 μl of Emeral™ accelerator to each well. Cover plate and count in a TopCount scintillation counter (Packard, Inc.) for 0.01 minutes/well.

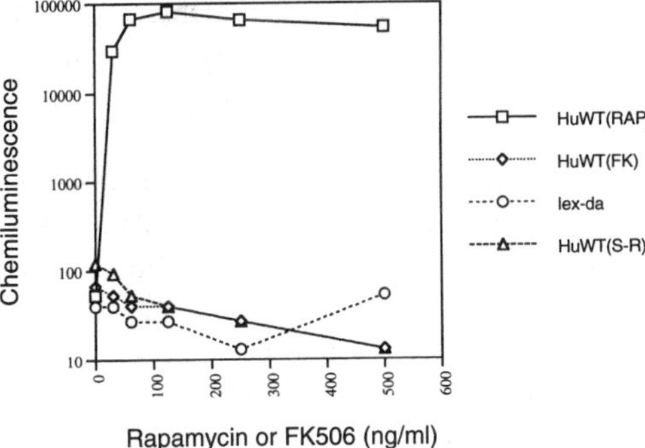

Fig. 15-3. Quantitative chemiluminescence assay to measure drug-mediated protein-protein interaction. Expression of the *lacZ* reporter gene is a measure of drug-mediated interaction. The former is detected using a substrate of ß-galactosidase that produces a chemiluminescent signal when cleaved. Rapamycin (squares), but not FK506 (diamonds), mediates interaction between human FKBP12 (bait) and the Rapt1 binding domain in the LexA/VP16 two-hybrid system. No complex formation occurs when the Rapt1 serine-2035 is mutated to arginine (triangles) or if the LexA-FKBP12 bait is substituted with the LexA-da control bait (circles).

Using this quantitative chemiluminescence assay, we measured the interaction of FKBP12 with the Rapt1 binding domain as a function of drug concentration. Addition of rapamycin from from 0 to 100 ng/ml increased ß-galactosidase activity approximately 1000-fold. The effect was specific for rapamycin; FK506 over the same concentration range did not increase ß-galactosidase activity significantly over background levels. If a Rapt1 mutant is used (serine-2035 to arginine) rather than wild type or if LexA-da, a control bait, is substituted for the LexA-FKBP12 bait, rapamycin-mediated complex formation is obliterated (Figure 15-3). The sensitivity and specificity of this microtiter assay are ideal for conducting high-throughput screens.

Discussion

Using the methods described here, we adapted the two-hybrid system to identify cellular protein targets of small molecule ligands. In reconstruction experiments to test the approach, FK506 mediated interaction between FKBP12 and calcineurin in a dose-dependent fashion. The FKBP12 baits in yeast strains deleted for the FKBP12 gene conferred sensitivity to rapamycin, indicating that the fusions were capable of binding ligand and interacting with the endogenous yeast target. These preliminary experiments established the optimal conditions for conducting two-hybrid screens in the presence of rapamycin.

Using human FKBP12 as a bait, we conducted two-hybrid screens with rapamycin to identify proteins that interact with the FKBP12/rapamycin complex. From these screens, we isolated the mammalian homolog of yeast TOR (Heitman et al. 1991; Kunz et al. 1993; Cafferkey et al. 1993; Helliwell et al. 1994), designated RAPT1 (Chiu et al. 1994). The identical protein (called RAFT1 or FRAP) was isolated as the target of the FKBP12/rapamycin complex using biochemical approaches (Sabatini et al. 1994; Brown et al. 1994). In the two-hybrid screens, we used a random-primed cDNA library with an average insert size of 1 kb (provided by S. Hollenberg; Vojtek et al. 1993), which permitted the identification of a small region of 133 amino acids within the 289 kD RAPT1 that is sufficient for interaction with the FKBP12/rapamycin complex (Chiu et al. 1994).

Certain protocols for conducting two-hybrid screens recommend plating primary transformants directly onto media that selects for growth of transformants expressing the reporter genes. When conducting screens to identify ligand-dependent interactors, it is advantageous to first pool library transformants prior to plating them onto selective media. The pooled transformants can then be subjected to different selection conditions, such as various concentrations of ligand or 3-AT, to optimize the conditions for isolating ligand-dependent protein interactors. The same pooled transformants can also be screened to identify cellular proteins that interact with the receptor in the absence of ligand.

In addition to its use in screens to identify targets of ligand-receptor complexes, the two-hybrid system can be adapted to a microtiter format for performing high-throughput screens to identify novel ligands that mediate protein-protein interaction. A chemiluminescent substrate of ß-galactosidase provides a sensitive method for detecting expression of *lacZ*, one of two reporters used in most two-hybrid systems. Establishing the conditions under which interaction between the bait and activation domain hybrid is minimal in the absence of ligand will facilitate the detection of true positives in the assay. Ligands identified in high-throughput screens using the two-hybrid system should be tested to determine if their effect on reporter gene expression occurs via interaction with the bait or activation domain hybrid or by some other mechanism. The last scenario, although unlikely, is a formal possibility given the induction of the *HIS3* reporter by rapamycin in certain two-hybrid strains. One concern in using yeast for drug screens is the possibility of a high false-negative rate due to poor intracellular accumulation of certain classes of compounds. However, the sensitivity and specificity of the microtiter assay for monitoring protein interaction mediated by rapamycin indicate that the two-hybrid system can be utilized for high-throughput screens for novel molecules that mediate protein interaction.

References

Bartel, P.L., Chien, C.-T., Sternglanz, R., and Fields, S. (1993). Using the two-hybrid system to detect protein-protein interactions. In *Cellular Interactions in Development: A Practical Approach*. Hartley, D.A., (ed.). Oxford: Oxford University Press. pp. 153–179.

Berlin, V., and Chiu, M.I. (1995). Identification of novel cell cycle targets using small molecule ligands. In *Cell Cycle: Materials and Methods*, Pagano, M., ed. New York, Springer-Verlag. pp. 145–156.

Bierer, B.E., Mattila, P.S., Standaert, R.F., Herzenberg, L.A., Burakoff, S.J., Crabtree, G., and Schreiber, S.L. (1990). Two distinct signal transmission pathways in T lymphocytes are inhibited by complexes formed between an immunophilin and either FK506 or rapamycin. Proc. Natl. Acad. Sci. USA 87: 9231–9235.

Brown, E.J., Albers, M.W., Shin, T.B., Ichikawa, K., Keith, C.T., Lane, W.S. and Schreiber, S.L. A mammlian protein targeted by G1-arresting rapamycin-receptor complex (1994). Nature 369: 756-758.

Cafferkey, R., Young, P.R., McLaughlin, M.M., Bergsma, D.J., Koltin, Y., Sathe, G.M., Faucette, L., Eng, W.-K., Johnson, R.K., and Livi, G.P. (1993). Dominant missense mutations in a novel yeast protein related to mammalian phosphatidylinositol 3-kinase and VP34 abrogate rapamycin cytotoxicity. Mol. Cell. Biol. 13:6012–6023.

Chevray, P.M., and Nathans, D. (1992). Protein interaction cloning in yeast: Identification of mammalian proteins that react with the leucine zipper of Jun. Proc. Natl. Acad. Sci. USA 89:5789–5793.

Chiu, M.I., Katz, H. and Berlin, V. (1994). RAPT1, a mammalian homolog of yeast TOR, interacts with the FKBP12/rapamycin complex. Proc. Natl. Acad. Sci. USA 91:12574–12578.

Clipstone, N. A., and Crabtree, G.R. (1992). Identification of calcineurin as a key signalling enzyme in T-lymphocyte activation. Nature 357: 695–697.

Fields, S., and Song, O.-K. (1989). A novel genetic system to detect protein-protein interactions. Nature 340:245–246.

Flanagan, W.M., Corthesy, B., Bram, R.J., and Crabtree,G.R. (1991). Molecular cloning and over expression of the human FK506-binding protein FKBP. Nature 352:803–807.

Frantz, B., Nordby, E.C., Bren, G., Steffan, N., Paya, C.V., Kincaid, R.L., Tocci, M.J., O'Keefe, S.J., and O'Neill, E.A. (1994). Calcineurin acts in synergy with PMA to inactivate IkB/MAD3, an inhibitor of NF-kB. EMBO J. 13:861–870.

Fruman, D.A., Klee, C.B., Bierer, B.E., and Burakoff, S.J. (1992). Calcineurin phosphatase activity in T lymphocytes is inhibited by FK506 and cyclosporin A. Proc. Natl. Acad. Sci. USA 89:3686–3690.

Gyuris, J., Golemis, E., Chertkov, H., and Brent, R. (1993). CDil, a human G1 and S phase protein phosphatase that associates with Cdk2. Cell 75:791–803.

Handschumacher, R.E., Harding, M.W., Rice, J., Drugge, R.J., and Speicher, D.W. (1984). Cyclophilin: a specific cytosolic binding protein for cyclosporin. Science 226:544–546.

Hannon, G.J., Demetrik, D., and Beach, D. (1993). Isolation of the Rb-related p130 through its interaction with CDK2 and cyclins. Genes Dev. 7:2378–2391.

Harding, M.W., Galat, A., Uehling, D.E., and Schreiber, S.L. (1989). A receptor for the immunosuppressant FK506 is a *cis-trans* peptidyl-prolyl isomerase. Nature 341: 758–760.

Heitman, J., Movva, N.R., and Hall, M.N. (1991). Targets for cell cycle arrest by the immunosuppressant rapamycin in yeast. Science 253:905–909.

Helliwell, S., Wagner, P, Kunz, J, Deuter-Reinhard, M, Henriquez, R., and Hall, M.N. (1994). TOR1 and TOR2 are structurally and functionally similar but not identical phosphatidylinositol kinase homologues in yeast. Mol. Biol. Cell 5:105–118.

Hoffman, C.S., and Winston, F. (1987) A ten-minute DNA preparation from yeast efficiently releases autonomous plasmids for transformation of *Escherichia coli*. Gene 57:267–272.

Kunz, J., Henriquez, R., Schneider, U., Deuter-Reinhard, M., Movva, N.R., and Hall, M.N. (1993). Target of rapamycin in yeast, TOR2, is an essential phosphatidylinositol kinase homolog required for G1 progression. Cell 73:585–596.

Lee, W.L., Ryan, F., Swaffield, J.C., Johnston, S.A., and Moore, D.D. (1995). Interaction of thyroid-hormone receptor with a conserved transcriptional mediator. Nature 374:91–94.

Liu, J., Albers, M.W., Wandless, T.J., Luan, S., Alberg, D.G., Belshaw, P.J., Cohen, P., MacKintosh, C., Klee, C.B., and Schreiber, S.L. (1992). Inhibition of T cell signaling by immunophilin-ligand complexes correlates with loss of calcineurin phosphatase activity. Biochemistry 31:3896–3901.

Liu, J., Farmer, J.D., Lane, W.S., Friedman, J., Weissman, I., and Schreiber, S.L. (1991). Calcineurin is a common target of cyclophilin-cyclosporin A and FKBP-FK506 complexes. Cell 66:807–815.

Mattila, P.S., Ullman, K.S., Fiering, S., Emmel, E.A., McCutcheon, M., Crabtree, G.R., and Herzenberg, L.A. (1990). The actions of cyclosporin-A and FK506 suggest a novel step in the activation of T lymphocytes. EMBO J. 9:4425–4431.

McCaffrey, P.G., Perrino, B.A., Soderling, T.R., and Rao, A. (1993). NF-ATp, a T lymphocyte DNA-binding protein that is a target for calcineurin and immunosuppressive drugs. J. Biol. Chem. 268:3747–3752.

Rothstein, R. (1991). Targeting, disruption, replacement, and allele rescue: integrative DNA transformation in yeast. Methods Enzymol. 194:281–301.

Sabatini, D.M., Erdjument-Bromage, H., Lui, M., Tempst, P. and Snyder, S.H. (1994). RAFT1: A mammalian protein that binds to FKBP12 in a rapamycin-dependent fashion and is homologous to yeast TORs Cell 78:35–43.

Sherman, F. (1991). Getting started with yeast. In *Guide to Yeast Genetics and Molecular Biology*. Guthrie, C., and Fink, G. R., eds. Methods Enzymol. 194:3–20.

Siekierka, J.J., Hung, S.H.Y., Poe, M., Lin, C.S., and Sigal, N.H. (1989). A cytosolic binding protein for the immunosuppressant FK506 has peptidyl-prolyl isomerase activity but is distinct form cyclophilin. Nature 341:758–760.

Tocci, M.J., Matkovich, D.A., Collier, K.A., Kwok, P., Dumont, F. Lin, S. Degudicibus, S., Siekierka, J.J., Chin, J., and Hutchinson, N.I. (1989). The immunosuppressant FK506 selectively inhibits expression of early T cell activation genes. J. Immunol. 143:718–726.

Vojtek, A.B., Hollenberg, S.M., and Cooper, J.A. (1993). Mammalian ras interacts directly with the serine/threonine kinase raf. Cell 74:205–214.

16

Use of a Combinatorial Peptide Library in the Two-Hybrid Assay

Meijia Yang

Protein-peptide interactions, such as the recognition of a continuous epitope by an antibody or the binding of a peptide hormone to its receptor, play important roles in many biological processes. Additionally, a short peptide sequence, embedded and constrained within a protein scaffold, may mediate a protein-protein interaction in a similar manner as a ligand-receptor interaction. The goal of designing biologically relevant peptidomimetics has led pharmaceutical research efforts to develop simple assays for protein-peptide interactions. Advances in this area include the isolation of peptide ligands from libraries displayed on the surface of filamentous bacteriophage (for review, see Gallop et al. 1994). We have used the two-hybrid system to detect protein-peptide interactions in vivo and have demonstrated that this assay is sensitive and specific for analyzing randomly generated peptide ligands (Yang et al. 1995). Our approach is complementary to the phage display method, as it allows various genetic manipulations. In this chapter, I discuss issues relevant to the use of a random oligonucleotide-encoded peptide library in the two-hybrid assay. A detailed protocol for construction of this library in an activation domain vector is included (see also chapter 5). Also, I discuss the identification and characterization of peptide ligands for the human retinoblastoma protein (pRb) that were obtained from such a library.

General Considerations

Before using the two-hybrid system to identify peptide ligands, the following issues that relate to combinatorial peptide library approaches need to be addressed: (1) the selection system must allow the detection of protein-peptide interactions that usually have lower affinity than protein-protein interactions; (2) the complexity of the peptide library should be within the screening capacity of yeast transformation protocols; (3) a quantitative assay for affinity should be included in the selection system to facilitate the analysis of the physicochemical properties involved in binding; (4) nonspecific protein-ligand interactions should not be detected or should be easily eliminated. Compared to either the phage (Cwirla et al. 1990; Scott and Smith, 1990; Devlin et al. 1990) or the lacI plasmid (Cull et al. 1992) methods to display random peptides, the two-hybrid system offers some advantages with respect to sensitivity and quantitation. Conversely, the two-hybrid system has a limited capacity to deal with highly complex libraries as compared to the other peptide display systems.

We determined that weak protein-peptide interactions with an apparent K_d of greater than 70 μM can be detected in the two-hybrid assay (Yang et al. 1995), confirming earlier experiments that indicated that this method is more sensitive than coimmunoprecipitation (Li and Fields, 1993). The sensitivity of the two-hybrid system appears to be due to an amplification of an initial signal by multiple rounds of transcription and translation, and the production of a stable reporter enzyme such as ß-galactosidase, which can catalyze a chromogenic reaction. Thus, the two-hybrid system may allow the detection of protein-peptide interactions of an affinity comparable to those detected by phage display (1 to 1000 μM, Gallop et al. 1994).

The two-hybrid system can be used as a quantitative assay to analyze the relative affinities of protein-protein interactions. Affinities measured by co-immunoprecipitation correlated with the transcriptional signal in the yeast assay (Li and Fields 1993). This feature allowed the identification of amino acid residues involved in the binding of SV40 large T antigen to the tumor suppressor protein p53 (Li and Fields 1993). However, a second independent affinity assay, such as BIAcore or coimmunoprecipitation assays, should be carried out to confirm the relative affinities determined by the two-hybrid assay. It is necessary to confirm, for example, by Western blot, that the hybrid proteins being compared are expressed at similar levels. A newer generation of two-hybrid vectors with features such as conditional expression of the hybrid proteins (to avoid toxiciy problems) and lower plasmid copy number per cell (to reduce fluctuation of copy number) may help determine relative affinities more accurately.

In combinatorial peptide library approaches using phage display, the phage produce the peptides as fusions to the major or the minor coat protein. These peptide libraries can be highly complex, as 10^9 different peptides can be screened (for review, see Gallop et al. 1994). A similarly complex peptide library for use in the two-hybrid system would require the screening of billions of yeast transformants. The current yeast transformation protocols can be conveniently performed to screen up to ~10^7 different clones. Therefore, the complexity that can be screened in the yeast assay is far less than that of phage display. To increase the number of short peptides that can be screened in the two-hybrid system, the length of the random peptides can be increased by synthesizing longer oligonucleotides, such that each peptide of the library contains several shorter peptides of desired length.

Two-hybrid searches result in the identification of false positives, defined as activation domain fusions that can induce expression of reporter genes in the presence of DNA-binding domain fusions unrelated to the target protein (Harper et al. 1993; Bartel et al. 1993). False positives activate transcription of the two-hybrid reporter genes by unknown mechanisms. Such false positives were not found in the peptide library described here (M. Yang, unpublished data), suggesting that they may require larger protein domains. These domains are encoded only by the larger-sized inserts used in the construction of genomic or cDNA libraries.

Construction of a Random Peptide Library

We constructed a library of random 16mer peptides as fusions to the C-terminus of the yeast Gal4p transcription activation domain by cloning double-strand oligonucleotides into the vector pGAD.GH (Figure 16-1). Alternatively, the peptide could be located N-terminal to the activation domain or within a folded protein that allows the display of conformationally-constrained peptides. The random oligonucleotide sequences were flanked by two constant regions, each containing a unique restriction site, *Bam*HI or *Eco*RI. During the synthesis of single-stranded oligonucleotides, a random codon was defined by equal incorporation of all four nucleotides at the first and the second position and by equal incorporation of G or T at the third position. For each random codon, the $4 \times 4 \times 2$ (= 32) possible combinations can encode all 20 amino acids and the stop codon TAG. In order to encode all possible pentamer combinations, the total number of independent clones should exceed 32^5 or 33.5 million, a number not readily screened by current yeast transformation protocols. One solution to this problem is to design longer random peptides, which are a repetition of shorter peptides. A 16mer peptide has 12 combinations of continuous pentamers, and therefore a total of ~3 million 16mer peptides is required to encompass all possible pentamers. Our random 16mer peptide library contained 10 million different clones, which encompasses all possible pentamers and 10% of all possible hexamers.

Protocols 1 through 3 describe procedures for constructing the peptide library using an activation domain vector.

a) Random oligonucleotides and primers for PCR:

5'-GAACTAGTGGATCCC(NNK)$_{16}$TAGGAATTCGGCCGC-3'

BamHI EcoRI

+

5'-GAACTAGTGGATCCC (5' primer)

3'-ATCCTTAAGCCGGCG (3' primer)

N: equal incorporation of all four nucleotides;
K: equal incorporation of G or T.

b) The reading frame of the random peptides in the pGAD.GH.

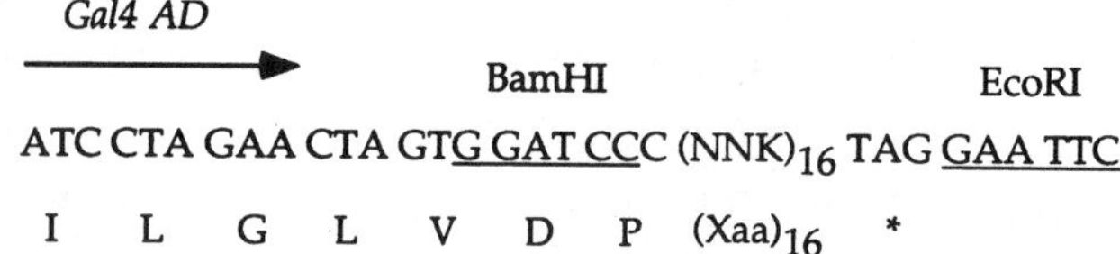

Fig. 16.1. Construction of a peptide library by cloning radom oligonucleotides into the activation domain vector pGAD.GH (Hannon et al. 1993).

Protocol 1: Preparation of Double-Stranded Oligonucleotides

1. Make the following mixture for PCR amplification.

20 pmole	single-stranded oligonucleotides
6 nmole	5' primer
6 nmole	3' primer
100 μl	10 × PCR buffer (from supplier)
10 μl	25 mM deoxynucleotide triphosphate (dNTP) mixture

   Adjust the volume with sterile water to 990 μl and add 10 units Taq DNA polymerase (final volume of 1000 μl).
2. Aliquot the PCR mixture into 100 μl reaction volume for amplification and cover the mixture with 50 μl mineral oil.
3. Perform PCR using the following parameters:

5 cycles	30 seconds	95°C
	90 seconds	35°C
	40 seconds	72°C
20 cycles	30 seconds	94°C
	30 seconds	45°C
	40 seconds	72°C

4. Extract DNA with chloroform.
5. Extract DNA with phenol and then with chloroform.
6. Precipitate DNA with ethanol.
7. Dissolve DNA pellet in 50 µl water. Remove 5 µl for controls.
8. Digest the DNA with 2 µl (20 units) of *Eco*RI and *Bam*HI enzymes for 2 hours, in a final volume of 60 µl using *Eco*RI buffer (final concentration: 100 mM Tris-HCl, pH 7.5, 50 mM NaCl, 10 mM $MgCl_2$, 0.025% Triton X-100). Note: Digest 2.5 µl of the DNA from step 7 with *Eco*RI or *Bam*HI as a control.
9. Isolate the digested DNA by electrophoresis using a 20% polyacrylamide gel (acrylamide:bisacrylamide, 29:1) buffered with TBE (90 mM Tris-borate, 2.5 mM EDTA). Load uncut DNA from step 7 and singly cut DNA from step 8 to check for complete digestion. Stain the gel in ethidium bromide and cut out gel slices containing the DNA bands and transfer to a microfuge tube. Crush the gel slice with sterile pipette tips.
10. Add 500 µl TE buffer (10 mM TrisCl pH 7.4, 1 mM EDTA) per gel slice and 5 µl 10% SDS; shake the tube at 37°C overnight or at 55°C for 3 to 4 hours.
11. Add 50 µl of 5 M potassium acetate, pH 5.2, and vortex for 5 seconds. Spin the tube at 10,000 rpm for 10 minutes to remove the SDS and gel debris. Transfer the supernatant to a sterile tube.
12. Extract the DNA with phenol and then with chloroform.
13. Add 10 µg polyacrylamide as carrier. (Preparation of the polyacrylamide solution: Dissolve 20 mg acrylamide in 1 ml sterile water. Add 1 µl TEMED and 10 µl 10% ammonium persulfate and allow polymerization for 1 hour. Dilute the viscous solution to 1 mg/ml final concentration with sterile water. Use a small amount of dithiothreitol or ß-mercaptoethanol to quench the free radicals.) Precipitate the DNA with 2 volumes of ethanol and wash the DNA once with 70% ethanol. Dissolve the DNA in 50 µl sterile water.

Protocol 2: Preparation of the Vector DNA for Ligation

Success in peptide library construction depends largely on the ligation efficiency, which is determined by many factors. First, high-quality vector DNA is required. Preparation of the vector for ligation involves restriction digestion and phosphatase treatment, and conditions for both steps should be carefully optimized in pilot experiments. I found that DNA prepared by CsCl gradient is the most amenable for both restriction digestion and phosphatase treatment. Although oligonucleotide inserts are readily available from the PCR amplification and restriction digestion, it is important that digestion of both ends of the inserts is complete. As a rule, quantitation of the yield or the efficiency for each treatment step during the vector processing is needed for optimization or troubleshooting.

1. In order to determine conditions for complete restriction digestion, perform two separate restriction digestions with 50 µg purified plasmid DNA each. One digestion is to be incubated with 20 units of *Eco*RI and the other with 20 units of *Bam*HI, in a final volume of 60 µl using *Eco*RI buffer (see Protocol 1). After the addition of the enzymes, remove small aliquots of plasmid DNA at time points of 5, 10, 20, and 40 minutes and analyze the plamid DNA by agarose

gel electrophoresis. As judged by DNA band pattern, the digestion for each of the enzymes should be complete after 5 minutes of incubation. If it is not, try to lower the DNA amount used for digestion or use another batch of the plasmid, since the most common cause for incomplete digestion is contaminants in the plasmid DNA preparation. Do not excessively increase the amount of the restriction enzyme because an excess of enzyme storage buffer can cause star activity. Once the restriction digestion conditions are optimized, repeat the digestion with both enzymes in a single digestion and incubate for 30 minutes for complete digestion.

2. Add 6.6 µl of 10X phosphatase buffer (0.5 M Tris Cl, pH 8.7, 50 mM $MgCl_2$) and 1 unit of calf intestine phosphatase to the digested DNA. Incubate the mixture at 37°C for 5 to 10 minutes. (Longer incubation is not necessary and often leads to degradation of the DNA.)

3. Stop the phosphatase treatment by phenol/chloroform extraction. Precipitate the DNA by adding 6 µl 3 M sodium acetate, pH 5.2, 10 µg polyacrylamide carrier (see Protocol 1), and 180 µl ethanol. Dissolve the DNA precipitate in 50 µl sterile water and estimate the yield by analyzing 5 µl of the DNA sample by agarose gel electrophoresis. The completion of the double digestion can be assessed as follows. Use 1 unit T4 polynucleotide kinase to treat 1 µl cut, recovered DNA in a 10 µl reaction in kinase buffer. After the kinase reaction, use 1 µl of the kinase reaction mixture to perform a ligation in a 10 µl reaction in ligase buffer. This treatment should recircularize singly cut plasmid DNA. Next, transform the ligated DNA into *Escherichia coli* using a similar amount of the uncut plasmid as a control. The number of *E. coli* colonies transformed by ligated DNA represents uncut or singly cut plasmid DNA and should be at least 1/1000 or less as compared to the number with undigested DNA.

Protocol 3: Ligation and Transformation

A standard ligation protocol can be used. The ligated DNA should be precipitated using yeast tRNA as a carrier. The precipitation is an important step for removal of salt from the ligation mixture, which can prevent high efficiency *E. coli* transformation by electroporation.

1. Use an estimated amount of DNA equivalent to a 3:1 insert/vector ratio for ligation. As a rule, use 1 Weiss unit of T4 DNA ligase for 1 µg vector DNA in a 10 µl ligation reaction. Keep in mind that ligation conditions may vary and need to be tested for each combination of plasmid and insert. Generally, the volume of ligation reaction in each tube should not exceed 10 µl, and the reaction is carried out at 15°C for 12 hours.
2. After the ligation, add 40 µl sterile water and 10 µl or less of 10 µg/µl yeast tRNA to a 10 µl ligation mixture. Use 120 µl ethanol to precipitate the DNA. Recover ligated plasmids in 20 µl ultra pure water.
3. Use 1 µl of the recovered DNA to transform *E. coli* by electroporation according to the procedures suggested by the manufacturer. Estimate the titer of the colony forming units by plating out a dilution series of the *E. coli* suspension on LB ampicillin plates. Repeat electroporation until the expected complexity of the library is reached. Plasmid DNA for the library can be directly prepared from the cells scraped from the plates; do not amplify the library by growth in liquid culture.

Identification and Analysis of Peptides that Bind to the Retinoblastoma Protein (pRB)

A number of cellular or viral proteins have been found to interact with pRb (reviewed by Weinberg, 1995). Mammalian D-type cyclins or oncogene products of DNA tumor viruses, such as adenovirus E1A, SV40 large T antigen, and human papilloma virus (HPV) E7, contain a five amino acid motif Leu-X-Cys-X-Glu (LXCXE) that mediates their binding to pRb (Dowdy et al. 1993; Ewen et al. 1993; DeCaprio et al. 1989; Dyson et al. 1989; Moran, 1988). Hypophosphorylated form(s) of pRb bind to and repress at least three E2F transcription factors (E2Fs 1, 2, and 3), and the inhibition of the E2Fs is abrogated upon hyperphosphorylation of pRb by CDK2 and/or CDK4 complexes (for review, see Weinberg 1995). The oncogene products E1A, E7, and large T antigen bind to pRb with stronger affinities than cellular proteins. Thus, by targeting pRb, the viral oncogene proteins abrogate the cell-cycle-dependent regulation of E2F transcription factors (Bandara and La Thange 1991; Helin et al. 1992; Kaelin et al. 1992; Dowdy et al. 1993; Ewen et al. 1993). However, the physical or chemical properties that differentiate the affinities of the various pRb-interacting proteins containing the LXCXE motif remained unclear. Mutational analysis of the large T antigen and E7 proteins suggested that the LXCXE motif is necessary for binding, whereas other amino acids in the region of the motif contribute to the affinity (DeCaprio 1988; Munger et al. 1989; Gage et al. 1990; Heck et al. 1992).

To understand how the LXCXE motif might be recognized by pRb, we searched for pRb-binding peptide ligands using a pRb fragment (amino acids 301–928) cloned into the DNA-binding domain vector pAS1 (Durfee et al. 1993). The pRb fragment contains region A (amino acids 393–572) and region B (amino acids 646–772). Together they constitute the AB-pocket, a minimum binding domain required for the recognition of two types of pRb interactors, the LXCXE motif-containing proteins and the E2F transcription factors (Hu et al. 1990; Huang et al. 1990; Kaelin et al. 1990, Helin et al. 1992; Kaelin et al. 1992). A screen of three million clones of the peptide library yielded seven peptides that can specifically bind to pRb. All seven contain the LXCXE motif (Table 16-1), demonstrating that the two-hybrid system can specifically identify peptide ligands present in a random library. Because earlier experiments showed that two-hybrid signals can be correlated with the affinities between interacting proteins (Li and Fields, 1993), the expression of the reporter gene *GAL1-lacZ* was used to determine the relative affinities of the peptide ligands for pRb. Both the LXCXE motif and other amino acids adjacent to this motif contributed to these affinities.

In order to determine which residues were most critical for binding, we mutagenized the p1 and p7 peptides by suboptimal PCR (Leung et al. 1989)

Table 16-1 Peptides identified from a library screen interact with the pRb in the two-hybrid assay. (Adapted from Yang et al. 1995.)

Peptide	Sequence[a]	ß-galactosidase activity[b]
P1	YGLWI**L** W**C**D**E**EGLDLG	45
P2	NQLLGDV**L** A**C**Y **E**QEVE	6.9
P3	WTEL**L** F**C**F **E**QVYGDPF	6.8
P4	EGGD**L** G**C**D **E**SWSEGYT	6.0
P5	CDG**L** L**C**T **E**TLL	5.4
P6	GGCPGAN**L** C**C**F **E**KSLD	2.5
P7	TTWRER**L** R**C**E **E**NGLGV	1.1

[a]Peptide sequence are aligned by the conserved residues (indicated in bold).

[b]Values for liquid ß-galactosidase activity represent the mean of at least four assays, and the standard errors were generally 10 to 20% of the mean.

and identified presumptively stronger or weaker binding derivatives (Table 16-2), based on increased or decreased ß-galactosidase activity, respectively. Derivatives of p7 with increased activity lost a positively charged arginine, and all derivatives of p1 with decreased activity had fewer charged aspartic acid or glutamic acid residues. P1 derivatives with unaltered activities did not gain or lose any charged residue. Furthermore, the two-hybrid signal for the highly active peptide P1 and its derivative P1-a1 correlated with biochemical affinities determined by a BIAcore binding assay; p1 bound to pRb with an apparent K_d of 13 to 23 μM, and P1-a1 with an apparent K_d of 61 to

Table 16-2 P1 and P7 peptide derivatives and their binding to the pRb in the two-hybrid assay. (Adapted from Yang et al. 1995.)

Peptide	Sequence	ß-galactosidase activity[a]
P1	Y G L W I **L** W **C** D **E** E G L D L G	45
P1-r	- - - - T - - - - - - - - - - -	45
P1-a	- - - - - - - - E - - - - - - -	42
P1-h	- - - R - - - - - - - - - - - -	27
P1-1	- - - - - - R - - - - - - - - -	22
P1-i	- - - - - - - - - - - - - G - -	18
P1-20	- - - - - - - - G - - - - - - -	16
P1-K	- - - - - - - - - - G - - - - -	12
P1-a1	- - R - - - - - - - - - - - - -	11
P7	T T W R E R **L** R **C** E **E** N G L G V	2.2
P7-7	- - - G - - - - - - - - - - - -	4.4
P7-14	- - - E - - - - - - - - - - - -	7.3

[a]Values for liquid ß-galactosidase activity represent the mean of at least four assays, and the standard errors were generally 10 to 20% of the mean.

76 μM. Although high-affinity pRb-binding peptides apparently must contain negatively charged residues, as charged residues can be found at nearly every position in the region near the LXCXE motif, the mechanism for this effect cannot be easily attributed to specific electrostatic interactions between pRb and the side chains of the peptide.

Figge et al. (1993) analyzed the CD spectrum of a 23-amino acid T antigen peptide containing the LXCXE motif and found 40% net alpha-helical content for the peptide in solution. Consistent with this structure, proline substitutions in a 15-amino acid T antigen peptide that changed the motif sequence from LFCSE to LPCPE abolished its binding to pRb. Our screen for pRb ligands did not yield any peptide containing the LXCXE motif at or near the C-terminus of the peptide, although the complexity of the library predicted the existence of such peptides. It is known that the formation of secondary structure is unfavorable at the end of flexible, short peptides. Possibly, the LXCXE motif forms a secondary structure that places the leucine, cysteine, and glutamic acid residues in a spatial arrangement that is recognized by pRb. The formation of this secondary structure in the LXCXE-containing peptides might be favored by the presence of the negative charges. The effect of charges on secondary structure formation is not unprecedented; it has been observed that the stability of an alpha-helix formed by a peptide can be strongly influenced by both the presence and the distribution of charged residues (Shoemaker et al. 1987). However, it should be noted that the viral proteins are tightly folded compared to the peptides, and that the protein scaffold contributes significantly to the folding of the LXCXE motif important for high-affinity binding to pRb. Interestingly, the LXCXE motif contained in D-type cyclins is at the N-terminus of the proteins; there are only two amino acid residues between the N-terminal methionine and the motif (Dowdy et al. 1993).

One question relevant to the usefulness of the random peptide library approach is whether differences in the affinities of the pRb-binding peptides is related to differences in affinities among various cellular and viral pRb-binding proteins. Though a definitive answer to this question remains elusive, data from earlier mutational analyses of large T antigen and HPV E7 proteins are consistent with the analysis derived from the peptide library. First, DeCaprio et al. (1988) identified five T antigen mutants that are deficient in binding to pRb and in promoting focus formation and soft agar growth by rodent cells. Among the five T antigen mutants, four had an altered LXCXE motif, whereas the remaining mutant (K7) had a change at the glutamic acid residue at position +1 and serine at the position +5 (relative to the LXCXE motif) to a lysine and an asparagine residue, respectively. Thus, the K7 mutant had an increase of two positive charges in the region recognized by pRb while the LXCXE motif remained intact. Similarly, HPV E7 proteins from the benign, "low-risk" strains HPV-6b and HPV-11 bind to pRb with much reduced affinity as compared to E7 proteins from the "high-risk" strains HPV-16 and HPV-18. The latter HPVs are found in genital tract lesions that often progress to malignant cancers. Sequence comparison revealed that the region containing the sequence GLXCXE in the E7 proteins of "low-risk" HPVs is changed to DLXCXE in "high-risk" HPVs (Munger et al. 1989; Gage et al. 1990; Heck et al. 1992). Additionally, substitution of the glycine residue in the GLHCYE with an aspartic acid residue increased binding of the "low-risk" E7 protein to pRb by 5-fold, and the mutant HPV-6 with this change gained significant transformation activity when cotransfected into primary baby kidney cells along with an activated *ras* oncogene (Heck et al. 1992).

Consistent with the mutational analysis of the viral proteins, our results suggest an explanation for why the viral oncogene products are able to deprive cells of the regulatory function of pRb. During the early G1 phase of the cell cycle, pRb exists in a underphosphorylated form that binds to and inactivates the E2F transcription factors (for review, see Weinberg 1995). Repression of E2F function by pRb is abrogated by hyperphosphorylation of pRb by CDK complexes (Ewen et al. 1993; Dowdy et al. 1993) or by interaction with viral proteins (Bandara and La Thange 1991; Chittenden 1991; Helin et al. 1992; Kaelin et al. 1992). Intriguingly, the regions containing the LXCXE motif in the viral oncoproteins are highly negatively charged (Figge et al. 1993). Thus, the observation that the overall net charge determines the affinities of the pRb ligands might explain the high affinities of the SV40 T-antigen and HPV E7 proteins. Our mutagenesis analysis of the pRb-binding peptides supports the idea that net negative charge of the sequence surrounding the LXCXE motif region is important for the large T antigen or E7 proteins to decouple pRb from its regulation of the E2F transcription factor.

Summary and Future Prospects

During two-hybrid searches for peptide ligands, we encountered several problems with this approach. First, we failed in searches using such targets as the human tumor suppressor protein p53, the p53 (R173H) mutant, and an oncogenic Ras protein. The limitation to this approach could be due to several reasons. For example, Lane and coworkers discovered that short peptides derived from the C-terminus of p53 can activate DNA-binding by the full-length protein (Hupp et al. 1995). The shortest peptide that activated DNA-binding contained twelve residues. The peptide library used in our screening was of limited complexity and only encompassed all possible pentamers. To increase the size of the library is not advantageous unless yeast transformation efficiency can be increased significantly. A more practical approach might be to construct a biased activation domain library using fragments of proteins known to interact with the target protein. Although this approach would require construction of a peptide library for each protein target, the probability of finding peptide ligands may be significantly increased.

The random peptides described here possess free C-termini and thus are fully flexible to adopt a myriad of conformations. This flexibility may also have reduced the probability of finding peptide ligands. It is possible to constrain the random peptides by displaying them as loops in a tightly folded protein (Colas et al. 1996). Conformationally constrained peptides can exhibit higher affinities than linear, unconstrained peptides; O'Neil and coworkers (1992) identified cyclic peptide ligands for integrin that contain the tripeptide RGD and were more reactive in the cyclic form.

Concerns have arisen over whether some two-hybrid interactions with K_d values in the μM range are spurious because such interactions might not occur in vivo (Allen et al. 1995). Generally, biochemical binding assays should be carried out to confirm two-hybrid interactions. It is also possible to use a less sensitive reporter strain and to limit the expression of both hybrid proteins in order to increase the threshold for the detection. However, we found that pRb-peptide interactions with affinities in the μM range can be specifically detected. The high specificity and low affinity probably reflect additional requirements, such as a folded structure of peptides, that are needed for the detection of protein-peptide interactions. As discussed earlier, recognition of peptides with the LXCXE motif by pRb likely involves formation of a secondary structure and cannot be attributed solely to electrostatic and/or hydrophobic interactions.

Yeast cells bearing a peptide ligand and the target protein have a growth advantage in the appropriate strain background, and both the ligand and the target protein plasmids are easily retrievable for retesting and for sequence analysis. Given that the transcription signal may be proportional to the affinity between an interacting ligand and the target protein, the two-hybrid search permits the identification of strong or weak peptide ligands in a single round of screening. Obviously, such a comparison requires that the peptide fusions be expressed at similar levels, which can be facilitated by controlling the expression of the hybrid proteins with an inducible promotor and by using centromere-based plasmids that do not vary in copy number per cell. Our results suggest that difference in the affinities of peptide ligands may be useful to guide secondary approaches, such as the design of peptidomimetics with increased affinity. In addition, the determination of NMR structures for the identified peptide ligands can complement the affinity analysis. Binding in the current two-hybrid system is multivalent, that is, there are numerous sites for the DNA-binding domain hybrid in the promoter of the reporter gene. Multivalent ligand selection usually increases the sensitivity of the affinity selection, allowing the detection of weak interactions, while a monovalent approach selects for high-affinity interactions (Lowman et al. 1991). It should be possible to engineer a two-hybrid reporter gene with a single DNA-binding site for a protein that binds DNA as a monomer, such that only strong interactions are detected. Thus, with a mutant pool derived from a weak binding peptide obtained from a screen in a sensitive reporter strain, strong binding peptides may be obtained by using a less sensitive reporter strain.

Unlike phage display, the two-hybrid system allows the introduction of an additional component in the affinity selections (see chapter 7). Peptide ligands could be used as such a component, to modulate pre-existing protein-protein interactions, similar to the disruption of the binding between pRb and E7 by peptides containing the LXCXE motif (Jones et al. 1990). Overexpression of the competing peptide might be required in order to suppress sufficiently the expression of the reporter gene. The two-hybird counter-selection scheme (see chapter 7), combined with a combinatorial peptide library, could be used to select peptides that disrupt protein-protein interactions. A peptide library constructed in a vector lacking a transcription activation domain is required for such a selection strategy. This combinatorial approach might be useful to identify peptides that disrupt pharmaceutically important interactions.

ACKNOWLEDGMENTS I thank Drs. Gunter Bernhard and Xuemei Cao for advice on construction of the peptide library, Dr. Stanley Fields for support and advice, and the members of the Fields laboratory for discussions and help throughout this work. I also thank Drs. B. Aronson, T. Miller, and P. Tegtmeger for comments on the manuscript.

References

Allen, J.B., Walberg, M.W., Edwards, M.C., and Elledge, S.J. (1995). Finding prospective partners in the library: the two-hybrid system and phage display find a match. Trends Biochem. Sci. 20:511–516.

Bandara, L.R., and La Thange, N.B. (1991). Adenovirus E1a prevents the retinoblastoma gene product from complexing with a cellular transcription factor. Nature 351:494–497.

Bartel, P., Chien, C.-T., Sternglanz, R., and Fields, S. (1993). Elimination of false positives that arise in using the two-hybrid system. BioTechniques 14:920–924.

Chien, C.-T., Bartel, P.L., Sternglanz, R., and Fields, S. (1991). The two-hybrid system: A method to identify and clone genes for proteins that interact with a protein of interest. Proc. Natl. Acad. Sci. U.S.A. 88:9578–9582.

Chittenden, T., Livingston, D.M., and Kaelin, Jr., W.G. (1991). The T/E1A-binding domain of the retinoblastoma product can interact selectively with a sequence-specific DNA-binding protein. Cell 65:1073–1082.

Colas, P., Cohen, B., Jessen, T., Grishina, I., McCoy, J., and Brent, R. (1996). Genetic selection of peptide aptamers that recognize and inhibit cyclin-dependent Kinase 2. Nature 380:548–550.

Cull, M.G., Miller, J.F., and Schatz, P.J. (1992). Screening for receptor ligands using large libraries of peptides linked to the C terminus of the lac repressor. Proc. Natl. Acad. Sci. USA 89:1865–1869.

Cwirla, S.E., Peters, E.A., Barrett, R.W., and Dower, W.J. (1990). Peptides on phage: A vast library of peptides for identifying ligands. Proc. Natl. Acad. Sci. USA 87:6378–6382.

DeCaprio, J.A., Ludlow, J.W., Figge, J., Shew, J.-Y., Huang, C.-M., Lee, W.-H., Marsilio, E., Paucha, E., and Livingston, D.M. (1988). SV40 large tumor antigen forms a specific complex with the product of the retinoblastoma susceptibility gene. Cell 54:275–283.

Devlin, J.J., Panganiban, L.C., and Devlin, P.E. (1990). Random peptide libraries: A source of specific protein binding molecules. Science 249:404–406.

Dowdy, S.F., Hinds, P.W., Louie, K., Reed, S.I., Arnold, A., and Weinberg, R.A. (1993). Physical interaction of the retinoblastoma protein with human D cyclins. Cell 73:499–511.

Durfee, T., Becherer, K., Chen, P.-L., Yeh, S.H., Yang, Y., Kilburn, A.E., Lee, W.-H., and Elledge, S.J. (1993). The retinoblastoma protein associates with the protein phosphatase type 1 catalytic subunit. Genes Dev. 7:555–569.

Dyson, N., Howley, P.M., Munger, K., and Harlow, E. (1989). The human papilloma virus-16 E7 oncoproteins is able to bind to the retinoblastoma gene product. Science 243:934–937.

Ewen, M.E., Sluss, H.K., Sherr, C.J., Matsushime, H., Kato, J., and Livingston, D.M. (1993). Functional interactions of the retinoblastoma protein with mammalian D-type cyclins. Cell 73:487–497.

Fields, S., and Song, O.-K. (1989). A novel genetic system to detect protein-protein interactions. Nature 340:245–246.

Figge, J., Breese, K., Vajda, S., Zhu, Q.-L., Eisele, L., Andersen, T.T., McColl, R., Friedrich, T., and Smith, T.F. (1993). The binding domain structure of retinoblastoma-binding proteins. Prot. Sci. 2:155–164.

Gage, J.R., Meyers, C., and Wettstein, F. O. (1990). The E7 proteins of the nononcogenic human papillomavirus type 6b (HPV-6b) and the oncogenic HPV-16 differ in retinoblastoma protein binding and other properties. J. Virol. 64:723–730.

Gallop, M.A., Barrett, R.W., Dower, W.J., Fodor, S.P., and Gordon, E.M. (1994). Applications of combinatorial technologies to drug discovery. 1. Background and peptide combinatorial libraries. J. Med. Chem. 37:1233–1251.

Hannon, G.J., Demetrick, D., and Beach, D. (1993). Isolation of the Rb-related p130 through its interaction with CDK2 and cyclins. Genes Dev. 7:2378–2391.

Harper, J.W., Adami, G.R., Wei, N., Keyomarsi, K., and Elledge, S.J. (1993). The p21 Cdk-interacting protein Cip1 is a potent inhibitor of G1 cyclin-dependent kinases. Cell 75:805–816.

Heck, D.V., Yee, C.L., Howley, P.M., and Munger, K. (1992). Efficiency of binding the retinoblastoma protein correlates with the transforming capacity of the E7 oncoproteins of the human papillomaviruses. Proc. Natl. Acad. Sci. USA 89:4442–4446.

Helin, K., Lees, J.A., Vidal, M., Dyson, N., Harlow, E., and Fattaey, A. (1992). A cDNA encoding a pRB-binding protein with properties of the transcription factor E2F. Cell 70:337–350.

Hu, Q., Dyson, N., and Harlow, E. (1990). The regions of the retinoblastoma protein needed for binding to adenovirus E1A or SV40 large T antigen are common sites for mutations. EMBO J. 9:1147–1155.

Huang, S., Wang, N.-P., Tseng, B.-Y., Lee, W.-H., and Lee, E.H. (1990). Two distinct and frequently mutated regions of retinoblastoma protein are required for binding to SV40 T antigen. EMBO J. 9:1815–1822.

Hupp, T.R., Sparks, A., and Lane, D.P. (1995). Small peptides activate the latent sequence-specific DNA binding function of p53. Cell 83:237–245.

Jones, R.E., Wegrzyn, R.J., Patrick, D.R., Balishin, N.L., Vuocolo, G.A., Riemen, M.W., Defeo-Jones, D., Garsky, V.M., Heimbrook, D.C., and Oliff, A. (1990). Identification of HPV-16 E7 peptides that are potent antagonists of E7 binding to the retinoblastoma suppressor protein. J. Biol. Chem. 265:12782–12785.

Kaelin, Jr., W.G. Ewen, M.E., and Livingston, D.M. (1990). Definition of the minimal Simian virus 40 large T antigen- and adenovirus E1A-binding domain in the retinoblastoma gene product. Mol. Cell. Biol. 10:3761–3769.

Kaelin, Jr., W.G., Krek, W., Sellers, W.R., DeCaprio, J.A., Ajchenbaum, F., Fuchs, C.S., Chittenden, T., Li, Y., Farnham, P.J., Blanar, M.A., Livingston, D.M., and Flemington, E.K. (1992). Expression cloning of a cDNA encoding a retinoblastoma-binding protein with E2F-like properties. Cell 70:351–364.

Leung, D.W., Chen, E., and Goeddel, D.V. (1989). A method for random mutagenesis of a defined DNA segment using a modified polymerase chain reaction. Technique 1:11–15.

Li, B. and Fields, S. (1993). Identification of mutations in p53 that affect its binding to SV40 large T antigen by using the yeast two-hybrid system. FASEB J. 7:957–963.

Lowman, H.B., Bass, S.H., Simpson, N., and Wells, J.A. (1991). Selecting high-affinity binding proteins by monovalent phage display. Biochemistry 30:10832–10838.

Moran, E. (1988). A region of SV40 large T-antigen can substitute for a transforming domain of the adenovirus E1A products. Nature 334:168–170.

Munger, K., Werness, B.A., Dyson, N., Phelps, W.C., Harlow, E., and Howley, P.M. (1989). Complex formation of human papillomavirus E7 proteins with the retinoblastoma tumor suppressor gene product. EMBO J. 8:4099–4105.

O'Neil, K.T., Hoess, R.H., Jackson, S.A., Ramachandran, N.S., Mousa, S.A., and DeGrado, W.F. (1992). Identification of novel peptide antagonists for GPIIb/IIIa from a conformationally constrained phage peptide library. Proteins 14:509–515.

Scott, J.K. and Smith, G.P. (1990). Searching for peptide ligands with an epitope library. Science 249:386–390.

SenGupta, D. J., Zhang, B. Kraemer, B., Pochart, P., Fields, S., and Wickens, M. (1996). A three-hybrid system to detect RNA-protein interactions in vivo. Proc. Natl. Acad. Sci. USA 93:8496–8501.

Shoemaker, K.R., Kim, P.S., York, E.J., Stewart, J.M., and Baldwin, R.L. (1987). Tests of the helix dipole model for stabilization of a-helices. Nature 326:563–567.

Weinberg, R.A. (1995). The retinoblastoma protein and cell cycle control. Cell 81:323–330.

Whyte, P., Buchkovich, K.J., Horowitz, J.M., Friend, S.H., Raybuck, M., Weinberg, R.A., and Harlow, E. (1988). Association between an oncogene and an anti-oncogene: the adenovirus E1A proteins bind to the retinoblastoma gene product. Nature 334:124–129.

Yang, M., Wu, Z., and Fields, S. (1995). Protein-peptide interactions analyzed with the yeast two-hybrid system. Nucl. Acids Res. 23:1152–1156.

Part IV

RELATED METHODOLOGIES

Although protein-protein interactions play an important role in cellular function, proteins also interact with other macromolecules. Two methods that are direct descendents of two-hybrid technology are the one- and three-hybrid systems. The one-hybrid system, described by Chong and Mandel in chapter 17, has proven useful in the identification of proteins that bind to particular DNA sequences. The three-hybrid system, discussed by Zhang, Kraemer, SenGupta, Fields, and Wickens in chapter 18, is a recent development to detect protein-RNA interactions.

The one-, two-, and three-hybrid systems all rely on the generation of a transcriptional signal to indicate particular macromolecular interactions. In chapter 19, Johnsson and Varshavsky describe an alternative system for the detection of protein-protein interactions, the ubiquitin-based split-protein sensor. Interactions occurring in the yeast cytosol can be detected by proteolytic cleavage of a reporter protein, an approach that complements nuclear-based assays.

17

Isolation of DNA-Binding Proteins Using One-Hybrid Genetic Screens

Jayhong A. Chong
Gail Mandel

Several techniques are available for cloning cDNAs coding for sequence-specific DNA-binding proteins. One successful approach is the purification of transcription factors by DNA-affinity chromatography, followed by subsequent cloning of the cDNA using probes deduced from protein sequence (Kadonga and Tjian 1986). This technique has several limitations. It requires ample starting material as well as the formation of a stable in vitro complex between the DNA and the binding factor. Another approach, "recognition site cloning," involves screening bacteriophage expression cDNA libraries with a radiolabeled DNA sequence corresponding to the known binding site for the transcription factor (Singh et al. 1988). A disadvantage of this approach is that it requires renaturation of fusion proteins on filters, which may not allow the binding of the protein to the target DNA sequence. Additionally, both of these approaches are time and labor intensive.

More recently, a new methodology based upon the yeast two-hybrid system has been developed. This methodology, a one-hybrid system, has been used to clone cDNAs encoding both yeast and mammalian transcription factors. It has worked in at least one case where recognition site cloning failed (Wang and Reed 1993). Several different versions of the one-hybrid approach have been described (Wang and Reed 1993; Inouye et al. 1994; Li and Herskowitz 1993; Dowell et al. 1994; Chong et al. 1995) that rely on histidine prototroph selection and/or a screen for the *lacZ* gene product for identification of the DNA-binding domain of the desired transcriptional factor. This chapter will describe the general strategy for performing the one-hybrid screen. The reader may refer to specific references noted in the text (or in other chapters in this book) for the detailed descriptions of additional yeast strains, vectors, and Gal4p activation domain (GAD) cDNA libraries that might be used.

Conceptual Basis for One-Hybrid Genetic Screen

The one-hybrid genetic screen is a modification of the two-hybrid system (Fields and Song 1989), and similar conceptually to an in vivo approach used to identify target DNA sequences for known transcriptional factors (Wilson et al. 1991). In contrast to the latter, the one-hybrid approach is used to identify cDNAs coding for proteins that bind a specific DNA element. The better this binding site is characterized, the better the chance of success in cloning the desired cDNA. For example, smaller target sequences will help prevent the occurrence of spurious interactions of proteins with the DNA in vivo. Additionally, mutated DNA elements that no longer bind the desired transcriptional factor in mobility shift assays are valuable for eliminating interactions with nonspecific DNA-binding proteins (as follows).

The one-hybrid approach requires the introduction of reporter plasmids into yeast. These plasmids carry a marker gene for selection after transformation. In addition, the reporter plasmids contain either a multimerized DNA element or multiple distinct DNA elements (see, for example, Dowell et al. 1994) cloned upstream of a yeast promoter. The foreign DNA elements replace the upstream activator sequence (UAS) of the yeast promoter. The yeast promoter drives expression of a reporter gene such as *HIS3* or *lacZ* in the presence of transcriptional activators or activation domains that are anchored in place by binding to the foreign target DNA sequences. Specialized activation domain-tagged cDNA libraries are then introduced into the yeast reporter strains. In-frame fusions of the Gal4p activation domain (GAD) with DNA-binding domains that interact with the multimerized cognate binding sites are identified through histidine prototroph selection and/or by blue reaction products in the ß-galactosidase colony color screen. Because of the potential for false positives due to interactions with nonspecific DNA-binding domains, it is advisable to generate parallel yeast strains containing mutated DNA elements that do not bind the desired protein in a mobility shift assay, upstream of the same yeast promoters.

Step-by-Step Procedure for the One-Hybrid Screen

Generation of the Yeast Indicator Strains

Examples of reporter plasmids used in one-hybrid screening are shown in Figure 17-1. The shuttle vectors contain the col E1 origin of replication and the gene coding for ampicillin resistance for growth and selection in bacteria. Two to four copies of the desired binding site are synthesized as complementary oligonucleotides, annealed, and then subcloned upstream of truncated yeast promoters such as those for the *GAL1* and *CYC1* genes that lack the upstream activation sequences (UAS) for these promoters (see Figure 17-1). The yeast promoters drive expression of two different reporter genes, *HIS3* and *lacZ*. Often, the multimerized DNA elements are unstable in bacteria so that it is desirable to perform the subcloning using strains deficient in recombination pathways (e.g., SURE cells, Stratagene, Inc.). Often, it is not known how important the orientation and spacing of the multimerized DNA elements are for an interaction with a DNA-binding domain. Therefore, the conservative approach is to clone the reiterated DNA elements in a head-to-tail fashion in the orientation that they assume in the native gene.

After the number and orientation of the elements are confirmed by sequence analysis, the reporter plasmids are either introduced separately into yeast to create two distinct yeast reporter strains (see Inouye et al. 1994) or introduced sequentially into the same yeast strain (Chong et al. 1995). Several strains (W303-1a, YM955, GGY1, and derivatives of PCY2) have been used for this purpose. For the transformations, the yeast vectors are linearized at a restriction site for targeted integration at a selected locus. For example, the reporter plasmid pTH containing the *HIS3* reporter gene (Figure 17-1) is restricted at the *Pvu*II site in the *LYS2* gene fragment for integration at the *LYS2* locus. It is prudent to select several yeast colonies after transformation and confirm the site of integration of the reporter plasmids by Southern blot analysis using probes for the DNA element and reporter gene. Subsequently, the *lacZ* reporter plasmid pJAC13 is restricted with *Xba*I for integration at the *TRP1* locus (Figure 17-1) into one of the Ura^+ colonies from the pTH transformation. Southern blotting of DNA from several colonies is used to confirm integration of the second reporter plasmid (note that the order of integration of the reporter plasmids is dependent upon the presence of the selectable markers on the individual plasmids). The reporter strain is maintained under selection for the *URA3* and *TRP1* markers. At this point, growth of the reporter strain should be checked on -His selective media to ensure that the yeast do not contain a transcription factor that binds to the test DNA element and activates transcription of the reporter gene. The entire process should be repeated to generate a yeast reporter strain that contains a mutated version of the target DNA elements. Putative GAD-DNA-binding domain proteins obtained from the initial selection and screen in the wild-type reporter strain should be tested to iden-

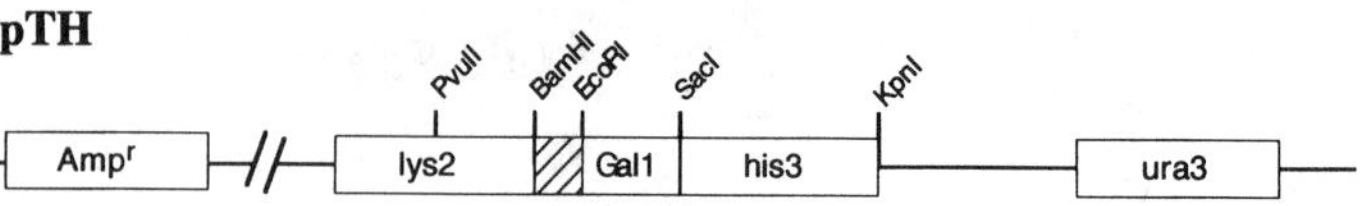

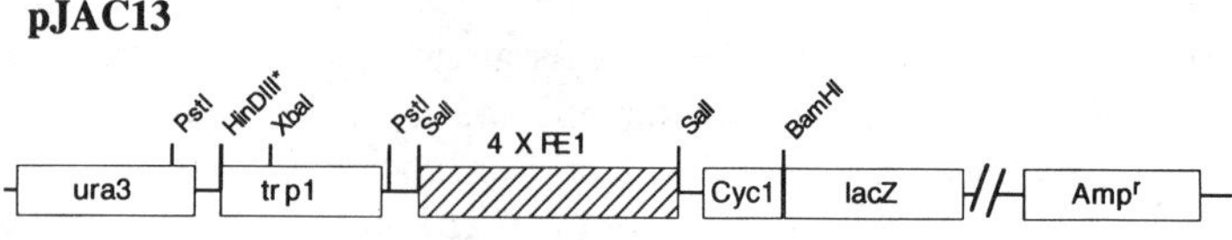

▨ = Cloning site for multimerized target DNA sequence

*The indicated HinDIII site is far from unique

Fig. 17-1. Vectors used to generate yeast reporter strains. Plasmid pTH (Flick and Johnston 1990; Flieg et al. 1986, Johnston and Davis 1984) contains the *HIS3* auxotrophic marker driven by a truncated *GAL1* promoter. Multimerized binding sites should be cloned directly upstream of the *GAL1* promoter using the unique *Eco*RI restriction site. pTH also contains the selectable *URA3* marker, and a fragment of the *LYS2* gene which can be used, after restriction with *Xba*I, for integration into yeast via homologous recombination. The *lacZ* reporter gene plasmid pJAC13 (derived from a plasmid, pCZi3GAL; Lue and Kornberg 1987; Lue et al. 1989) contains four copies of the type II silencer element RE1 (Chong et al. 1995) cloned into a unique *Sal*I site upstream of the yeast *CYC1* promoter. The RE1-binding sites can be replaced with other multimerized sites by digestion with the *Sal*I restiction endonuclease. In addition to the *URA3* selectable marker, pJAC13 also contains the *TRP1* marker. Integration through homologous recombination in the *TRP1* gene may be facilitated by restriction with *Xba*I. Because pJAC13 contains both *URA3* and *TRP1* selectable markers, while pTH contains only *URA3*. pTH-derived plasmids should be introduced into the yeast strain first. Select Ura+ colonies can then be transfected with pJAC13 and selected for tryptophan prototrophy.

tify those that do not activate reporter gene expression in the mutated yeast strain (Figure 17-2).

Transformation of the Reporter Strain with a GAD cDNA Library

These libraries, prepared from a wide variety of tissue types and species, are now available commercially and from several investigators. The same libraries that are used for the yeast two-hybrid system can also be used for

Yeast one-hybrid screen for REST

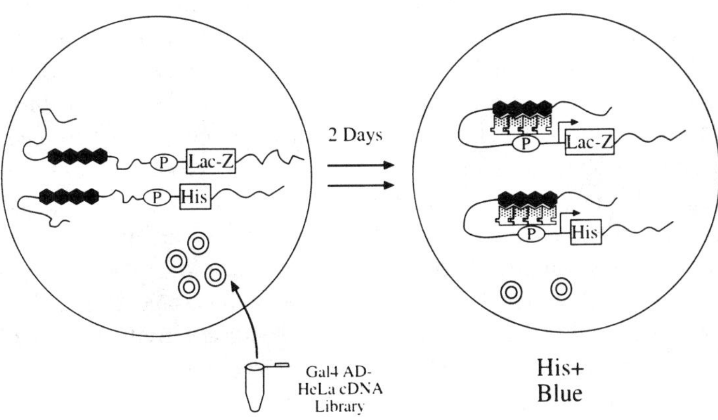

Putative REST fusion protein does not bind mutated Type II silencer elements

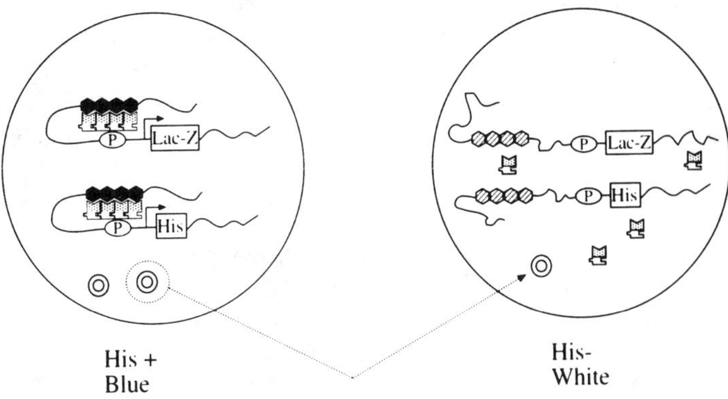

Fig. 17-2. One hybrid strategy. A GAD-cDNA library is introduced into a yeast reporter strain harboring the appropriate multimerized binding sites upstream of the *HIS3* and *lacZ* reporter plasmids (top panel). Complementary DNA clones are isolated from all His$^+$, blue yeast colonies and reintroduced into yeast reporter strains bearing mutated versions of the binding sites (bottom panel). The phenotype should now be His$^-$ and lacZ$^-$.

the one-hybrid system. The libraries are generally available as frozen bacterial cultures which must be amplified by standard techniques before the plasmid library DNA is recovered. The plasmid library is transformed using standard protocols (see chapters 2, 3, and 4). For convenience, 50 μg of library DNA is introduced into yeast. Transformation efficiency is monitored by plating an aliquot of the transformants onto -Leu selective medium (if the library vector bears the *LEU2* marker gene) and counting the number of colonies.

Between 20,000 and 100,000 yeast are plated onto each of ten 150 mm plates of -Leu, -His selective medium and incubated at 30°C. Generally, three types of colonies are observed. First, all of the yeast in the transformation will grow a little because of the histidine and leucine in the media used to grow the yeast prior to transformation. This limited growth is seen as a hazy background on the plates. Second, numerous distinct colonies will slowly begin to appear on top of the hazy background. These colonies likely represent yeast that have received the library plasmids and thus are Leu^+. However, because they are His^-, they will eventually stop growing. Third, colonies that grow to several mm in diameter over the next week are Leu^+ and His^+. Individual colonies are picked at this stage, streaked for single colonies, and then screened for ß-galactosidase activity using standard techniques (chapters 2, 3, and 4). Alternatively, we have found it convenient to test the His^+ yeast colonies as they grow up by using a sterile toothpick to transfer a small amount of a colony to a grid on a Whatman filter. The filter is frozen and thawed several times in liquid nitrogen and then exposed to the chromogenic substrate. Yeast that turn blue are then streaked for single colonies on -His selective medium and retested for β-galactosidase activity. *LacZ* reporter gene expression may be present in only a subset of the His^+ colonies. The His^+ background is unexplained but may represent, at least in part, revertants of the *HIS3* gene, as well as proteins that bind to sequences other than the test DNA elements.

Criteria for Showing that the His^+, $lacZ^+$ Phenotype of a Yeast Colony Is Due to the Presence of a cDNA from the Library Transformation

The putative positive yeast colonies are streaked for single colonies. Two complementary strategies can then be used on single colonies to ensure that the His^+, $lacZ^+$ phenotype is conferred by the library plasmid. First, the yeast colonies can be cured of the plasmid as follows. The yeast are grown for several generations in complete liquid media and then are plated onto rich medium. The resulting colonies are replica-plated separately onto selective -Leu medium and -His medium. If the His^+ phenotype is the result of the presence of the library plasmid, then growth on the -His selective plates should be limited to the yeast colonies that also grow on the -Leu plates. Second, the candidate yeast colonies can be grown in -His selective media and the library plasmids recovered by standard techniques (see chapters 2 and 4) and introduced by electroporation into *Escherichia coli* for amplification. The plasmids are then introduced into the original yeast reporter cells and tested for the ability to confer the His^+ $lacZ^+$ phenotype to the transformed yeast.

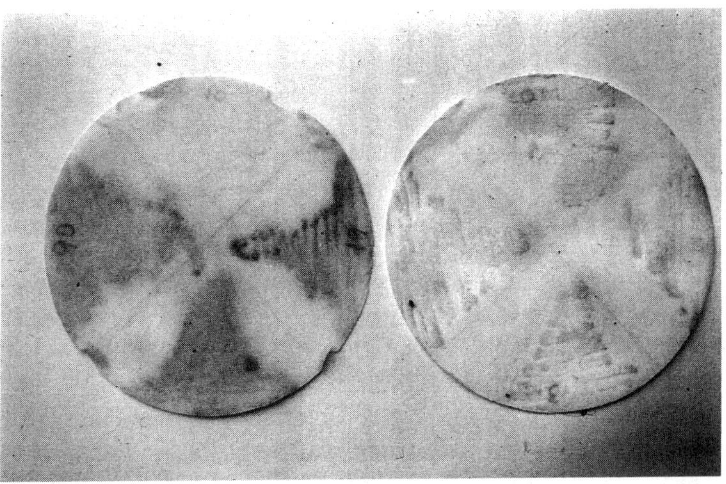

Fig. 17-3. Colorimetric assay for β-galactosidase activity distinguishes specific from nonspecific binding of DNA-binding proteins identified in the one-hybrid screen. The filter sections represent streaks of 4 individual His$^+$ yeast colonies that were examined for *lacZ* gene expression as described in the text. Three blue colonies and 1 white colony are shown in the left panel. The white colony represents a nonspecific DNA-binding protein that turned blue after 6 hours, while the other colonies turned blue after approximately 1 hour. The same 4 cDNAs introduced into the yeast strain carrying mutated versions of the binding site (right panel) did not turn blue when the assay was continued for the same amount of time as for the wild-type reporter strain.

Verification of the Specificity of Interaction between the Cloned cDNA Encoding the DNA-Binding Domain and the Cognate DNA Elements

The library plasmids recovered from the His$^+$ yeast transformants should be introduced into the yeast strain harboring the mutated DNA elements. These yeast are replica-plated onto both -Leu medium and -His medium. The yeast should be Leu$^+$ (the plasmid selectable marker) and His$^-$, indicating that there is no interaction between the identified DNA-binding domain and the mutated target DNA sequence. Further evidence is provided by testing the yeast colonies that grew on the -Leu plate for β-galactosidase activity, which should be absent. A comparison of yeast colony β-galactosidase activity in strains containing wild-type and mutated versions of a target DNA element is shown in Figure 17-3. It should be noted that the time of the assay is important. If the β-galactosidase assay is incubated long enough, even the mutated strain may turn blue, presumably due to weak binding of the factor to the mutated target DNA sequence.

Authentication of Cloned cDNAs

Criteria need to be established for determining that the cloned cDNA encodes the true protein of interest. The plasmids recovered from the His+, lacZ+ yeast are electroporated into bacteria for plasmid amplification and DNA sequence analysis. Homologies to known DNA-binding domains, deduced from the DNA sequence, are obviously encouraging. Full-length cDNAs can be generated by screening cDNA libraries with the cDNA obtained from the yeast screen. The full-length cDNAs can be introduced into cells along with reporter genes bearing the DNA elements to assay whether the identified factor has the correct function (for example, transcriptional activation or repression). Antibodies can be generated against the protein and used in mobility shift assays to determine whether the identified factor is present in nuclear extracts of the correct cell types.

Potential Problems

If no positive yeast colonies result from the library transformations, it may be due to low representation of the transcription factor cDNA in the library. Several different cDNA libraries should be screened to help alleviate this problem. If the target binding site is recognized by a protein endogenous to yeast, a His+ strain may result. In such a case, a different yeast strain may be tried.

References

Chong, J.A., Tapia-Ramirez, J., Kim, S., Toledo-Aral, J.J., Zheng, Y., Boutros, M.C., Altshuler, Y.M., Frohman, M.A., Kraner, S.D., and Mandel, G. (1995). REST: A mammalian silencer protein that restricts sodium channel gene expression to neurons. Cell 80:949–957.

Dowell, S.J., Romanowski, P., and Diffley, J.F. (1994). Interaction of Dbf4, the Cdc7 protein kinase regulatory subunit, with yeast replication origins in vivo. Science. 265: 1243–1246.

Fields, S., and Song, O. (1989). A novel genetic system to detect protein-protein interactions. Nature 340:245–246.

Flick, J.S., and Johnston, M. (1990). Two systems of glucose repression of the GAL1 prmoter in *Saccharomyces cerevisiae*. Mol. Cell. Biol. 10:4757–4769.

Fleig, U.N., Pridmore, R.D., and Philippseu, P. (1986). Construction of LYS2 cartridges for use in genetic manipulations of *Saccharomyces cerevisiae*. Gene 46:237–245.

Inouye, C., Remondelli, P., Karin, M., and Elledge, S. (1994). Isolation of a cDNA encoding a metal response element binding protein using a novel expression cloning procedure: the one hybrid system. DNA and Cell Biol. 13:731–742.

Johnston, M., and Davis, R.W. (1984). Sequences that regulate the divergent GAL1-GAL10 promoter in *Saccharomyces cerevisiae*. Mol. Cell Biol. 4:1440–1448.

Kadonga, J.T. and Tjian, K.R. (1986). Affinity purification of sequence-specific DNA binding proteins. Proc. Natl. Acad. Sci. USA 83:5889–5893.

Li, J.J., and Herskowitz, I. (1993). Isolation of ORC6, a component of the yeast origin recognition complex by a one-hybrid system. Science 262: 1870–1874.

Lue, N.F., and Kornberg, R.D. (1987). Accurate initiation at RNA polymerase II promoters in extracts from *Saccharomyces cerevisiae*. Proc. Natl. Acad. Sci. USA 84:8839–8843.

Lue, N.F., Buchman, A.R., and Kornberg, R.D. (1989). Activation of yeast RNA polymerase II transcription by a thymidine-rich upstream element in vitro. Proc. Natl. Acad. Sci USA 86:486–490.

Singh, H., Lebowitz, J.H., Baldwin, J.A.S., and Sharp, P.A. (1988). Molecular cloning of an enhancer binding-protein: isolation by screening of an expression library with a recognition site DNA. Cell 52:415–423.

Wang, M.M., and Reed, R.R. (1993). Molecular cloning of the olfactory neuronal transcription factor Olf-1 by genetic selection in yeast. Nature 364:121–126.

Wilson, T.E., Fahrner, T.J., Johnston, M., and Milbrandt, J. (1991). Identification of the DNA binding site for NFGI-B by genetic selection in yeast. Science 252:1296–1300.

18

A Three-Hybrid System to Detect and Analyze RNA-Protein Interactions in Vivo

Beilin Zhang
Brian Kraemer
Dhruba SenGupta
Stanley Fields
Marvin Wickens

RNA-protein interactions are critical in diverse cellular processes, such as splicing, translation, early development, and infection by RNA viruses. For example, the infectious cycle of human immunodeficiency virus (HIV-1) relies on the interactions between reverse transcriptase and the RNA genome to generate viral DNA, between Tat and TAR to produce viral mRNA, between Rev and RRE to export viral RNA, and between viral proteins and the RNA genome to assemble viral particles.

Few general methods are available to detect and analyze RNA-protein interactions in vivo. In vitro, RNA-protein interactions can be detected using the same collection of biochemical methods as are commonly employed to study DNA-protein interactions, including nitrocellulose filter binding, electrophoretic mobility shifts, and UV cross-linking. In vivo, specific combinations of RNA and protein can be studied by analyzing the normal biological function of that interaction. However, such assays are idiosyncratic and can be complicated by the fact that mere binding may be insufficient for the biological function.

Several systems have been described that may permit the detection and analysis of a wide range of RNA-protein interactions. For example, phage display can be exploited to identify specificity determinants of RNA-binding proteins in vitro (Laird-Offringa and Belasco 1995). Alternatively, RNA-protein interactions can be assayed in vivo by placing an RNA bind-

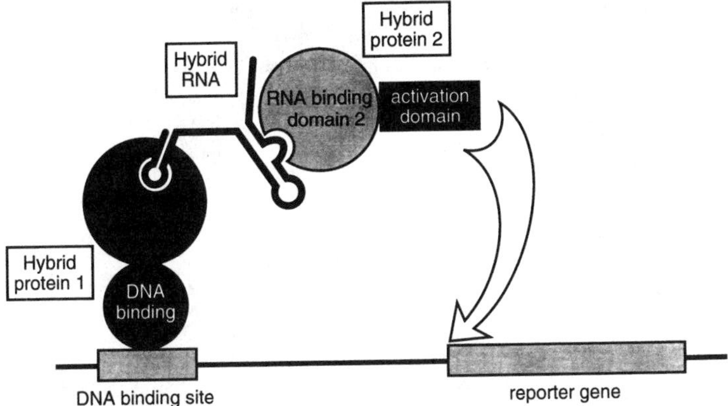

Fig. 18-1. Diagram of the three-hybrid system. Adapted from SenGupta et al. (1996).

ing site in such a location that it represses translation of a specific mRNA; binding of the cognate protein represses translation of the mRNA, which is readily detectable (MacWilliams et al. 1993; Stripecke et al. 1994). An anti-termination system involving the RNA-binding N protein of bacteriophage lambda also has been adapted to analysis of RNA-protein interactions in bacterial cells (Harada et al. 1996).

In this chapter, we describe a three-hybrid system to detect and analyze RNA-protein interactions in vivo (SenGupta et al. 1996). It relies on the physical, not biological, properties of the RNA and protein molecules. As a result, a wide variety of RNA-protein interactions can be detected and analyzed. It possesses, in principle, many of the same strengths and limitations of the two-hybrid system for dissecting protein-protein interactions. Because the three-hybrid system is in an early stage of development, we emphasize the principles of the system and its potential applications.

Strategy of the Three-Hybrid System

In the two-hybrid system, a protein-protein interaction brings two hybrid proteins together and results in the activation of a reporter gene. In the three-hybrid system (SenGupta et al. 1996), diagrammed in Figure 18-1, a hybrid RNA molecule functions as the bridge between two hybrid proteins. Hybrid protein 1 contains RNA-binding domain 1 fused to a DNA-binding domain, while hybrid protein 2 contains a different RNA-binding domain, 2, fused to a transcription activation domain. The hybrid RNA contains recognition sites for the two RNA-binding domains. The interaction of this

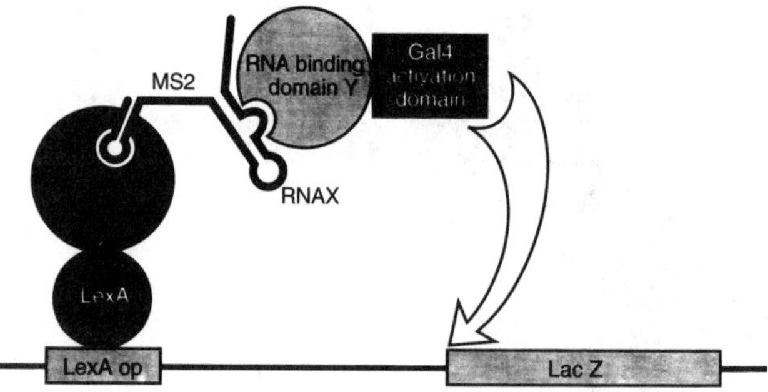

Fig. 18-2. Specific components of the three-hybrid system: a common arrangement. Both *lacZ* and *HIS3* have been used as reporters, and are present in strain L40-coat. See text for details.

RNA with the two hybrid proteins is required for transcription of the reporter gene.

The specific configuration that has been analyzed in the greatest detail is depicted in Figure 18-2. In this arrangement, hybrid protein 1 consists of the LexA DNA-binding protein fused to the coat protein of bacteriophage MS2. Hybrid protein 2 consists of the Gal4p transcription activation domain linked to the RNA-binding domain, Y. The hybrid RNA consists of two MS2 RNA binding sites linked to RNA sequence X. The interaction between RNAX and RNA-binding domain Y brings the activation domain to the promoter and results in the transcription of the reporter gene. Hybrid protein 1, the presence of MS2 sites in the hybrid RNA, and the Gal4p activation domain are fixed; RNAX and RNA-binding domain Y vary among experiments.

Figure 18-2 presents a simplified view of the complexes that may form at the promoter. Multiple LexA operators are present (four in the *HIS3* promoter and eight in the *lacZ* promoter), and LexA protein binds as a dimer; thus multiple hybrid protein 1 molecules may be tethered to the promoter via their DNA-binding domains. Moreover, MS2 coat protein also might oligomerize, perhaps resulting in recruitment of non-DNA-bound hybrid protein 1 molecules to the promoter region. The hybrid RNA contains two MS2 binding sites, and MS2 coat protein binds as a dimer to a single site. Thus one RNA molecule could interact simultaneously with DNA-bound and non-DNA-bound LexA/MS2 coat protein molecules. In sum, these attributes of the system may make it particularly sensitive in detecting the second RNA-protein interaction: multiple LexA/MS2 coat protein molecules may reside at the promoter, and only one need be occupied by a hybrid RNA to activate transcription.

Table 18-1 Examples of known RNA-protein interactions detected in the three-hybrid assay

RNA	Protein	Affinity in vitro (~K_d)
IRE	IRP1	10^{-10} to 10^{-11}
U1	U1-70K	-
TAR (HIV)	Tat	10^{-8}
poly(A)	PAB	10^{-8}
IRES (EMCV)	PTB	10^{-7} to 10^{-8}

Each of the indicated interactions has been tested in the three-hybrid system. Controls for specificity include, in each case, tests with mismatched combinations of RNA and protein. Citations for the in vitro affinities of the interactons are as follows. For IRE/IRP1, Barton et al. (1990) and Haile et al. (1989); for HIV-1 TAR/Tat, Slice et al. (1992); for poly(A)/PAB, Gorlach et al. (1994) and Sachs et al. (1987; for IRES/PTB, Witherell et al. (1993).

General Considerations

Several different RNA-protein interactions have been assayed successfully using the system (Table 18-1). The sequence specificity of the interactions typically has been assessed using mismatched pairs of RNA and protein. As indicated in Table 18-1, the affinity of the interactions detected ranges from K_ds of approximately 10^{-10} to 10^{-8}M. These data strongly suggest that other specific interactions of comparable affinity likely will be detectable. However, affinity is not the only issue that will influence whether a given interaction is detectable. The abundance, the conformation, and the cellular location of the hybrid RNAs can all influence transcriptional activation in the system, and may vary among RNAs.

The secondary and tertiary structures of RNAs often are critical in their interactions with proteins. Most of the RNAs tested to date in the three-hybrid system form relatively stable structures in vitro. Ostensibly "unstructured" RNAs might be more prone to the formation of unproductive, alternative conformations when present in the context of a hybrid RNA. This potential problem is minimized by the fact that the vector portions of the hybrid RNA either form stable structures themselves (RNaseP RNA leader, MS2 binding sites) or are AU-rich (RNaseP RNA terminator); as a result, their interference with the structure of the inserted sequences may be minimal. Furthermore, it is not necessary that every molecule in the RNA population fold correctly; proper folding of only a fraction of the molecules is likely to be sufficient to lead to transcriptional activation.

The three-hybrid assay requires that the RNAs be nuclear. RNAs may be nuclear either because they are derived from RNaseP RNA and its promoter (since RNaseP RNA is not known to enter the cytoplasm), or because after transport to the cytoplasm, the RNAs are returned to the nucleus by binding to their cognate hybrid proteins, which then translocate to the nucleus. The second hybrid protein carries a nuclear localization signal.

The use of an RNA polymerase III promoter to produce the hybrid RNA may restrict the range of sequences that can be analyzed. Since RNA polymerase III often terminates at runs of four or more consecutive uridine residues (Bogenhagen and Brown 1981; Geiduschek and Tocchini 1988), it may prove beneficial to use alternative RNA polymerases, such as those from bacteriophages T7 and SP6. The effect of potential polymerase III terminators in an RNA sequence of interest can be assessed directly by Northern blotting, or by inserting the entire RNA sequence between MS2 and IRE elements and then assaying the IRE/IRP1 interaction in the three-hybrid system.

This same approach permits an assessment of the length of RNA that may be inserted. Although this point has not been investigated systematically, insertion of as much as 500 nucleotides of RNA sequence between IRE and MS2 sites has had little effect on the IRE/IRP1 interaction.

The minimal affinity required to yield detectable transcriptional activation in the three-hybrid system has not been determined. In the two-hybrid system, however, protein-protein interactions with Kd values greater than 1 μM can be detected (Estojak et al. 1995; Yang et al. 1995); as similar reporter strains are used in the two assays, it may be possible to detect RNA-protein interactions of relatively low affinity. Complexes involving an RNA and multiple proteins—such as the spliceosome and ribosome — often rely on the additive effects of weak interactions to achieve stability. To reconstitute such interactions in the three-hybrid system, it may be necessary to provide additional proteins in yeast via transformation with additional plasmids.

Several families of RNA-binding proteins have been identified based on shared sequence motifs. One common class of RNA-binding proteins—those containing an RNA Recognition Motif (RRM) (Burd and Dreyfuss 1994)—stimulate transcription in the three-hybrid system and do so only with their appropriate targets [for example, PAB, PTB, and U1-70K (Table 18-1)].

Potential Applications

The three-hybrid system described here provides a rapid and potentially versatile method to detect RNA-protein interactions in vivo, toward a variety of ends. The potential applications mirror those of the two-hybrid system, but provide access to a new constellation of biological and clinical problems.

- *Dissecting known RNA-protein interactions.* The method should be capable of defining domains, as well as single amino acid residues or nucleotides, that are necessary in vivo for RNA-protein interactions that have been previously characterized. The ability of the system to detect several different interactions strongly suggests that this approach will be generally useful. The relationship between Kd and transcriptional activity will be critical in such analyses.
- *Identifying and cloning new proteins that bind to specific RNA sequences.* The method should be useful for identifying and cloning the genes for RNA-bind-

ing proteins that recognize biologically important RNA sequences. Such screens will be analogous to those performed in the two-hybrid system, but with an RNA rather than a protein molecule as bait. This application is discussed in greater detail below.

- *Identifying natural RNA targets.* By preparing a library of hybrid RNAs, each of which carries a different cellular RNA sequence fused to coat protein binding sites, it may be possible to identify specific mRNAs that bind to a defined protein.
- *Artificial RNA ligands.* The three-hybrid system may provide an in vivo method to identify or assay synthetic RNA oligonucleotides with selective affinity for defined proteins, analogous to in vitro approaches that exploit iterative selections.
- *Inhibitors of RNA-protein interactions.* The three-hybrid system may permit the rapid screening for inhibitors of a known RNA-protein interaction, such as those involved in viral replication, transcription, and assembly. In such assays, it likely would prove useful to use a counter-selectable reporter (such as *URA3*).
- *RNA-RNA interactions.* Through simple adaptations of the three-hybrid system, it may be possible to extend the method to generate a four-hybrid system for the analysis of RNA-RNA interactions in vivo, using two fixed protein hybrids and two different hybrid RNAs. This would permit the analysis of novel RNA-RNA interactions, as well as a possible in vivo selection for nucleic acid hybridization.

Hybrid RNA Molecules: General Features

The hybrid RNAs have the general structure depicted in Figure 18-3. They consist of the yeast RNaseP RNA leader sequence, two tandem MS2 coat protein binding sites, and the RNA sequence of interest, X. The MS2 and RNAX sequences can be in either position relative to one another. Two MS2 coat protein binding sites are used because binding to adjacent sites are cooperative (Bardwell and Wickens 1990; Witherell et al. 1990). Furthermore, the sites contain a nucleotide change that enhances binding to the coat protein (Lowary and Uhlenbeck 1987). The RNAs are transcribed from the yeast RNaseP RNA (*RPR1*) promoter, which is recognized by RNA polymerase III. This drives the synthesis of small RNA molecules, up to several thousand molecules per cell (Good and Engelke 1994). In addition, transcripts from the RNaseP promoter presumably do not enter the pre-mRNA processing pathway, which may facilitate the assay.

Constructing Plasmids that Encode Hybrid RNA Molecules

Two Routes We have created four plasmids for general use in producing hybrid RNAs, pMS2-1, pMS2-2, pIII/MS2-1, and pIII/MS2-2 (diagrammed in Figure 18-4). They provide two alternative routes for preparing identical plasmids encoding identical hybrid RNAs.

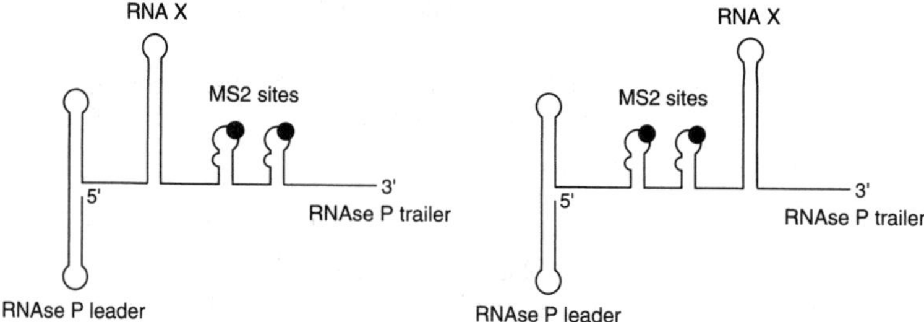

Fig. 18-3. Schematic diagram of the secondary structure of the hybrid RNA molecules. The RNA sequence of interest, X, is depicted as a stem-loop structure.

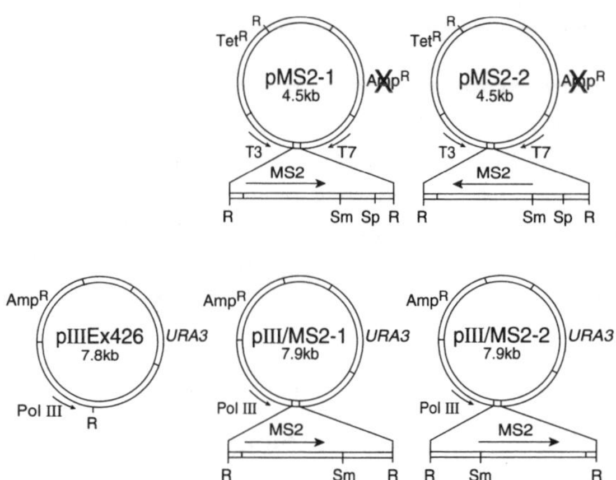

Fig. 18-4. Plasmid vectors for creating hybrid RNAs. Arrows labeled MS2: two tandem MS2 coat protein-binding sites, as described in the text and depicted in Fig. 18-2 (polarity of arrow is 5' to 3' in the sense-strand); T7 and T3: bacteriophage T7 and T3 RNA polymerase promoters, respectively (arrows indicate direction of transcription); pol III: RNA polymerase III promoter, including RNase P RNA sequences (arrows indicate direction of transcription); Tet^R: gene encoding tetracycline resistance in *E. coli;* Amp^R: gene encoding ampicillin resistance in *E. coli* (eliminated by small deletions in pMS2-1 and pMS2-2); *URA3:* the yeast *URA3* gene. Restriction sites are abbreviated as follows. R, *Eco*RI; Sm, *Sma*I; Sp, *Spe*I. In pMS2-1 and pMS2-2, the *Sma*I and *Spe*I sites are unique, and can be used to insert a sequence of interest; in pIII/MS2-1 and pIII/MS2-2, only the *Sma*I site is unique. Origins of replication are not depicted. pIII/MS2-1, pIII/MS2-2 and pIIIEx426RPR are multicopy yeast plasmids. pMS2-1 and pMS2-2 are bacterial plasmids and do not propagate in yeast.

We will term the DNA encoding the RNA sequence of interest, RNAX. RNAX can be cloned into the appropriate vector by either of two routes, resulting in identical hybrid RNAs.

Route 1: RNAX can be cloned into the unique *Sma*I site of pIII/MS2-1 or pIII/MS2-2.

Route 2: RNAX can be cloned into the unique *Sma*I or *Spe*I sites of pMS2-1 or pMS2-2 (see maps that follow). *Eco*RI fragments containing either 5' MS2-RNAX 3' or 5' RNAX-MS2 3' can then be moved into *Eco*RI-cleaved pIIIEx426RPR.

Route 1 saves one cloning step, and so may often be the more desirable approach. However, it may be easier to determine the orientation and sequence of the inserted sequence in the hybrid RNA using Route 2 for two reasons. First, the pMS2 plasmids have more restriction sites near the inserted RNA sequence, making it easier to analyze orientation (particularly with short inserts). Second, inserts in the pMS2-1 and pMS2-2 plasmids can be sequenced using conventional T7 and T3 primers.

pIII/MS2-1 and pIII/MS2-2 Vectors Cloning to generate hybrid RNAs using these vectors is summarized diagrammatically in Figure 18-5. pIII/MS2-1 and pIII/MS2-2 are yeast/*E. coli* shuttle vectors in which an RNA polymerase III promoter directs transcription of an RNA containing two tandem MS2 sites. Sequences to be tested in the three-hybrid assay are inserted directly into the unique *Sma*I restriction site (the only unique insertion site for such sequences). In pIII/MS2-1, the *Sma*I site is 3' of the tandem MS2 coat protein binding sites; in pIII/MS2-2, the *Sma*I site is 5' of the MS2 sites. Both plasmids are multicopy in yeast, and carry a yeast *URA3* marker. In *E. coli*, the plasmids confer ampicillin resistance.

pMS2-1 and pMS2-2 Vectors pMS2-1 and pMS2-2 provide an alternative means of generating the same hybrid RNA plasmids. They are *E. coli* plasmids derived from pBluescriptII KS(+), and do not carry an RNA polymerase III promoter or any other yeast sequences. They confer tetracycline resistance in *E. coli.* They carry two tandem MS2 coat protein binding sites flanked by bacteriophage T7 and T3 RNA polymerase promoters. Unique *Sma*I and *Spe*I sites can be used to insert a sequence of interest either 5' (pMS2-2) or 3' of MS2 sequences (pMS2-1). The sequence of any inserted DNA can be determined using T7 and T3 promoter-specific primers. After cloning RNAX sequence into either of these vectors, the region of these plasmids encoding the hybrid RNAs can be excised as an *Eco*RI fragment. This *Eco*RI fragment can then be inserted into the unique *Eco*RI site of the vector, pIIIEx426RPR (Good and Engelke 1994; Figure 18-4) to generate a plasmid that can be used to produce the desired hybrid RNA in yeast. [pIIIEx426RPR, constructed by Good and Engelke (1994), carries an RNA polymerase III promoter immediately adjacent to a unique *Eco*RI insertion site, and was used in generating the pIII/MS2-1 and pIII/MS2-2 vectors.]

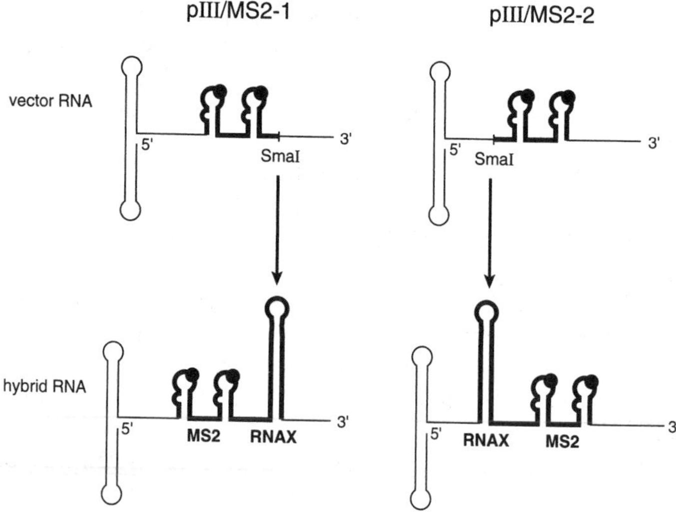

Fig. 18-5. Schematic diagram of the method of cloning hybrid RNAs using *pIII/MS2-1* and *pIII/MS2-2* vectors. In the example shown, sequence RNA X, a stem-loop, has been inserted into either pIII/MS2-1 or pIII/MS2-2. Thin line, RNaseP RNA sequences; bold lines, RNAX or MS2 binding-site sequences; black dot, point mutation in the loop of the MS2 binding site.

Such plasmids will carry the yeast *URA3* gene and confer ampicillin resistance in *E. coli*, and be identical to those generated using the pIII/MS2-1 and pIII/MS2-2 vectors.

Deciding on the Order of Sites The relative positions of RNAX and MS2 sites can make a difference in the extent of transcriptional activation observed. For example, in the IRE/IRP1 interaction, placing the IRE 5' of the MS2 sites results in 2- to 3-fold more transcription than when the IRE is placed 3' of the MS2 sites. We do not know whether this is a general feature of hybrid RNAs or a peculiarity of IRE-containing RNAs.

Three main considerations influence which polarity (5' RNAX-MS2 3' vs. 5' MS2-RNAX 3') is preferable.

- *Potential RNA polymerase III terminators.* A series of four or more Ts in succession can function as an RNA polymerase III terminator, depending on sequence context. If the Ts are at the 3' end of RNAX, it may be preferable to use the 5' MS2-RNAX 3' polarity. If the Ts are in the middle of RNAX, then it may be necessary to assess whether they present a problem. One way to do this is to clone RNAX between the IRE and MS2 sites and check whether the resulting RNA still works in the IRE/IRP1 assay. (There is a unique *Sma*I site in pIII/IRE-MS2 that can be used for this purpose.) If the IRE/IRP1 interaction still occurs and activates transcription, then the Ts in RNAX probably will not be a problem; if the IRE/IRP1 interaction cannot be detected, then it may be necessary to mutate the candidate terminator. Direct analysis of the RNA, by Northern blotting or nuclease protection for example, can also be used to determine the location of the RNA's 3' terminus. Hybrid RNAs typically are very abundant, facilitating such analyses.

- *RNA folding.* The folding of the hybrid RNAs may differ in the two orientations. Although computer predictions of RNA structure are inconclusive, you may wish to use an RNA folding program to see whether one arrangement appears to be more likely to succeed than the other.
- *Position of the RNA element in its native context.* For certain RNA-protein interactions, the location of the RNA sequence relative to the 3' or 5' end of the RNA molecule is critical. For example, the interaction of certain proteins with tRNAs requires not only specific sequences in the tRNA, but also that the tRNA be at or near the 3' end of the molecule. In such cases, hybrid RNAs should be constructed to mimic the natural context.

In the absence of any other considerations, based on our limited tests, we recommend placing RNAX 5' of the MS2 sites.

Reporter Strain

The yeast strain L40-*ura3* (a gift of T. Triolo and R. Sternglanz, SUNY Stony Brook) is used in the system. The strain has the genotype *MATa, ura3-52, leu2-3,112, his3Δ200, trp1Δ1, ade2, LYS2::(LexA-op)$_4$ -HIS3, ura3::(LexA-op)$_8$ -lacZ.* The LexA-MS2 plasmid, the RNA plasmid and the activation domain plasmid can be introduced by transformation into this strain. The interaction between the three hybrid molecules results in the transcription of both the *HIS3* and *lacZ* genes, both of which have been placed under control of LexA binding sites.

In many applications of the system, the LexA-MS2 coat protein is fixed. We therefore constructed a strain with a LexA-MS2 coat protein fusion gene integrated into the chromosome. This strain, called L40-coat, is a derivative of L40-*ura3* and has the genotype *MATa, ura3-52, leu2-3,112, his3Δ200, trp1Δ1, ade2, LYS2::(LexA-op)$_4$ -HIS3, ura3::(LexA-op)$_8$ -lacZ, LexA-MS2 coat protein (TRP1).* Two plasmids, one encoding the hybrid RNA and the other encoding the RNA-binding domain Y linked to the activation domain, need to be transformed into this strain to detect the RNA-protein interaction.

The Activation Domain Plasmid

The plasmid encoding hybrid protein 2 carries the Gal4p activation domain fused to RNA-binding domain 2. We commonly use the vector, pACTII (S.

Elledge, Baylor College of Medicine) for this purpose. It carries a *LEU2* marker.

Analyzing Known RNA-Protein Interactions

The protocol to test and analyze a known RNA-protein interaction is similar to that of the two-hybrid assay. The RNA sequence of interest is cloned into a hybrid RNA vector, as previously described. The RNA-binding domain is cloned into an activation domain vector carrying a *LEU2* marker, such as pACTII.

The following is a sample protocol and results obtained analyzing the interaction between the iron response element (IRE) and a protein to which it binds tightly (IRP1). (This example is drawn from SenGupta et al. 1996.)

1. *Construct the plasmids.* An *Ava*I fragment containing the IRE from rat ferritin light chain (Eisenstein et al. 1993) was cloned into the *Sma*I site of both pIII/MS2-1 and pIII/MS2-2, generating pIII/IRE-MS2 and pIII/MS2-IRE. The rabbit IRP1 gene was cloned into pACTII to make pAD-IRP1.
2. *Transform the plasmids.* Yeast strain L40-*ura3* was transformed with pLexA-MS2, pAD-IRP plus either pIII/IRE-MS2 or pIII/MS2-IRE, and the transformants were plated out on SD-Trp-Leu-Ura media. Alternatively, yeast strain L40-coat can be transformed with pAD-IRP plus one of the RNA plasmids and plated out on SD-Leu-Ura plates.
3. *Assay the reporter gene.* Assays of transcriptional activity were the same as are used in the two-hybrid system (chapters 2, 3 and 4). Both *lacZ* and *HIS3* can be used as reporters.

Results obtained with the IRE/IRP interaction are depicted as examples in Figures 18-6 and 18-7. Figure 18-6 presents results using the *lacZ* gene as reporter to *screen* for the RNA-protein interaction. This experiment was performed as a series of triple transformations with plasmids encoding each of the three hybrid molecules. Rows 1 and 2 are the complete system, and yield elevated levels of ß-galactosidase. Rows 3 to 10 are controls, in which one component of the system is missing or altered. None yield significant levels of ß-galactosidase.

Figure 18-7 presents results using the *HIS3* gene as a reporter to *select* for the RNA-protein interaction. In this case, double transformations were performed into the strain L40-coat, which carries an integrated copy of the LexA-MS2 coat protein fusion gene. Again, only the two combinations that should activate (Sectors 1 and 2) do so.

If activation does not occur, it may be prudent to analyze the RNA and protein components directly. The level and length of hybrid RNA can be determined directly (for example, by Northern blotting). Levels of transcription from the RNA plasmid are generally high, so that detection of hybrid RNAs should be facile. Problems may arise due to RNA instability or RNA polymerase III terminators that prevent synthesis of the necessary regions of the hybrid RNA. To analyze the structure and expression of hybrid

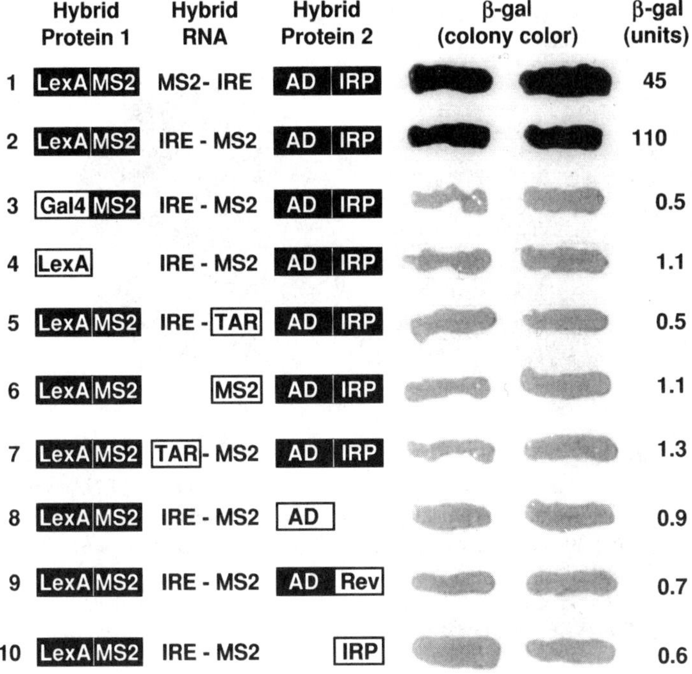

Fig. 18-6. The IRE/IRP1 interaction monitored by activation of ß-galactosidase. [Adapted from Sengupta et al. (1996)]. Three plasmids, encoding the proteins and RNAs indicated at the left of the figure, were introduced into the yeast strain L40-*ura3* by transformation, plating on media lacking tryptophan, leucine, and uracil. Yeast triple transformants were assayed for ß-galactosidase activity either by the colony color assay, or by direct measurement of enzyme activity. Duplicates of the colony color assay are shown. The two components present in each hybrid factor are depicted in the proper polarity, either N- to C-terminal, or 5' to 3'. Proteins are indicated in boxes, each box corresponding to a domain. Black boxes indicate domains present in the complete, functional system. White boxes indicate domains that are not present in the complete, functional system. MS2 in black box, MS2 coat protein; IRE, Iron Response Element; MS2, two MS2 RNA binding sites; TAR, stem-loop at 5' end of HIV-1 mRNAs; IRP1, Iron Response Protein [also called Iron Response Element Binding Protein (IRE-BP) and Iron Response Factor (IRF)]; AD, the activation domain of the Gal4p transcription factor of *S. cerevisiae;* Rev, an HIV-1 transactivator.

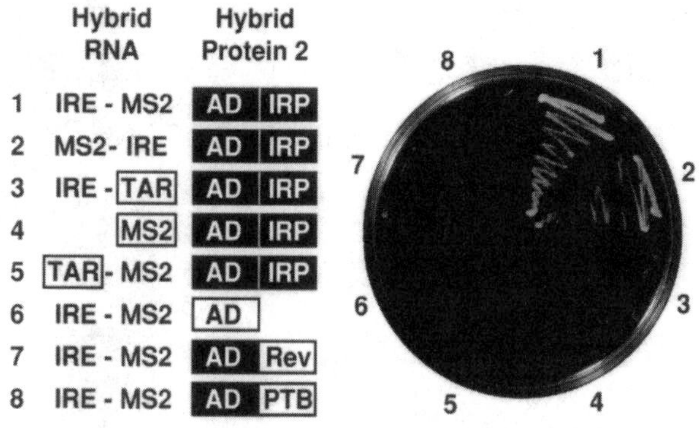

Fig. 18-7. The IRE/IRP1 interaction monitored by selection: activation of *HIS3*. Two plasmids, encoding the indicated hybrid RNAs and activation domain fusions, were introduced by transformation into strain L40-coat. After selecting transformants for the presence of the plasmids, colonies were restreaked onto media selecting for expression of *HIS3*. Only the two combinations that should lead to IRE/IRP1 interactions yield growth.

protein 2, sequence the region in which the fusion between the activation domain and the RNA-binding domain occurred, to ensure that the reading frame was maintained. If pACTII has been used as the expression vector, expression of the protein can be monitored directly by Western blotting, using antibodies directed against either the Gal4p activation domain or the influenza hemagglutinin epitope.

Screening for New Proteins with a Known RNA Sequence: General Strategy and Considerations

One common application of the three-hybrid system may lie in the identification and cloning of proteins that bind to a known RNA sequence. Such a screen is exactly analogous to those in the two-hybrid system, except that the "bait" is an RNA rather than a protein (Figure 18-8). The three-hybrid system is designed in such a way that existing activation domain libraries in *LEU2*-containing vectors can be used directly in three-hybrid screening. Using the strain L40-coat, both *lacZ* and *HIS3* markers can be monitored as reporters.

A protocol describing such a screen is given in Figure 18-9. The protocol described was actually used to isolate IRP1/AD plasmids from a "spiked" liver cDNA/AD library. In this reconstruction experiment, the IRP1/AD

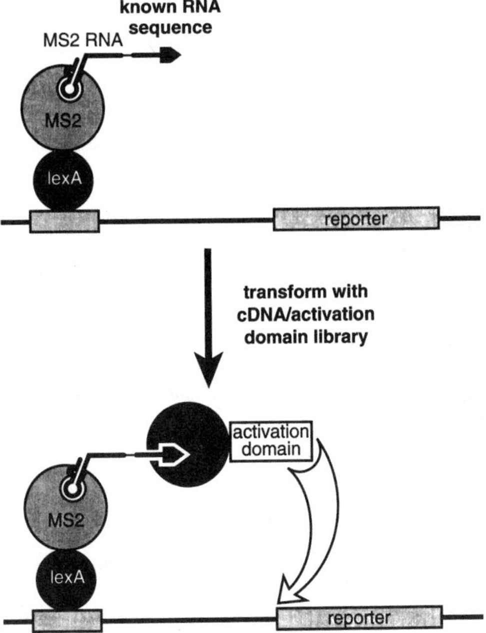

Fig. 18-8. General principle of a three-hybrid screen for a protein that interacts with a known RNA sequence. See text for details.

clone was diluted 1 part in 10,000 into a liver cDNA/AD library, and successfully recovered.

In the first steps of the screen (steps 1 and 2 in Figure 18-9), the library is introduced into strain L40-coat carrying the relevant RNA plasmid. Positives are identified by their ability to activate the *HIS3* reporter. 3-aminotriazole (3-AT), a competitive inhibitor of the *HIS3*-encoded enzyme, can be included at this step. Inclusion of 3-AT will be essential if the presence of the hybrid RNA itself activates transcription. High-affinity, stable interactions between a protein and the hybrid RNA will generate strains resistant to higher concentrations of 3-AT; to detect low-affinity interactions, it may be necessary to use little or no 3-AT.

In performing a three-hybrid screen of this type, "false" positives will be obtained that potentiate transcription without interacting with the hybrid RNA. Such "RNA-independent positives" include, for example, proteins that interact directly with the LexA/MS2 coat protein hybrid. RNA-independent positives can be more frequent than those that are RNA-dependent, and so should be identified and discarded. In the protocol in Figure 18-9, they are removed in two ways. In step 2, the only selection is for ac-

PROTOCOL	COMMENTS
1 Transform with RNA plasmid, selecting for *URA3*	Use strain L40-coat, which carries a LexA-MS2 coat protein gene on the chromosome, and has both *LacZ* and *HIS3* under the control of the three-hybrid system RNA plasmid carries *URA3* marker
2 Transform Activation Domain library in *LEU2* vector, selecting for *LEU2* and for activation of *HIS3* gene; do not select for RNA plasmid	Many standard AD libraries are in *LEU2* vectors, and so can be used directly Cells that do not require the RNA plasmid to activate *HIS3* expression can lose the RNA plasmid Concentration of 3AT, which monitors level of *HIS3* expression, is important at this step
3 Test positives (*HIS3*+) for presence of RNA plasmid by selecting for *URA3*	Keep transformants that have maintained the RNA plasmid
4 Select against the RNA plasmid on 5-FOA plates	Generates transformants that lack the RNA plasmid
5 Assay β-galactosidase on filters	Keep white colonies; these are positives that apparently require the RNA plasmid to activate
6 Mate with a series of tester strains carrying either wild type or mutant hybrid RNAs	Use strain R40-coat, which is identical to L40-coat but of opposite mating type. The larger the collection of RNA mutants analyzed, the better
7 Assay β-galactosidase on filters and pick positives with appropriate sequence specificity	Keep colonies carrying plasmids that activate only with the "wild type" RNAs, and not with binding-defective mutant RNAs
8 Functional tests of positives with correct binding specificity	Such tests are essential

Fig. 18-9. A protocol for screening an activation domain library using a hybrid RNA as bait. See text for details.

tivation of *HIS3*, not for maintenance of the RNA plasmid; as a result, cells that can activate *HIS3* without the RNA will often lose the RNA plasmid. In step 3, cells are first tested directly for the presence of the RNA plasmid. In step 4, the RNA plasmid is counterselected by growing colonies on media containing 0.1% 5-FOA (fluoroorotic acid). In step 5, cells that grow in the presence of 5-FOA (and hence have lost the RNA plasmid) are assayed again for *lacZ* expression. In step 6, the RNA plasmid is reintroduced. The reporter genes should again be activated. Candidates satisfying these criteria are studied further.

Additional plasmids and strains are being developed to facilitate the rapid identification and elimination of RNA-independent positives.

False positives in the initial screening can arise if the two proteins interact through a bridging cellular RNA (as opposed to the hybrid RNA). In this fashion, RNA-binding proteins may sometimes be identified that do not actually bind to the desired RNA sequence. Indeed, two-hybrid screens in which the "bait" is an RNA binding protein sometimes yield another RNA-binding protein that likely interacts through an endogenous yeast RNA. Importantly, in a three-hybrid screen of the type diagrammed in Figure 18-9, positives of this type (that is, involving an endogenous yeast RNA) will be eliminated, as they will score as RNA-independent. For most applications of the three-hybrid system, this situation, therefore, is not a problem.

Subsequent screens focus on assessing the sequence specificity of the interaction. Ideally, a series of hybrid RNAs, each one carrying a mutation that affects the RNA-protein interaction in a predictable way, are reintroduced into the strain from which the RNA plasmid has been cured. Reintroduction of wild-type and mutant RNAs is facilitated by using the strain R40-coat (see Figure 18-9). This strain is identical to L40-coat, but of opposite mating type. Thus mating (rather than transformation) can be used to reintroduce wild-type or mutant hybrid RNAs into candidate positive L40-coat strains. The levels of expression of *lacZ* and *HIS3* are monitored in the resulting diploids.

Clones that satisfy all of the screening criteria above should be analyzed further. In particular, recombinant protein produced from the identified activation domain plasmid should be tested for in vitro binding to the specific RNA sequence.

Ideally, a three-hybrid screen of the type diagrammed in Figure 18-9 identifies clones with the appropriate RNA-binding specificity. As in the two-hybrid system, the screen alone can not ensure that those clones are physiologically relevant. Functional tests are required.

Perspective

The three-hybrid assay potentially provides the same versatility to analysis of RNA-protein interactions as the two-hybrid system does for protein-protein interactions. Thus the assay may enable the development of related approaches to identify compounds that disrupt an RNA-protein interaction; to incorporate modifying activities that alter the structure or activity of the RNA or protein; and to attempt genome-wide approaches to identify RNA-protein interactions.

ACKNOWLEDGMENTS This work was supported by a grant from the National Science Foundation Center for Molecular Biotechnology and by grants from the National Institutes of Health.

References

Bardwell, V. J., and Wickens, M. (1990). Purification of RNA and RNA-protein complexes by an R17 coat protein affinity method. Nucl. Acid Res. 18:6587–6594.

Barton, H. A., Eisenstein, R. S., Bomford, A., and Munro, H. N. (1990). Determinants of the interaction between the iron-responsive element-binding protein and its binding site in rat L-ferritin mRNA. J. Biol. Chem. 265:7000–7008.

Bogenhagen, D. F., and Brown, D. D. (1981). Nucleotide sequences in Xenopus 5S DNA required for transcription termination. Cell 24:261–270.

Burd, C. G., and Dreyfuss, G. (1994). Conserved structures and diversity of functions of RNA-binding proteins. Science 265:615–621.

Clever, J., Sassetti, C., and Parslow, T. G. (1995). RNA secondary structure and binding sites for gag gene products in the 5' packaging signal of human immunodeficiency virus type 1. J. Virol. 69:2101–2109.

Eisenstein, R. S., Tuazon, P. T., Schalinske, K. L., Anderson, S. A., and Traugh, J. A. (1993). Iron-responsive element-binding protein. Phosphorylation by protein kinase C. J. Biol. Chem. 268:27363–27370.

Estojak, J., Brent, R., and Golemis, E. A. (1995). Correlation of two-hybrid affinity data with in vitro measurements. Mol. Cell. Biol. 15:5820–5829.

Geiduschek, E. P., and Tocchini, V. G. (1988). Transcription by RNA polymerase III. Annu. Rev. Biochem. 57:873–914.

Good, P. D., and Engelke, D. R. (1994). Yeast expression vectors using RNA polymerase III promoters. Gene 151:209–214.

Gorlach, M., Burd, C. G., and Dreyfuss, G. (1994). The mRNA poly(A)-binding protein: localization, abundance, and RNA-binding specificity. Exp. Cell Res. 211:400–407

Haile, D. J., Hentze, M. W., Rouault, T. A., Harford, J. B., and Klausner, R. D. (1989). Regulation of interaction of the iron-responsive element binding protein with iron-responsive RNA elements. Mol. Cell. Biol. 9:5055–5061.

Harada, K., Martin, S. S., and Frankel, A. D. (1996). Selection of RNA-binding peptides in vivo. Nature 380:175–179.

Laird-Offringa, I. A., and Belasco, J. G. (1995). Analysis of RNA-binding proteins by in vitro genetic selection—identification of an amino acid residue important for locking U1A onto its RNA target. Proc. Natl. Acad. Sci. USA 92:11859–11863.

Lowary, P. T., and Uhlenbeck, O. C. (1987). An RNA mutation that increases the affinity of an RNA-protein interaction. Nucl. Acid Res. 15:10483–10493.

MacWilliams, M. P., Celander, D. W., and Gardner, J. F. (1993). Direct genetic selection for a specific RNA-protein interaction. Nucl. Acids Res. 21:5754–5760.

Sachs, A. B., Davis, R. W., and Kornberg, R. D. (1987). A single domain of yeast poly(A)-binding protein is necessary and sufficient for RNA binding and cell viability. Mol. Cell. Biol. 7:3268–3276.

SenGupta, D. J., Zhang, B., Kraemer, B., Pochart, P., Fields, S., and Wickens, M. (1996). A three-hybrid system to detect RNA-protein interactions in vivo. Proc. Natl. Acad. Sci. USA 93:8496–8501.

Slice, L. W., Codner, E., Antelman, D., Holly, M., Wegrzynski, B., Wang, J., Toome, V., Hsu, M. C., and Nalin, C. M. (1992). Characterization of recombinant HIV-1 Tat and its interaction with TAR RNA. Biochemistry 31:12062–12068.

Stripecke, R., Oliveira, C. C., McCarthy, J. E., and Hentze, M. W. (1994). Proteins binding to 5' untranslated region sites: a general mechanism for translational regulation of mRNAs in human and yeast cells. Mol. Cell. Biol. 14:5898–5909.

Witherell, G. W., Gil, A., and Wimmer, E. (1993). Interaction of polypyrimidine tract binding protein with the encephalomyocarditis virus mRNA internal ribosomal entry site. Biochemistry 32:8268–8275.

Witherell, G. W., Wu, H. N., and Uhlenbeck, O. C. (1990). Cooperative binding of R17 coat protein to RNA. Biochemistry 29:11051–11057.

Yang, M., Wu, Z., and Fields, S. (1995). Protein-peptide interactions analyzed with the yeast two-hybrid system. Nucl. Acid Res. 23:1152–1156.

19

Split Ubiquitin
A Sensor of Protein Interactions In Vivo

Nils Johnsson
Alexander Varshavsky

Multiprotein complexes mediate the bulk of biological processes (Alberts 1984, Creighton 1992). The knowledge of these complexes is extensive for oligomeric proteins whose subunit interactions are strong enough to withstand in vitro conditions. However, many oligomeric assemblies, while relevant physiologically, are transient in vivo or unstable in vitro. The understanding of in vivo protein interactions (and especially of their temporal aspects) is still fragmentary and largely qualitative, the limitations of existing in vivo methods being a major reason. Assays for in vivo protein interactions include cross-linking of proteins with cell-penetrating reagents such as, for example, formaldehyde (Solomon et al. 1988; Creighton 1992) and use of resonance energy transfer between dye-coupled proteins microinjected into cells (Adams et al. 1991). Genetic analyses of protein interactions include searches for synthetic lethal or extragenic suppressor mutations (Guarente 1993a), which occur in genes whose products are at least functionally (and often physically) associated with a protein of interest. Another approach, the two-hybrid technique (Fields and Song 1989; Chien et al. 1991; Gyuris et al. 1993; Guarente 1993b; Phizicky and Fields 1995), is based on expressing one protein as a fusion to a DNA-binding domain of a transcriptional activator, and expressing another protein as a fusion to a transcriptional activation domain. If the test proteins interact in vivo, a transcriptional activator is reconstituted, resulting in the induction of a reporter gene. Applications of the two-hybrid technique to the discovery of specific protein interactions in vivo have been, and continue to be, immensely fruitful (reviewed by Phizicky and Fields 1995). However, this otherwise powerful method cannot address temporal aspects of a protein-protein interaction. In addition, the two-hybrid technique constrains the set of detectable protein interactions to those that occur (or can be "reproduced") in the nucleus, in proximity to the reporter gene.

Described below is another method, the *u*biquitin-based *s*plit-*p*rotein *s*ensor (USPS, Figure 19-1), which makes it possible to detect and monitor a protein-protein interaction as a function of time, at the natural sites of this interaction in a living cell (Johnsson and Varshavsky 1994a). The USPS technique stems from the following properties of ubiquitin (Ub), a 76-residue, single-domain protein (Figure 19-1A). Protein fusions containing Ub are rapidly cleaved in vivo by Ub-specific proteases (UBPs), which recognize the folded conformation of Ub (Bachmair et al. 1986; Johnsson and Varshavsky 1994b; Varshavsky 1992, 1996). When a C-terminal fragment of Ub (C_{ub}) is expressed as a fusion to a reporter protein, the fusion is cleaved only if an N-terminal fragment of Ub (N_{ub}) is also expressed in the same cell. This reconstitution of native Ub from its fragments, detectable by the in vivo cleavage assay, is not observed with a mutationally altered N_{ub}. However, if C_{ub} and the altered N_{ub} are each linked to polypeptides that interact in vivo, the cleavage of the fusion containing C_{ub} is restored, yielding a generally applicable assay for kinetic and equilibrium aspects of in vivo protein interactions (Johnsson and Varshavsky 1994a; Figure 19-1).

The use of USPS is illustrated below with test proteins containing a homodimerizing leucine zipper; we also mention a more recent study utilizing constructs that allow the detection of a transient in vivo interaction between a signal sequence and components of the translocation apparatus at the endoplasmic reticulum (ER) membrane. At present, USPS can be used to test for and explore an in vivo interaction between two or more predetermined proteins in a cell. USPS can also be employed as a protein-interaction assay in vitro (in a cell-free system). A USPS-based screen for genes whose products interact with a protein of interest should be feasible as well.

Materials and Methods

Strains, Media, Pulse-Chase, and Immunoblotting

All experiments used the YPH500 strain of the yeast *Saccharomyces cerevisiae* (*MATα ura3 lys2 trp1 ade2 his3 leu2*) (Sikorski and Hieter 1989; Johnsson and Varshavsky 1994b) grown at 30°C to A_{600} of ~1 in a synthetic (SD) medium (Sherman et al. 1986) containing 0.1 mM $CuSO_4$. Pulse-chase experiments, including the preparation of cell extracts in the presence of N-ethylmaleimide (to inhibit UBPs), immunoprecipitation with a monoclonal antibody to the "ha" epitope (Field et al. 1988), SDS-PAGE (12%),

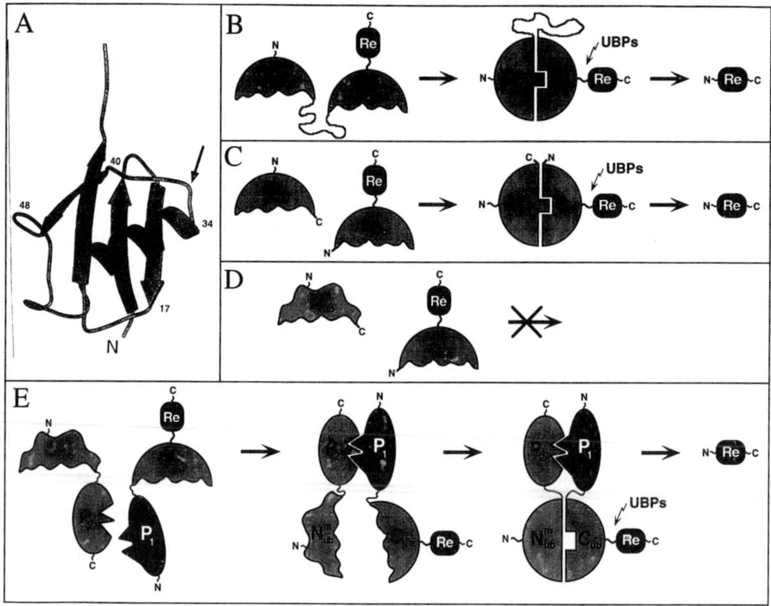

Fig. 19-1. Split ubiquitin as a proximity sensor in vivo (Johnsson and Varshavsky 1994a). (A) A ribbon diagram of Ub (Vijay-Kumar et al. 1987). An arrow denotes the site of either a 68-residue insertion or a cut between the subdomains of Ub. Some of the residue numbers are indicated. Ile-13, the site of mutations analyzed in this work, is in the second strand of the b-sheet, where it interacts with the hydrophobic face of the a-helix (Vijay-Kumar et al. 1987). (B) A newly formed Ub moiety bearing an insertion (wavy line) between its N-terminal (N_{ub}) and C-terminal (C_{ub}) subdomains, and linked to a reporter protein (Re; black oval). The insertion did not detectably interfere with the Ub folding, which was required for the in vivo cleavage of the fusion by Ub-specific proteases (UBPs; lightning arrow), yielding the free reporter. (C) When N_{ub} and C_{ub} were coexpressed as separate fragments, with C_{ub} still linked to the reporter, significant in vivo reconstitution of a quasi-native (recognizable by UBPs) Ub moiety was observed. (D) In vivo reconstitution of Ub from its separate, coexpressed fragments did not occur with a mutant N_{ub} fragment, denoted as N^{m}_{ub}, that bore a single-residue replacement at position 13. Conformational destabilization of N^{m}_{ub} relative to its wild-type counterpart N_{ub} is indicated by the altered shape of the N^{m}_{ub} subdomain. (E) *U*b-based *s*plit-*p*rotein *s*ensor (USPS). N^{m}_{ub}, an altered Ub fragment that failed to reconstitute Ub in the presence of C_{ub}, did support reconstitution if the two Ub fragments were linked to polypeptides P_1 and P_2 that interacted in vivo. Reduced conformational stability of Ub that has been reconstituted with N^{m}_{ub} instead of N_{ub} is denoted by a gap between the Ub subdomains.

and fluorography, were carried out as described (Johnsson and Varshavsky 1994b; Johnson et al. 1992, 1995), except that zero-time samples were withdrawn and processed 1 minute after the addition of a chase medium (Johnsson and Varshavsky 1994b). Immunoblotting with the monoclonal anti-ha antibody was performed using the ECL detection system (Amersham).

Test Proteins

The final constructs [verified by DNA sequencing (Ausubel et al. 1992)] resided in the *S. cerevisiae/Escherichia coli* plasmid vectors pRS314 or pRS316 (Sikorski and Hieter 1989), and were expressed in yeast from the induced P_{CUP1} promoter. The *S. cerevisiae* Ub gene was amplified, using polymerase chain reaction (PCR) (Ausubel et al. 1992), from the previously engineered *Sal*I site immediately upstream of the Ub start codon (Johnsson and Varshavsky 1994b) to the first C of codon 37, and from codon 35 to codon 76. The primers were constructed in a way that yielded, after ligation of the two amplified fragments, a *Bam*HI site between codons 35 and 37 in the Ub open reading frame (ORF). Ligation of this ORF to a fragment encoding dha (DHFR-ha) (Johnsson and Varshavsky 1994b) yielded an ORF encoding Ub-dha (Figure 19-2-I), which contained the sequence Met-His-Arg-Ser-Gly-Ile-Met between Gly-76 of Ub and Val-1 of DHFR. Constructs II–IV (Figure 19-2) were produced by replacing the *Sal*I-*Bst*XI fragment in construct I with appropriately designed double-stranded oligonucleotides. Construct V (Figure 19-2) was produced using PCR, *S. cerevisiae* genomic DNA, and primers designed to amplify the region of *STE6* (McGrath and Varshavsky 1989; Kuchler et al. 1989) from codon 196 to codon 262. The resulting fragment was inserted into the *Bam*HI site between the Ub codons 35 and 37 in constructs I–IV, yielding constructs V–VIII. In the construct IX (Figure 19-2), the residue 35 of Ub was preceded by a 32-residue linker all of whose residues except the N-terminal Met-Gly-Gly were specified by codons 234 to 262 of *STE6*. The z_1-C_{ub} portion of construct XIV contained the above Ste6p-derived sequence preceding the C_{ub} moiety, the leucine zipper region of *S. cerevisiae* Gcn4p (residues 235–281, denoted as z_1) (Vinson et al. 1989; Hinnebusch 1984; O'Shea et al. 1991; Pu and Struhl 1993), the construction-generated N-terminal Met, and the sequence Gly-Glu-Ile-Ser-Thr. Constructs X–XIII (Figure 19-2), derived from construct XIV and constructs V–VIII, encoded Gly-Glu-Ile-Ser-Thr-Leu-Glu C-terminally to z_1, and Gly-Gly-Ser-Thr-Met between z_1 and N_{ub}. The z_1 motif in N_{ub}^{wt}-z_1 and its derivatives but not in z_1C_{ub}-dha bore a Met-250 → Thr-250 replacement (residue numbers of Gcn4p), which occurred during construction; this replacement would be expected to weaken the interaction between z_1 domains (Pu and Struhl 1993; O'Shea et al. 1991).

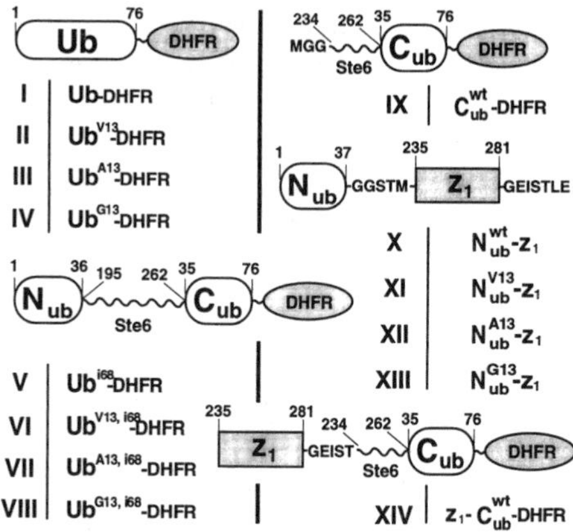

Fig. 19-2. Fusion constructs. Fusions of this work contained some of the following elements: (1) a Ub moiety, either wild-type (construct I) or bearing single-residue replacements at position 13 (constructs II-IV). (2) A Ub moiety containing the 68-residue insertion (denoted as "Ste6") derived from the cytosolic region of *S. cerevisiae* Ste6p between its transmembrane segments 4 and 5 (McGrath and Varshavsky 1989; Kuchler et al. 1989). The insertion was positioned after residue 36 of Ub (construct V). (3) A Ub moiety bearing both the above insertion and a single-residue replacement at position 13 (constructs VI–VIII). (4) A C-terminal fragment of Ub (C_{ub}, residues 35–76) bearing the 32-residue, Ste6p-derived sequence at its N-terminus (construct IX). (5) The same fusion whose N-terminus was extended, via the linker sequence Gly-Glu-Ile-Ser-Thr, with the 47-residue homodimerization motif ("leucine zipper", or z_1) of *S. cerevisiae* Gcn4p (Vinson et al. 1989; Hinnebusch 1984; O'Shea et al. 1991; Pu and Struhl 1993) (residues 235–281 of Gcn4) (construct XIV). (6) An N-terminal fragment of Ub (N_{ub}, residues 1–37) bearing the wild-type Ub sequence or a single-residue replacement at position 13, and a C-terminal extension containing the linker sequence Gly-Gly-Ser-Thr-Met followed by the z_1 leucine zipper of Gcn4p (constructs X–XIII). (7) Mouse dihydrofolate reductase (DHFR) bearing a C-terminal ha epitope (Field et al. 1988) (denoted as "DHFR" in the diagram and as "dha" in the text).

The USPS Technique

In Vivo Folding of Ubiquitin Containing an Insertion and/or a Single-Residue Replacement

Ub is a 76-residue, single-domain protein (Figure 19-1A) that is present in cells either free or covalently linked to other proteins. Ub plays a role in a number of processes, primarily through routes that involve protein degradation (Hershko and Ciechanover 1992; Varshavsky 1992, 1996). In eukaryotes, newly formed Ub fusions are rapidly cleaved by Ub-specific proteases (UBPs) after the last residue of Ub at the Ub-polypeptide junction (Papa and Hochstrasser 1993; Baker et al. 1992; Finley et al. 1989; Bachmair et al. 1986). It has been shown that the cleavage of a Ub fusion by UBPs requires the folded conformation of Ub (Johnsson and Varshavsky 1994b).

In the initial constructs used to explore the USPS technique, Ub was joined to the N-terminus of the mouse dihydrofolate reductase (DHFR), whose C-terminus was extended with the "ha" epitope tag (Field et al. 1988), yielding a Ub-dha (Ub-DHFR-ha) test protein of 22 kD. The Ub moiety whose Ile-3 and Ile-13 residues [which are buried in the hydrophobic core of Ub (Vijay-Kumar et al. 1987)] have been replaced by Gly residues is a poor UBP substrate (Johnsson and Varshavsky 1994b). To make similar but less destabilizing alterations of Ub, only Ile-13 was replaced with either Val, Ala, or Gly. The resulting Ub fusions (Figure 19-2-II, III, VI) were completely cleaved in vivo by the end of either a 5-minute or 2-minute labeling with ^{35}S-methionine, as was Ub-dha, bearing wild-type Ub (Figure 19-3A, lanes *a–d*, and data not shown). We then asked whether a 68-residue insertion (denoted as i68) within a loop (residues 34–40) connecting the only a-helix of Ub to a ß-strand (Figure 19-1A, B) and a substitution at position 13 of Ub, if present together, result in a less efficient cleavage of a fusion by UBPs. The i68 insertion was derived from the cytosolic region of *S. cerevisiae* Ste6p between its transmembrane segments 4 and 5 (McGrath and Varshavsky 1989; Kuchler et al. 1989). This region was chosen because it was expected to be either flexible or folded in a way that positions its ends in proximity to each other. By the end of a 2-minute pulse, no uncleaved Ub^{i68}-dha, and at most traces of Ub$^{V13, i68}$-dha could be detected (Figures 19-2-V, VI, and 19-3B, lanes *a* and *b*). However, the cleavage of Ub$^{A13, i68}$-dha, and especially of Ub$^{G13, ¿68}$-dha was much slower, in that significant amounts of the uncleaved fusions were observed by the end of a 2-minute pulse (Figures 19-2-VII, VIII, and 19-3B, lanes *c* and *d*).

The i68 insertion placed the two "halves" of a nascent Ub farther apart (Figure 19-1A, B), and, therefore, was expected to retard the folding of Ub; this effect could be detected by the 2-minute pulse-cleavage assay if another destabilizing Ub alteration such as Ile-13 $\rightarrow$ Gly-13 (which by itself

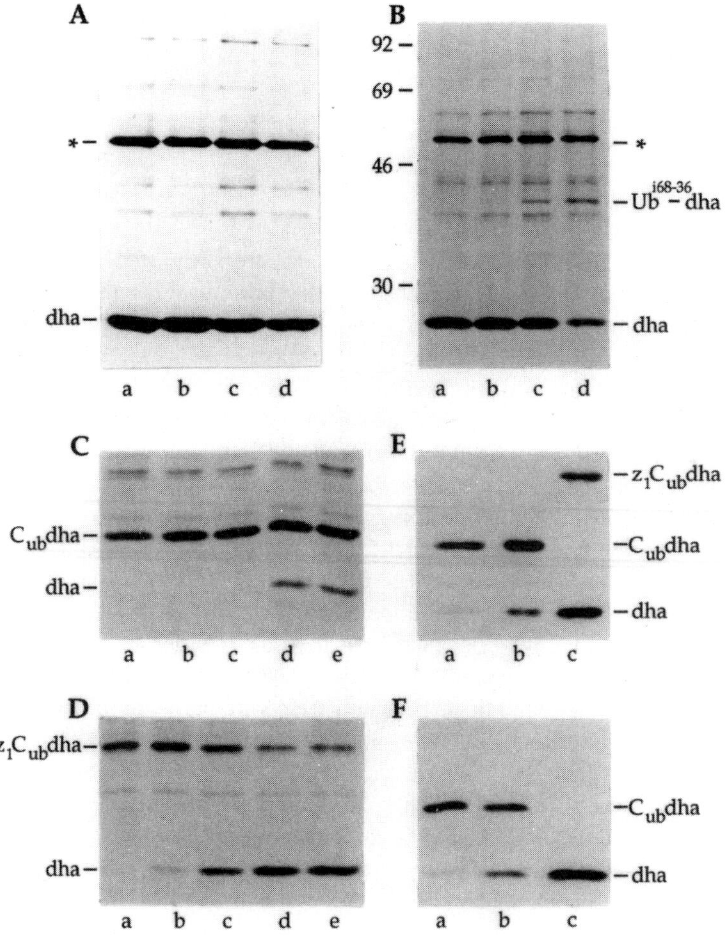

Fig. 19-3. Kinetic and equilibrium aspects of ubiquitin reconstitution in vivo. (A) Lane *a*, *S. cerevisiae* expressing Ub-dha (Fig. 19-2-I) were labeled for 5 minutes with ^{35}S-methionine, followed by extraction of proteins, immunoprecipitation with anti-ha antibody, and SDS-PAGE (see Materials and Methods). Lanes *b–d*, same as lane *a* but with the Ub moiety containing, respectively, Val, Ala or Gly instead of wild-type Ile at position 13 (Fig. 19-2-II, III, IV). (B) Lane *a*, same as lane *a* in A but with the labeling time of 2 minutes, and with *S. cerevisiae* expressing Ub^{i68}-dha (Fig. 19-2-V). Lanes *b–d*, same as lanes *b–d* in A but with the single-residue replacements at position 13 in the Ub^{i68} moiety (Fig. 19-2-VI, VII, VIII). (C) Lane *a*, same as lane *a* in B but with *S. cerevisiae* expressing C$_{ub}$-dha (Fig. 19-2-IX). Lane *b*, C$_{ub}$-dha was coexpressed with N_{ub}^{G13}-z_1 (Fig. 19-2-XIII). Lane *c*, C$_{ub}$-dha was coexpressed with N_{ub}^{A13}-z_1 (Fig. 19-2-XII). Lane *d*, C$_{ub}$-dha was coexpressed with N_{ub}^{V13}-z_1 (Fig. 19-2-XI). Lane *e*, C$_{ub}$-dha was coexpressed with N_{ub}^{wt}-z_1 (Fig. 19-3-XI). (D) Same as lane *a* in C but with *S. cerevisiae* expressing z_1C$_{ub}$-dha (Fig. 19-2-XIV). Lane *b*, z_1C$_{ub}$-dha was coexpressed with N_{ub}^{G13}-z_1. Lane *c*, z_1C$_{ub}$-dha was coexpressed with N_{ub}^{A13}-z_1. Lane *d*, z_1C$_{ub}$-dha was coexpressed with N_{ub}^{V13}-z_1. Lane *e*, z_1C$_{ub}$-dha was coexpressed with N_{ub}^{wt}-z_1. (E) Whole-cell extracts of *S. cerevisiae* expressing C$_{ub}$-dha (lane *a*), or C$_{ub}$-dha and N_{ub}^{G13}-z_1(lane *b*), or z_1C$_{ub}$-dha and N_{ub}^{G13}-z_1 (lane *c*) were fractionated by SDS-PAGE and analyzed by immunoblotting with anti-ha antibody. (F) Same as E but with *S. cerevisiae* expressing C$_{ub}$-dha (lane *a*), or C$_{ub}$-dha and N_{ub}^{A13}-z_1 (lane *b*), or z_1C$_{ub}$-dha and N_{ub}^{A13}-z_1 (lane *c*). Ub^{i68}-dha, free reporter (dha), C$_{ub}$-dha and z_1C$_{ub}$-dha are indicated. The asterisk denotes an unrelated *S. cerevisiae* protein that crossreacted with anti-ha antibody (Johnsson and Varshavsky 1994a, b; Johnson et al. 1995).

was insufficient to cause a detectable retardation of Ub folding) was present as well (Figure 19-3A, B). These results can be interpreted within the diffusion-collision model of protein folding (Creighton 1992; Kim and Baldwin 1982; Karplus and Weaver 1976), in which marginally stable units of isolated secondary structure form early and then coalesce into the native conformation. Indeed, in the native Ub, its first 34 residues are folded into an a-helix interacting with a double-stranded antiparallel b-sheet (Figure 19-1A) (Vijay-Kumar et al. 1987). Thus, the i68 insertion retards Ub folding primarily through a reduction in the frequency of collisions between the N-terminal and C-terminal subdomains of Ub, whereas the effect of substitutions at position 13 of Ub is a decreased conformational stability of its N-terminal subdomain (Figure 19-1A). This in vivo evidence for Ub subdomains (see also next section) is in agreement with the results of circular dichroism and NMR analyses of Ub and its fragments in a methanol-water solvent at low pH (Cox et al. 1993; Stockman et al. 1993).

In Vivo Reconstitution of Ubiquitin from Its Fragments

The relative insensitivity of Ub folding to a large insertion within the 34–40 loop (Figure 19-1A, B) suggested that separate, coexpressed fragments of Ub (produced by a cut within the 34–40 loop) may be able to reconstitute native Ub. In a test of this conjecture, C_{ub} (residues 35–76) was expressed as a fusion to the dha reporter (C_{ub}-dha), while an N-terminal fragment of wild-type Ub (residues 1–37, denoted as N_{ub}^{wt}) was expressed as a fusion to the "leucine zipper" homodimerization domain of the yeast Gcn4 protein (denoted as z_1) (Figure 19-2-IX, X) (O'Shea et al. 1991; Pu and Struhl 1993). (The reason for linking N_{ub}^{wt} to z_1 will become clear later.) When expressed by itself, C_{ub}-dha remained largely uncleaved; instead, the entire fusion was slowly degraded (Figure 19-4A, lanes *a–c*). However, coexpression of C_{ub}-dha and N_{ub}^{wt}-z_1 resulted in the cleavage of C_{ub}-dha, yielding dha, which accumulated during a 30-minute chase (Figure 19-4A, lanes *d-f*). This cleavage was slow in comparison to the cleavage of a fusion containing wild-type Ub (Figure 19-3A, lane *a*). The relatively low overall rate of the C_{ub}-dha cleavage apparently resulted from a low overall rate of "productive" in vivo collisions between the C_{ub} and N_{ub} fragments, in comparison to the rate of analogous collisions between the same subdomains of Ub when they are linked within a single polypeptide (Figure 19-1A, B).

We concluded that the cleavage of C_{ub}-dha requires the presence of N_{ub}, and is the consequence of an in vivo association between the C_{ub} and N_{ub} fragments (Johnsson and Varshavsky 1994a). This association results in at least transient formation of a Ub moiety which is similar enough to native Ub to be a substrate of UBPs. That fragments of a protein can associate to form a functional, quasi-native species has been demonstrated for a variety of proteins other than Ub; the examples include ribonuclease A, staphylococcal nuclease and other proteins in vitro (Anfinsen et al. 1971; Richards

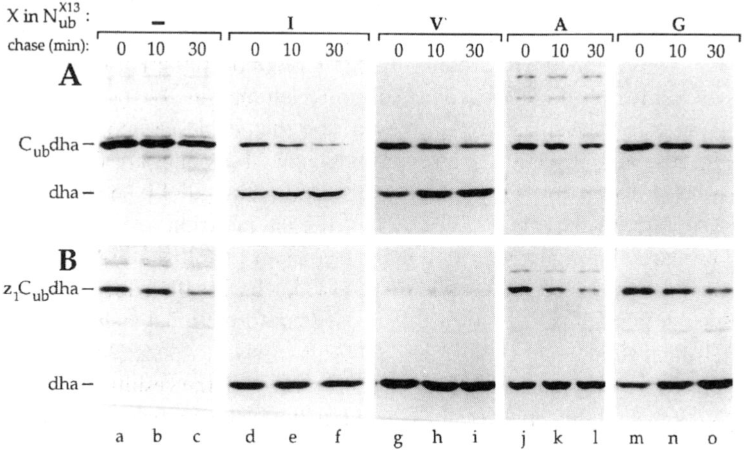

Fig. 19-4. Reconstitution of the folded ubiquitin from its coexpressed fragments: enhancement by *cis*-linked interacting polypeptides. (A) Lanes *a–c*, *S. cerevisiae* expressing C_{ub}-dha (Fig. 19-2-IX) were labeled for 5 minutes with ^{35}S-methionine, followed by a chase for 0, 10, and 30 minutes, extraction of proteins, immunoprecipitation with anti-ha antibody, and SDS-PAGE (see Materials and Methods). Note a slow degradation of C_{ub}-dha and the absence of the cleavage that yields dha. Lanes *d–f*, same as lanes *a–c* but C_{ub}-dha was coexpressed with N_{ub}^{wt}-z_1 (Fig. 19-2-X). Lanes *g–i*, C_{ub}-dha was coexpressed with N_{ub}^{V13}-z_1 (Fig. 19-2-XI), which bore Val instead of wild-type Ile at position 13 of Ub. Lanes *j–l*, C_{ub}-dha was coexpressed with N_{ub}^{A13}-z_1 (Fig. 19-2-XII). Lanes *m–o*, C_{ub}-dha was coexpressed with N_{ub}^{G13}-z_1 (Fig. 19-2-XIII). (B) Lanes *a–c*, same as lanes *a–c* in A but with z_1C_{ub}-dha (Fig. 19-2-XIV) instead of C_{ub}-dha. Lanes *d–f*, z_1C_{ub}-dha was coexpressed with N_{ub}^{wt}-z_1-. Lanes *g–i*, z_1C_{ub}-dha was coexpressed with N_{ub}^{V13}-z_1. Lanes *j–l*, z_1C_{ub}-dha was coexpressed with N_{ub}^{A13}-z_1. Lanes *m–o*, z_1C_{ub}-dha was coexpressed with N_{ub}^{G13}-z_1.

and Wyckoff 1971; Galakatos and Walsh 1987; Johnsson and Weber 1990), and also *S. cerevisiae* isoleucyl-tRNA synthetase, *E. coli* Lac permease and Tet protein in vivo (Landro and Schimmel 1993; Shiba and Schimmel 1992; Rubin and Levy 1991; Zen et al. 1994).

The efficiency of Ub reconstitution depended on the conformational stability of its N-terminal subdomain: coexpression of C_{ub}-dha with either N_{ub}^{G13}-z_1 or N_{ub}^{A13}-z_1 (Figure 19-2-IX, XII, XIII), bearing Gly or Ala instead of Ile at position 13, resulted in virtually no cleavage of C_{ub}-dha, in contrast to the results with either N_{ub}^{wt}-z_1 or N_{ub}^{V13}-z_1 (Figure 19-2-X, XI), which bore either wild-type Ile or Val, a hydrophobic residue larger than Ala and Gly, at position 13 (Figure 19-1C, D, and Figure 19-4A, lanes *j–o*; compare with lanes *d–i*). These results, together with the finding that the UBP-mediated cleavage of a Ub fusion requires folded Ub (Figure 19-3A, B) (Johnsson and Varshavsky 1994b) led to a new assay for in vivo protein interactions, as shown in the following section.

Split Ubiquitin as a Proximity Sensor

We asked whether the linking of two polypeptides that interact in vivo to N_{ub} and C_{ub} facilitates reconstitution of Ub. C_{ub}-dha was fused to a region of *S. cerevisiae* Gcn4p (residues 235–281) that contained the leucine zipper homodimerization domain (denoted as z_1) (Hinnebusch 1984; Pu and Struhl 1993). In the resulting z_1C_{ub}-dha, a 32-residue linker was inserted between z_1 and C_{ub} (Figure 19-2-XIV) to ensure that N_{ub} and C_{ub} subdomains could be spatially proximal within a z_1-mediated complex between z_1C_{ub}-dha and N_{ub}-z_1. When expressed by itself, z_1C_{ub}-dha remained uncleaved, and was slowly degraded during the chase (Figure 19-4B, lanes *a–c*). However, coexpression of z_1C_{ub}-dha and N_{ub}^{G13}-z_1, bearing Gly instead of Ile at position 13 of Ub, resulted in a significant cleavage of z_1C_{ub}-dha (yielding dha) in the course of a 30-min chase (Figures 19-2-XIII, XIV and 4B, lanes *m–o*). By contrast, no such cleavage was observed when N_{ub}^{G13}-z_1 was coexpressed with C_{ub}-dha, which lacked z_1 (Figure 19-4A, lanes *m–o*). Similar results (but with faster cleavage of z_1C_{ub}-dha) were obtained upon coexpression of z_1C_{ub}-dha and N_{ub}^{A13}-z_1 (Figure 19-4A, B, lanes *j–l*). Moreover, the enhancement of Ub reconstitution by z_1-z_1 interactions was observed even with pairs of Ub fragments that could yield Ub by themselves (in the absence of linked z_1). Specifically, whereas the coexpression of C_{ub}-dha and N_{ub}^{wt}-z_1 or N_{ub}^{V13}-z_1 resulted in detectable but slow cleavage of C_{ub}-dha that was still incomplete after 30 minutes of chase, coexpression of z_1C_{ub}-dha and N_{ub}^{wt}-z_1 or N_{ub}^{V13}-$z_1$resulted in the nearly complete cleavage of z_1C_{ub}-dha (yielding dha) by the end of a 5-minute pulse (Figure 19-4A, B, lanes *d–i*). The temporal resolution of this assay could be increased by shortening the labeling time from 5 to 2 minutes. For example, the fraction of z_1C_{ub}-dha cleaved by the end of a 2-minute pulse progressively increased when z_1C_{ub}-dha was coexpressed with N_{ub}^{G13}-z_1, N_{ub}^{G13}-z_1, N_{ub}^{G13}-z_1 or N_{ub}^{wt}-z_1 (Figure 19-3D, lanes *b–e*). By contrast, no cleavage of z_1C_{ub}-dha was observed when it was expressed by itself (Figure 19-3D, lane *a*), or when N_{ub}^{A13}-z_1 or N_{ub}^{G13}-z_1 were coexpressed with C_{ub}-dha, which lacked the z_1 zipper (Figure 19-3C, lanes *b* and *c*).

To determine steady-state levels of test proteins, cell extracts were analyzed by immunoblotting with anti-ha antibody (Figure 19-3E, F). When C_{ub}-dha was expressed by itself, it remained largely uncleaved (Figure 19-3E, F, lanes *a*). When N_{ub}^{A13}-z_1 was coexpressed with C_{ub}-dha, a fraction of C_{ub}-dha was cleaved to yield dha (Figure 19-3F, lane *b*). However, when N_{ub}^{A13}-z_1 was coexpressed with z_1C_{ub}-dha, virtually all of z_1C_{ub}-dha was cleaved to yield dha (Figure 19-3F, lane *c*). Similar results were obtained with N_{ub}^{G13}-z_1, except that a significant fraction of z_1C_{ub}-dha remained uncleaved in the presence of N_{ub}^{G13}-z_1 (Figure 19-3E lanes *b* and *c*).

Discussion

The USPS Technique

This chapter describes an assay for in vivo protein interactions, termed the USPS technique (*u*biquitin-based *s*plit-*p*rotein *s*ensor) (Figure 19-1). Protein fusions containing ubiquitin (Ub)—a 76-residue, single-domain protein—are rapidly cleaved in vivo by ubiquitin-specific proteases (UBPs), which recognize the folded conformation of Ub. When a C-terminal fragment of Ub (C_{ub}) is expressed as a fusion to a reporter protein, the fusion is cleaved only if an N-terminal fragment of Ub (N_{ub}) is also expressed in the same cell. This reconstitution of native Ub from its fragments, detectable by the in vivo cleavage assay, is not observed with a mutationally altered N_{ub}. However, if C_{ub} and the altered N_{ub} are each linked to polypeptides that interact in vivo, the cleavage of the fusion containing C_{ub} is restored, yielding a generally applicable assay for kinetic and equilibrium aspects of in vivo protein interactions (Johnsson and Varshavsky 1994a) (Figure 19-1E).

Features of USPS that distinguish it from the transcription-based two-hybrid technique include the possibility of monitoring a protein-protein interaction as a function of time, at the natural sites of this interaction in a living cell. Enhancement of Ub reconstitution by interacting polypeptides linked to fragments of Ub stems from a local increase in concentration of one Ub fragment in the vicinity of the other. This in turn increases the probability that the two Ub fragments coalesce to form a quasi-native Ub moiety, whose (at least) transient formation results in the irreversible cleavage of the fusion by UBPs (Figure 19-1D). This cleavage can be detected readily, and can be followed as a function of time (Figure 19-4) or at steady state (Figure 19-3E, F). Unlike the two-hybrid technique, which is based on the apposition of two structurally independent protein domains whose folding and functions do not require direct interactions between the domains, the USPS method involves reconstituting the conformation of a single-domain protein (Figure 19-1E).

USPS detects a spatial proximity of proteins but not necessarily their direct interaction. Therefore a USPS signal might also result from the binding of test proteins to a common ligand—another protein or a larger structure such as a microtubule or a ribosome. This feature of USPS should allow in vivo analyses of large protein complexes, including those that may be too unstable or too transient for detection by other methods. A separate, immunologically detectable reporter domain (Figure 19-1) is not an essential aspect of USPS: the cleavage of a C_{ub}-containing fusion can also be followed by measuring the enzymatic activity of a reporter protein that is inactive until released from a C_{ub}-containing fusion, or by using an antibody specific for a test protein linked to either the C-terminus or the N-terminus of C_{ub}. USPS also works with N_{ub} linked either to the N-terminus of a test protein (Figure 19-2) or to its C-terminus, or even with the N_{ub} moiety as an insert within a test protein (N. J and A. V., unpublished data).

Many aspects of USPS remain to be explored. For instance, given a delay between the association of two test proteins and the diagnostic cleavage event (Figure 19-1D), a USPS assay provides only an upper-limit estimate for the rate of association of the test proteins. The delay in cleavage is caused not only by a delay in the reconstitution of a quasi-native (cleavable) Ub moiety but also, possibly, by its lower than wild-type quality as a substrate of UBPs. In addition, the acceleration of Ub reconstitution by linked, interacting test proteins should be influenced by steric constraints on the spatial disposition of Ub fragments. Flexible linkers between these fragments and test proteins reduce but may not eliminate this source of ambiguity, whose actual extent remains to be determined. At the same time, these complications are likely to be negligible in a comparative USPS analysis, which utilizes identical linker arrangements and characterizes interactions between a protein of interest and a series of its structurally similar protein ligands (for example, mutant variants of a ligand). Another informative variation of USPS would be to position a UBP within a C_{ub}-containing fusion: the kinetics of a resulting *cis*-cleavage of the fusion upon the reconstitution of Ub could be compared to that of a *trans*-cleavage in the current version of USPS.

Other applications of USPS

Listed below are some of the USPS applications that remain to be explored.

1. USPS can be used to analyze protein interactions in vitro—in cell extracts or with purified fusions in the presence of a purified UBP such as yeast Ubp1p (Tobias and Varshavsky 1991). One advantage of in vitro USPS is the possibility of adding UBP at a desired time after the mixing of interacting species. (An approach to a temporal control of the UBP-mediated cleavage that would be applicable both in vitro and in vivo would be to identify pairs of Ub fragments whose reconstitution is temperature sensitive.) The kinetics of Ub reconstitution from its fragments in vitro can be compared with analogous reconstitutions in living cells to address the influences of translation, chaperonins, macromolecular crowding and other aspects of the in vivo condition. *In vitro* USPS can also be used to screen for peptide or nonpeptide ligands that interact with a ligand of interest.
2. Many proteins have more than one protein ligand. A version of USPS for a "many-body" interaction could be as follows. N_{ub}-P_1, a fusion between N_{ub} and a protein of interest P_1 is coexpressed with two other proteins: P_2C_{ub}-R_1, a fusion between C_{ub}, P_2 (a putative ligand of P_1) and a reporter R_1, and a similarly designed P_3C_{ub}-R_2 fusion containing a different reporter R_2, and P_3, another putative ligand of P_1. Comparing the kinetics of in vivo cleavages of P_2C_{ub}-R_1 and P_3C_{ub}-R_2 in the presence of N_{ub}-P_1 should provide information

about relative affinities of P_1 for P_2 and P_3, and about kinetic aspects of these interactions as well.

3. Lateral diffusion and interactions of membrane proteins underlie membrane-based reactions such as activation of receptors by hormones and adhesion between cells. In this version of USPS, the two Ub subdomains can be positioned within cytosol-exposed regions of two membrane proteins. Varying the positions of Ub subdomains within test proteins, using mutant Ub subdomains whose (facilitated) reconstitution is either temperature sensitive or constitutive, and monitoring the kinetics of diagnostic in vivo cleavages of the resulting fusions may yield insights into dynamic aspects of membrane-associated proteins that are difficult or impossible to obtain with existing methods.
4. A USPS-based screen for genes whose products interact with a protein of interest P_1 can employ a reporter R whose function is incompatible with an N-terminal extension such as P_1C_{ub}. The reporter would be activated upon Ub reconstitution (Figure 19-1E) and the cleavage of P_1C_{ub}-R that yields R. In this approach, cells expressing P_1C_{ub}-R are transformed with an expression library encoding random translational fusions to a mutant N_{ub} moiety and screened for cells expressing active reporter. Another approach to an USPS-based screen for genes whose products interact with a protein of interest P_1 can employ a conditionally active degradation signal, thereby bypassing the requirement for a cleavage-dependent activation of a reporter protein. Specifically, a reporter-containing P_1C_{ub}-d-R fusion of this screen would bear a destabilizing residue, such as, for example, Arg or Leu (denoted "d"), at the C_{ub}-R junction. Upon reconstitution of Ub at P_1C_{ub}-d-R, the UBP-mediated cleavage of this fusion at the C_{ub}-R junction would convert a relatively long-lived P_1C_{ub}-d-R into d-R, a short-lived protein that bears a destabilizing N-terminal residue and is degraded by the N-end rule pathway. The N-end rule relates the in vivo half-life of a protein to the identity of its N-terminal residue. The corresponding degradation signal, called the N-degron, comprises a destabilizing N-terminal residue and a specific internal lysine (or lysines) of a protein substrate (reviewed by Varshavsky 1992, 1996). Earlier work has shown that expression of an enzymatically active protein reporter such as, for example, the *S. cerevisiae* Ura3p, in *ura3Δ* cells confers the Ura^+ phenotype if the reporter is long-lived but the Ura^- phenotype if the reporter has been rendered short-lived through the activation of its N-degron (Dohmen et al. 1994). Conversion of the pre-N-degron of P_1C_{ub}-d-R into the N-degron of d-R through the UBP-mediated cleavage at the C_{ub}-R junction can therefore be employed to screen for ORFs that encode a mutant N_{ub} moiety linked to a protein that interacts with the protein of interest P_1. A USPS-based genetic screen remains to be implemented.

USPS was demonstrated above with homodimerizing polypeptides of the leucine zipper type (Figures 19-2 and 19-4). Note that a USPS assay with a homodimerization domain is expected to have a lower sensitivity than an otherwise similar assay in which a pair of interacting *hetero*dimerization domains is used. Indeed, z_1-mediated complexes among N_{ub}-z_1 and z_1C_{ub}-d_{ha} fusions (Figure 19-3) apparently included not only the "relevant" N_{ub}/C_{ub} complexes but also those of the N_{ub}/N_{ub} and C_{ub}/C_{ub} type.

More recently, we have used USPS to detect and analyze in vivo interactions between *S. cerevisiae* Sec62p and the signal sequences of either the *SUC2*-encoded invertase or the *MFα1*-encoded precursor of α-factor, a mating pheromone (Johnsson and Varshavsky, unpublished results). Sec62p is an integral membrane protein and essential component of the translocation complex which mediates the transport of proteins bearing signal sequences across the endoplasmic reticulum membrane (Deshaies and Schekman 1990). The USPS assay detected specific, transient interactions between Sec62p and the signal sequences; it also made possible a kinetic analysis of these in vivo interactions, which have previously been demonstrated using photocrosslinking in a cell-free system (Müsch et al. 1992). The transient proximity between a signal sequence and the membrane-embedded Sec62p could be detected using several N_{ub} and C_{ub} arrangements in Sec62p and signal sequence-bearing test proteins. Together with the results of the present work, these findings (Johnsson and Varshavsky, unpublished results) illustrate the versatility of USPS, its sensitivity to relatively weak and transient protein interactions, and the remarkable flexibility of "allowed" N_{ub} and C_{ub} configurations within test proteins.

Facilitated reconstitution of a single-domain protein from fragments that cannot reassociate by themselves is the concept of USPS that can be used to construct split-protein sensors other than split Ub. The previously described examples of protein reconstitution (Galakatos and Walsh 1987; Johnsson and Weber 1990; Landro and Schimmel 1993; Shiba and Shimmel 1992; Rubin and Levy 1991; Zen et al. 1994) suggest that many proteins can be manipulated to yield a set of fragments whose reassembly into a folded, single-domain species is facilitated by linked, interacting polypeptides or other ligands. The presently unique advantage of the Ub-based split-protein sensor stems from the in vivo cleavage assay whose dependence on Ub conformation makes it possible to monitor a protein-protein interaction as a function of time, at the natural sites of this interaction in a living cell (Figure 19-1E).

ACKNOWLEDGMENT These studies were supported by grants to A. V. from the National Institutes of Health (GM31530 and DK39520).

References

Adams, S. R., Harootunian, A. T., Buetchler, Y. J., Taylor, S. S., and Tsien, R. Y. (1991). Fluorescence ratio imaging of cyclic AMP in single cells. Nature 349:694–697.

Alberts, B. M. (1984). The DNA enzymology of protein machines. Cold Spring Harbor Symp. Quant. Biol. 49:1–12.

Anfinsen, C. B., Cuatrecasas, P., and Taniuchi, H. (1971). Staphylococcal nuclease: chemical properties and catalysis. Enzymes 4:177–204.

Ausubel, F. M., Brent, R., Kingston, R. E., Moore, D. D., Seidman, J. G., Smith, J. A. and Struhl, K., eds. (1992). *Current Protocols in Molecular Biology*. New York, Wiley.

Bachmair, A., Finley, D., and Varshavsky, A. (1986). In vivo half-life of a protein is a function of its amino-terminal residue. Science 234:179–186.

Baker, R. T., Tobias, J. W., and Varshavsky, A. (1992) Ubiquitin-specific proteases of *Saccharomyces cerevisiae*. J. Biol. Chem. 267:23364–23375.

Chien, C., Bartel, P. L., Sternglanz, R., and Fields, S. (1991). The two-hybrid system: a method to identify and clone genes for proteins that interact with a protein of interest. Proc. Natl. Acad. Sci. USA 88:9578–9582.

Cox, J. P. L., Evans, P. A., Packman, L. C., Williams, D. H., and Woolfson, D. N. (1993). Dissecting the structure of a partially folded protein. J. Mol. Biol. 234:483–492.

Creighton, T. E. (1992). *Proteins: Structures and Molecular Properties*. New York, W. H. Freeman.

Deshaies R. J., and Schekman, R. (1990). Structural and functional dissection of Sec62p, a membrane-bound component of the yeast endoplasmic reticulum protein import machinery. Mol. Cell. Biol. 10:6024–6035.

Dohmen, R. J., Wu, P., and Varshavsky, A. (1994). Heat-inducible degron: a method for constructing temperature-sensitive mutants. Science 263:1273–1276.

Field, J., Nikawa, J. I., Broek, D., MacDonald, B., Rodgers, L., Wilson, I. A., Lerner, R. A., and Wigler, M. (1988). Purification of a RAS-responsive adenylyl cyclase complex from *S. cerevisiae* by use of an epitope addition method. Mol. Cell. Biol. 8:2159–2165.

Fields, S., and Song, O.-K. (1989) A novel genetic system to detect protein-protein interactions. Nature 340:245–246.

Finley, D., Bartel, B., and Varshavsky, A. (1989). The tails of ubiquitin precursors are ribosomal proteins whose fusion to ubiquitin facilitates ribosome biogenesis. Nature 338:394–401.

Galakatos, N. G., and Walsh, C. T. (1987) Specific proteolysis of native alanine racemases from *Salmonella typhimurium*: identification of the cleavage site and characterization of the clipped two-domain proteins. Biochemistry 26:8475–8480.

Guarente, L. (1993a). Synthetic enhancement in gene interaction: a genetic tool come of age. Trends Genet. 9:362–366.

Guarente, L. (1993b). Strategies for the identification of interacting proteins. Proc. Natl. Acad. Sci. USA 90:1639–1641.

Gyuris, J., Golemis, E., Chertkov, H. and Brent, R. (1993). Cdi1, a human G1 and S phase protein phosphatase that associates with Cdk2. Cell 75:791–803.

Hershko, A., and Ciechanover, A. (1992). The ubiquitin system for protein degradation. Annu. Rev. Biochem. 61:761–807.

Hinnebusch, A. G. (1984). Evidence for translational regulation of the activator of general amino acid control in yeast. Proc. Natl. Acad. Sci. USA 81:6442–6446.

Johnson, E. S., Bartel, B., Seufert, W., and Varshavsky, A. (1992). Ubiquitin as a degradation signal. EMBO J. 11:497–505.

Johnson, E. S., Ma, P. C. M., Ota, I. M., and Varshavsky, A. (1995). A proteolytic pathway that recognizes ubiquitin as a degradation signal. J. Biol. Chem. 270:17442–17456.

Johnsson, N., and Varshavsky, A. (1994a). Split ubiquitin as a sensor of proteins interactions in vivo. Proc. Natl. Acad. Sci. USA 91:10340–10344.

Johnsson, N., and Varshavsky, A. (1994b). Ubiquitin-assisted analysis of protein transport across membranes. EMBO J. 13:2686–2698.

Johnsson, N., and Weber, K. (1990). Structural analysis of p36, a Ca^{2+}, lipid-binding protein of the annexin family, by proteolysis and chemical fragmentation. Eur. J. Biochem. 188:1–7.

Karplus, M., and Weaver, D. L. (1976). Protein-folding dynamics. Nature 260:404–409.

Kim, P. S., and Baldwin, R. L. (1982). Specific intermediates in the folding reactions of small proteins and the mechanism of protein folding. Annu. Rev. Biochem. 51:459–489.

Kuchler, K., Sterne, R. E., and Thorner, J. (1989). *Saccharomyces cerevisiae STE6* gene product: a novel pathway for protein export in eukaryotic cells. EMBO J. 8:3973–3984.

Landro, J. A., and Schimmel, P. (1993). Non-covalent protein assembly. Curr. Op. Str. Biol. 3:549–554.

McGrath, J. P., and Varshavsky, A. (1989). The yeast *STE6* gene encodes a homologue of the mammalian multidrug resistance P-glycoprotein. Nature 340:400–404.

Müsch, A., Wiedmann, M., and Rapoport, T. A. (1992). Yeast Sec proteins interact with polypeptides traversing the endoplasmic reticulum membrane. Cell 69:343–352.

O'Shea, E. K., Klemm, J. D., Kim, P. S., and Alber, T. (1991) X-ray structure of the GCN4 leucine zipper, a two-stranded, parallel coiled coil. Science 254:539–544.

Papa, F. R., and Hochstrasser, M. (1993). The yeast *DOA4* gene encodes a deubiquitinating enzyme related to a product of the human tre-2 oncogene. Nature 366:313–319.

Phizicky, E. M., and Fields, S. (1995) Protein-protein interactions: methods for detection and analysis. Microbiol. Rev. 59:94–123.

Pu, W. T., and Struhl, K. (1993). Dimerization of leucine zippers analyzed by random selection. Nucl. Acids Res. 21:4348–4355.

Richards, F. M., and Wyckoff, H. W. (1971). Bovine pancreatic ribonuclease. Enzymes 4:647–806.

Rubin, R. A., and Levy, S. B. (1991). Tet protein domains interact productively to mediate tetracycline resistance when present on separate polypeptides. J. Bact. 173:4503–4509.

Sherman, F., Fink, G. R., and Hicks, J. B. (1986). *Methods in Yeast Genetics*. New York, Cold Spring Harbor Laboratory Press.

Shiba, K., and Schimmel, P. (1992). Functional assembly of a randomly cleaved protein. Proc. Natl. Acad. Sci. USA 89:1880–1884.

Sikorski, R. S., and Hieter, P. (1989). A system of shuttle vectors and yeast host strains designed for efficient manipulation of DNA in *Saccharomyces cerevisiae*. Genetics 122:19–27.

Solomon, M. J., Larsen, P. L., and Varshavsky, A. (1988) Mapping protein-DNA interactions in vivo with formaldehyde: evidence that histone H4 is retained on a highly transcribed gene. Cell 53:937–947.

Stockman, B. J., Euvrard, A., and Scahill, T. A. (1993). Heteronuclear three-dimensional NMR spectroscopy of a partially denatured protein: the A-state of human ubiquitin. J. Biomol. NMR 3:285–296.

Tobias, J. W., and Varshavsky, A. (1991). Cloning and functional analysis of the ubiquitin-specific protease gene *UBP1* of *Saccharomyces cerevisiae*. J. Biol. Chem. 266:12021–12028.

Varshavsky, A. (1992). The N-end rule. Cell 69:725–735.

Varshavsky, A. (1996). The N-end rule. Cold Spring Harb. Symp. Quant. Biol. 60:461–478.

Vijay-Kumar, S., Bugg, C. E., and Cook, W. J. (1987). Structure of ubiquitin refined at 1.8 Å resolution. J. Mol. Biol. 194:531–544.

Vinson, C. R., Sigler, P. B., and McKnight, S. L. (1989). Scissor-grip model for DNA recognition by a family of leucine zipper proteins. Science 246:911–916.

Zen, K. H., McKenna, E., Bibi, E., Hardy, D., and Kabak, R. H. (1994). Expression of lactose permease in contiguous fragments as a probe for membrane-spanning domains. Biochemistry 33:8198–8206.

Index